New Trends in Optical Networks

New Trends in Optical Networks

Editors

Fabio Cavaliere
Luca Potì

MDPI

Basel • Beijing • Wuhan • Barcelona • Belgrade • Novi Sad • Cluj • Manchester

Editors
Fabio Cavaliere
Ericsson Research
Pisa, Italy

Luca Potì
Consorzio Nazionale
Interuniversitario per le
Telecomunicazioni - CNIT
Pisa, Italy

Editorial Office
MDPI
St. Alban-Anlage 66
4052 Basel, Switzerland

This is a reprint of articles from the Topical Collection published online in the open access journal *Applied Sciences* (ISSN 2076-3417) (available at: https://www.mdpi.com/journal/applsci/topical_collections/Optics_Networks).

For citation purposes, cite each article independently as indicated on the article page online and as indicated below:

Lastname, A.A.; Lastname, B.B. Article Title. *Journal Name* **Year**, *Volume Number*, Page Range.

ISBN 978-3-0365-8686-1 (Hbk)
ISBN 978-3-0365-8687-8 (PDF)
doi.org/10.3390/books978-3-0365-8687-8

Contents

About the Editors

Fabio Cavaliere

Fabio Cavaliere is an Expert in Photonic Systems and Technologies at Ericsson, and Rapporteur of ITU-T Question 6/15 (Characteristics of optical components, subsystems and systems for optical transport networks). Fabio is author of approximately 130 filed patent applications and more than 100 publications on optical networks, including the book "Photonics applications for radio systems and Networks (Artech House, Boston, USA). In 25 years of professional experience, his research activities have encompassed radio access networks, fiber access, high speed optical transmission, and integrated photonics. Fabio is in the Technical Program Committees of International Conferences on Optical Communications and Guest Editor of Applied Science's Topical Collection on Optical Networks and of the IEEE Communications Standards Magazine. He is in the Board of Stakeholders of Photonics 21 (the European technology platform on photonic technologies), in the Expert Advisory Board of NetworldEurope, and in the Strategic Advisory Board of the European Quantum Flagship.

Luca Potì

Luca Potì is Head of the Research Area "High Capacity and Secure Optical Communications" with the Interuniversity National Consortium for Telecommunications (CNIT) at the Photonic Networks and Technologies Lab in Pisa, Italy, and Full Professor at the Universitas Mercatorum in Roma, Italy. He has published one book, six book chapters, and more than 400 international journal papers, conference papers, and patents. He served as a coordinator and/or scientist within more than 50 Industrial and Institutional International Projects. He was chair, co-chair, and organiser for several International and National technical conferences and workshops. His research interests include optical communication and networks, quantum communications and computing, photonic integrated devices and subsystems, secure communications, optical steganography, authentication techniques, artificial intelligence applied to the optical physical layer, and advanced sensing techniques, including optical devices.

Preface

This book covers recent advances in optical communication networks with high impacts on the future digitized society and economy, encompassing all network layers from new optical components to smart network operations. Its contents include contributions from several worldwide-reputed experts in the sector. This book intends to provide a comprehensive view of all aspects involved in the design of modern optical networks and is addressed to both young and experienced researchers in academia and industry, so that they can pave the way to the future digital society.

Fabio Cavaliere and Luca Potì
Editors

applied sciences

MDPI

Article

Design and Implementation of Semi-Physical Platform for Label Based Frame Switching in Integrated Satellite Terrestrial Networks

Wei Zhou [1,2,†], Xing Jiang [1,*,†], Qingsong Luo [2,†], Shanguo Huang [3,†], Bingli Guo [3,†], Xiang Sun [4,†], Shaobo Li [5,6,†], Xiaochuan Tan [5,6,†], Mingyi Ma [5,6,†] and Tianwen Fu [1,†]

1 School of Information and Communication, Guilin University of Electronic Technology, Guilin 541004, China; 19021101008@mails.guet.edu.cn (W.Z.); 21022303025@mails.guet.edu.cn (T.F.)
2 The 34th Research Institute of China Electronics Technology Group Corporation, Guilin 541004, China; gioc34@sina.com
3 State Key Laboratory of Information Photonics and Optical Communications, Beijing University of Posts and Telecommunications, Beijing 100876, China; shghuang@bupt.edu.cn (S.H.); guobingli@bupt.edu.cn (B.G.)
4 Department of Electrical and Computer Engineering, University of New Mexico, Albuquerque, NM 87131, USA; sunxiang@unm.edu
5 The 54th Research Institute of China Electronics Technology Group Corporation, Shijiazhuang 050081, China; shaoboleigh@163.com (S.L.); tanxiaochuan333@163.com (X.T.); mamingyi2022@163.com (M.M)
6 Hebei Key Laboratory of Photonics Information Technology and Application, Shijiazhuang 050081, China
* Correspondence: jiang_x@guet.edu.cn
† These authors contributed equally to this work.

Citation: Zhou, W.; Jiang, X.; Luo, Q.; Huang, S.; Guo, B.; Sun, X.; Li, S.; Tan, X.; Ma, M.; Fu, T. Design and Implementation of Semi-Physical Platform for Label Based Frame Switching in Integrated Satellite Terrestrial Networks. *Appl. Sci.* **2022**, *12*, 6674. https://doi.org/10.3390/app12136674

Academic Editors: Luca Potì and Fabio Cavaliere

Received: 2 June 2022
Accepted: 28 June 2022
Published: 1 July 2022

Abstract: With the explosion of traffic demand in recent years, the integration of satellite optical networks and terrestrial networks (ISTN) creates a promising networking solution for future low-latency, high-rate, and high-capacity communications. Owing to the high cost of deploying and maintaining a satellite optical network, it is critical to carefully design and plan the network to ensure the performance of the network. Thus, a semi-physical simulation platform based on software-defined networks (SDNs) is developed to simulate a satellite optical network and evaluate the performance of the proposed label-based advanced orbiting system (AOS) frame switching method that adheres to the Consultative Committee for Space Data Systems' recommended standard (CCSDS). The semi-physical simulation platform has two major innovations: (1) adapting and integrating network protocols between the CCSDS and open system interconnect (OSI) reference models, particularly at the data link layer, and (2) the foundation for an SDN-based satellite optical network. In the control plane, real-time VxWorks Simulators serve as controllers to establish and manage various network protocols and the link manager protocol (LMP). Here, network protocols include open shortest path first (OSPF) for routing managing and controlling messages, constraint shortest path first–traffic engineering (CSPF-TE), and constraint-label distribution protocol (CR-LDP) for routing data services. LMP is used to assign and reserve satellite optical link resources. The performance of the architecture and protocols is evaluated via a semi-physical simulation platform.

Keywords: software-defined networks; integrated satellite–terrestrial network; semi-physical simulation platform; label-based CCSDS AOS frame switching

1. Introduction

With the exponential growth of traffic demands and data services over recent years, deploying satellite networks based on microwave communications has received much attention but has faced some challenges, including limited microwave frequency band resources, inaccurate transmission direction, and inadequate network capacity. On the other hand, free-space optics (FSO), which applies optical lasers to transmit data streams between two satellites, is becoming a substitute communications technology in satellite networks. As compared to microwave and other radio communications, FSO offers a higher frequency band, higher link capacity, lower power consumption, lower interference, and easier to identify interceptions or eavesdropping, and so FSO has been widely used in space-terrestrial broadband communications [1–5].

On the other hand, software-defined networks (SDNs) offer a novel architecture to achieve various network services, e.g., routing and load balancing, in an efficient and flexible way. SDN applies two basic strategies to facilitate flexible network management (1) SDN decouples the control plane of a network from its data plane, and (2) SDN conceptually concentrates on the entire viewing of the network intelligence (i.e., the control plane), and hence abstracts the underlying network architecture and improves the application optimization. In general, an SDN physically comprises controllers and a number of forwarding-rule-based white-box switches. The controllers make the routing decisions and forwarding rules based on the statistics sent from the switches, and the switches simply forward data based on forwarding rules from the controllers over the southbound interface, such as OpenFlow. Thus, SDN can potentially enhance the flexibility of network management through programmability and simplify network administration via abstracting the underlying network architecture. The existing satellite optical networks typically employ traditional network design, i.e., the control and data planes are dispersed in each switch, and thus it is challenging to develop an efficient routing strategy that can be adaptive to the network dynamics. In [6,7], SDN is applied in a satellite optical network, where a satellite–terrestrial station gateway may function as a controller to run the control plane that manages the whole satellite optical network and monitors the statistics of all the satellite nodes, and which only conducts routing and forwarding configurations according to the forwarding rules given by the controller. However, [6,7] suggest only an SDN-based end-to-end fragment-aware routing algorithm without considering the real-world on-satellite data link protocol, which is unique to the terrestrial network protocol.

Traffic congestion could severely degrade the network performance and is generally caused by imbalanced traffic load among different links and inadequate resource distribution [8]. The traditional routing protocols (e.g., open shortest path first (OSPF)) may lead to traffic congestion because they are mainly routed concerning the packets via the shortest path without considering the available resources and quality of services (QoS), resulting in label switching routers (LSRs) being overused for multiple label switching paths (LSPs) [9]. Traffic engineering (TE) is proposed to optimize network resource assignment via bandwidth allocation based on resource statistics. The optimized routing strategy can substantially balance the traffic load, increase the network throughput, and reduce network latency. In terms of routing algorithms, the current SDN-based routing algorithms can be classified into two categories: (1) Shortest routing method [10] employs OSPF-based congestion control (i.e., the Dijkstra algorithm) in the SDN-based low earth orbit (LEO) satellite network. However, the Dijkstra algorithm does not evaluate the network state and always picks the same forwarding route for the same source–destination pair, which may cause link congestion and throughput reduction. (2) The QoS routing method evaluates the QoS of different routing paths and selects the one that can provide the best service for a source–destination pair. However, the major issue with the QoS routing method is it is difficult to achieve multi-constraint routing, which is a well-known non-deterministic polynomial-hard (NP-hard) problem. If services are not distinguished, services with high QoS needs and high priority cannot be handled in time, resulting in network performance deterioration. Therefore, it is critical to design a routing algorithm that considers the dynamic features of satellite network structure, varied QoS needs of different traffic, and traffic load and capacity of different links [11]. On the other hand, the mentioned works only evaluate their proposed routing algorithms via simulations, which may not be feasible to be implemented in a real satellite network.

In order to bridge the gap between simulation and real satellite networks, an SDN-based semi-physical simulation platform is built to evaluate satellite networking and routing in this paper. The major contributions of the paper are summarized as follows:

1. A novel label-based Consultative Committee for Space Data Systems (CCSDS) advanced orbiting system (AOS) data link layer protocol, optimized OSPF protocol, constraint shortest path first–traffic engineering (CSPF-TE), and constraint-label distribution protocol (CR-LDP) protocol are designed for integrated satellite–terrestrial networks (ISTNs). In addition, a simulation platform is built to verify the performance of label-switching-based elastic ISTNs with no restrictions on simulated nodes based on a software environment, i.e., INTEL VxWorks, an embedded real-time operating system;
2. We develop the southbound interface of simulated nodes (act as controllers) in INTEL VxWorks to communicate with field-programmable gate array (FPGA) boards (which act as switches) to fully verify the performance of label-switching-based elastic ISTNs, which follows and expands the recommended standard of the CCSDS AOS protocol [12,13];
3. The programming codes running in the INTEL VxWorks simulated node could be easily applied to different micro-chip architectures according to the board support packages, e.g., Acorn RISC Machine (ARM), performance optimization with enhanced RISC–performance computing (PowerPC), and microprocessor without interlocked piped stages (MIPS), thus having high compatibility.

The remainder of the paper is structured as follows. In Section 2, principles and methods of SDN-based ISTNs are introduced. Sections 3 and 4 give an explicit description of the experimental results in terms of software simulation platform and semi-physical simulation platform, respectively. Finally, concluding remarks and future work are given in Section 5.

2. Principles and Methods

2.1. Principles of SDN-Based ISTNs

It is challenging to design an efficient networking solution for space networks due to the complexity and dynamical changes of network topology. Moreover, some studies have been working on optimizing the space network management, but they do not combine space networks and terrestrial networks together. Currently, some constellation satellite systems, such as the Starlink project, have been deployed, trying to compete with or even replace the existing terrestrial core networks. On the other hand, the integration of satellite networks and terrestrial networks can substantially increase the coverage and capacity of the Internet, but many challenges have not been addressed yet. In this paper, we propose a basic SDN-based ISTN system, as illustrated in Figure 1, that mainly includes a satellite optical network, satellite access users, and terrestrial user access networks with gateways. Here, the satellite optical network is the key part of the backbone network and comprises a number of high earth orbit (HEO) satellites, e.g., geostationary orbit (GEO) and inclined geosynchronous orbit (IGSO) satellites, and middle earth orbit (MEO) satellites. FSO is used to transmit data between two satellites in the satellite optical network. Each satellite in the satellite optical network can be considered as a label switching router with its own link-state database (LSDB). Different space access users, such as airplanes, unmanned aerial vehicles (UAVs), and LEO satellites, are interconnected together via the satellite optical network to achieve positioning, navigation, data transmission, and so on. Note that LEO satellites can also be part of the satellite optical network. In this case, each LEO satellite is viewed as a label switching router with its own LSDB.

Each HEO satellite is able to communicate with a terrestrial access network via a terrestrial gateway, which plays a vital role in realizing the integration between satellite and terrestrial networks through protocol conversion, e.g., between the Transmission Control Protocol (TCP)/IP protocol (used in the terrestrial access network) and the label-based AOS protocol (used in the satellite optical network). Note that only HEO satellites but not MEO satellites can communicate with terrestrial gateways due to the stability of HEO satellites, i.e., the locations of HEO satellites are more stable than MEO satellites in terms of maintaining FSO links that connect to terrestrial gateways. An SDN controller is deployed in the terrestrial network to control the behavior of satellites in the satellite optical network via the corresponding gateway to achieve constellation topology control and management, and allocate data channels for terminals in the user access network. In general, the purpose of the proposed SDN-based ISTN system is to interconnect space access users and terrestrial access networks, and achieve efficient inter-satellite routing and network operation administration and maintenance (OAM). This novel network architecture will pave the way to realizing an information-centric and/or content-centric networking Internet that offers network providers and users secure, high-performance, adaptable, and resilient services.

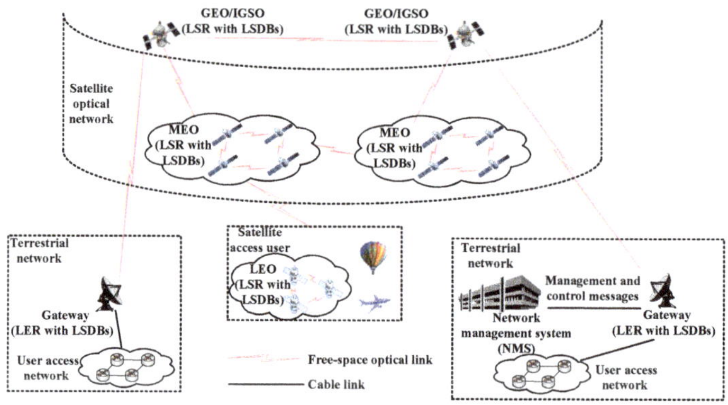

Figure 1. The typical application scenario of ISTNs.

Establishing and maintaining a satellite network is costly since (1) although the capital cost of the satellite decreases, launching a satellite to space is still expensive, and (2) repairing and changing the hardware settings of satellites is difficult owing to the harsh space environment. Hence, building a semi-physical satellite network platform to emulate various network services is necessary to ensure the correct network configurations before deploying a satellite network in space. Here, a semi-physical satellite network platform is a hardware and software co-designed simulator, where different FPGA boards are used to emulate the behaviors of satellites in a satellite optical network, and the whole communication network is built in a simulation environment. Satellite network designers may use the semi-physical platform to analyze and evaluate system performance, optimize network parameters, and significantly decrease the cost of maintaining the network. Semi-physical simulation, also known as hardware-in-the-loop simulation (HILS), is a type of simulation technology that incorporates some real hardware implementations into the simulation loop to better emulate the whole system. The method is regarded as one of the most important tools in the development of space systems in the aerospace field that can enhance the confidence level of the simulation results and address the system modeling challenges.

2.2. The Label Switching-Based ISTNs

2.2.1. Principles of Label Switching

In the data plane of ISTNs, various packets are categorized into different flows according to their information, e.g., source/destination IPv4 and/or IPv6 address. The packets that are transmitted over the same path are considered as a forwarding equivalence class (FEC). It may be feasible to employ a single label for a set of FECs, which is called link aggregation. A local unique fixed-length label will be appended to the packets in the same type of FEC. Once receiving an incoming packet, the label switching router extracts the in-label of the packet, replaces it with the out-label, and forwards the packet to the egress port according to the local label forwarding information base (LFIB) table, and thus transmits the packet to the neighbor label switching router. The packet is switched by different label switching routers until the egress label edge router is reached. In some ways, the ISTNs act similarly to a multi-protocol label switching (MPLS) network. The uni-directional data path, along which a data packet is transmitted in the MPLS domain, is defined as a label switching path. Currently, there are three typical methods of routing path calculation and label switching path setup:

1. Distributed routing and label switching path setup. All the label edge routers and label switching routers maintain their own link-state databases, and conduct routing computation and label switching path setup by themselves, without having terrestrial controller involvement;
2. Centralized routing, distributed label switching path setup. With the global knowledge of network status, a terrestrial controller is able to cause computer routing paths to meet multiple QoS requirements from the ingress label edge routers. Meanwhile, the ingress label edge router constructs a label switching path according to the related routing paths calculated by the terrestrial controller;
3. Centralized routing and label switching path setup. The terrestrial controller would calculate both routing paths and label switching paths and deliver them to routers.

Owing to the frequent topology changes of satellite networks, directly utilizing terrestrial distributed routing protocols (e.g., OSPF) would lead to constant route convergence, thus consuming valuable inter-satellite link (ISL) bandwidth. Hence, by following the idea of Method 1 above, we propose an optimized OSPF routing protocol and constraint label distribution protocol for a GEO/IGSO and MEO satellite network by considering the QoS requirements that are characterized as multiple dimensionless metrics. Here, QoS requirements can be measured by uni-directional link, cost, priority, duration time, forward/backward bandwidth and latency. In addition, link-state update flooding is restrained by periodic updates of the on-satellite local link-state database according to the network topology. The optimized OSPF routing algorithm is described in Algorithm 1. Furthermore, as a type of large-scale and heterogeneous network, whose network states dynamically vary in both time and space, any single node in the ISTNs will be capable of performing both the access function (e.g., multiple data granularity classifications and link aggregation) and the switching function (i.e., multi-layer data distribution, such as a traditional label edge router and label switching router). The aforementioned Method 1, therefore, is more capable of low overhead controlling and high scalability. Given the time-precise synchronization of the link-state database and label switching path, and a new signaling protocol needed for spreading labels among the label switching routers, Methods 2 and 3 may only be able to conduct a small number of label switching paths.

According to [14], the label distribution protocol (LDP) is designed to distribute labels to all the label edge routers and label switching routers along a label switching path. Currently, two typical protocols are specified by the Internet Engineering Task Force (IETF): (1) resource reservation protocol-traffic engineering (RSVP-TE) [15], which is a traffic engineering modification to the resource reservation protocol (RSVP) [16], and (2) constraint routing label distribution protocol [17], which is built based on LDP. The most significant distinction between the two protocols is how to share the control information inside the MPLS domain, namely: RSVP-TE employs a soft state method with update messages, while CR-LDP employs a hard state refreshing approach using TCP. For ease of implementation, we use CR-LDP and modify parts of its TCP function, and the algorithm of the optimized CR-LDP is revised in Algorithm 2. In contrast to the integrated services (IntServ), which use RSVP to provide end-to-end route reservations, differentiated services (DiffServ) are used in MPLS-based satellite networks to meet the QoS requirements. In the DiffServ method, all IP data are grouped into a behavior aggregate, which is tagged by a DiffServ code point (DSCP) inside each frame based on the same behavior request.

Algorithm 1 Optimized OSPF, Periodic Update of LSDB and Uni-directional Link Routing.

```
Initialize OSPF configure and periodically send out Hello message;
if message received then
    switch type of received message do
        case Hello:
            if without its own Router ID and in Full status then
                Switch to UNI_DIRECTION_SEND status and send uni-directional message;
            end if
            if with its own Router ID and in UNI_DIRECTION_SEND status then
                Switch to Full status;
            end if
            if in UNI_DIRECTION_RECEIVE status then
                Switch to Full status;
            end if
        case Database Description:
            Compare with link-state database, consistent with standard OSPF;
        case Link State Request:
            Organize and send link-state update message, consistent with standard OSPF;
        case Link State Update:
            if This update belongs to the link plan then
                Ignore this update message and restrain the flooding because the satellite update itself through the link plan;
            end if
            if This update does not belong to the link plan then
                Update link-state database and send link-state acknowledgment message, consistent with standard OSPF;
            end if
        case Link State Acknowledgment:
            Stop retransmitting link-state update message, consistent with standard OSPF;
        case Unidirectional:
            Switch to UNI_DIRECTION_RECEIVE status, organize link-state acknowledgment which includes uni-directional link and flood it;
        case NonUnidirectional:
            Switch to Down status, organize link-state acknowledgment which excludes uni-directional link and flood it;
end if
if Hello message not received before timer expires and in UNI_DIRECTION_SEND status then
    Switch to Down status and send NonUnidirectional message;
end if
```

Algorithm 2 CR-LDP, Constraint Label Distribution under the QoS Requirements.

```
Initialize CR-LDP configure;
if message received then
    switch type of received message do
        case LSP Build Request message:
            Reserve the bandwidth required by the forward label switching path, generate a label distribution request message and send it to the downstream node;
        case LDP Label Request message:
            switch View the displayed route type length value (TLV) and judge the node location do
                case Intermediate node:
                    Reserve bandwidth for the bi-directional label switching path, then update the label distribution request message and send it to the downstream node;
                case Tail node:
                    Allocate bandwidth directly for the backward label switching path, deliver the forwarding information base table to FPGA board in the data plane, generate a label distribution mapping message, and allocate label for the upstream node;
        case LDP Label Mapping message:
            switch View the displayed route TLV and judge the node location do
                case Intermediate node:
                    Allocate bandwidth for the bi-directional label switching path, deliver label forwarding information base tables to FPGA board in the data plane, generate a label distribution mapping message, and allocate label for the upstream node;
                case Tail node:
                    Allocate bandwidth for the forward label switching path, deliver forwarding information base table to FPGA board in the data plane;
end if
```

2.2.2. Principles of CCSDS AOS Frame Label Switching

CCSDS AOS allows the coexistence of ISTNs and multiple forms of data access. CCSDS AOS is able to transmit different types of data, such as voice, video, and images, in the form of a unified data flow over a spatial physical channel based on dynamic scheduling management of different virtual channels in the physical channel. Meanwhile, the flexible reuse mechanism and effective error correction/detection measures in AOS guarantee that the spatial physical channel has a high channel capacity, channel quality, and reliability to meet the needs of spacecraft data transmission and processing. The CCSDS AOS space data link protocol specifies the recommended procedures for transmitting various types of data using fixed-length packet units, i.e., transmission frames as shown in Figure 2a. However, in the intermediate node, AOS frames could not be switched directly, and all the traffic flows will be unpacked and forwarded to the upper-layer for the further process. As a result, the energy-resource-consuming frame unpackaging–packing–forwarding process cannot meet the growing demand for space data transmission and QoS routing performance. To address those issues, an extended AOS frame with **extended control field** is proposed to realize label-based frame switching directly in routers, thus avoiding the unnecessary unpacking–repacking process for bypassing traffic data, as shown in Figure 2b,c, respectively. In Figure 3, there is an explicit comparison between the open system interconnect (OSI) and extended CCSDS reference models, where the label switching sublayer is added between the data link protocol sublayer and the synchronization and channel coding sublayer, and an extended label-based AOS switching protocol is applied in the CCSDS protocol stack as follows:

- The **label** field is added in an AOS frame to forward traffic flows based on labels instead of IP-address-based forwarding;
- The **data communication network** field is used as an overhead for forwarding management and controlling messages for the control plane and the management plane;

- The **OAM** field, i.e., operation administration and maintenance, is used for link connection detection, and thus to realize make-before-break routing handover.

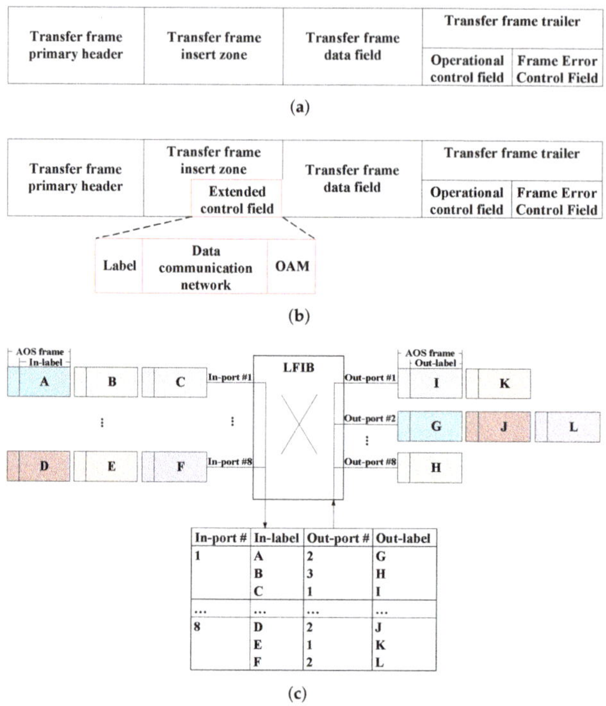

Figure 2. The label-based CCSDS AOS frame switching. (**a**) The CCSDS recommended AOS frame. (**b**) The proposed AOS frame with extended control field. (**c**) The principle of label swapping for the AOS frame according to LFIB.

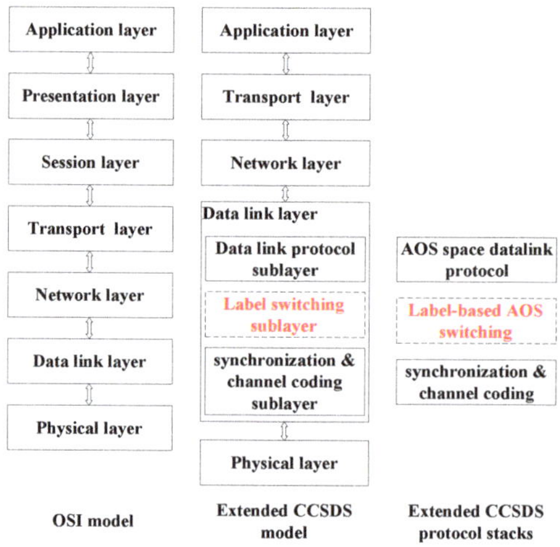

Figure 3. OSI reference model v.s. label-based extended CCSDS reference model.

2.3. The Design of SDN-Based ISTNs

2.3.1. The Functional Block Diagram of SDN-Based ISTNs

The network management system, controllers, and hardware modules are the three components of the proposed semi-physical simulation system. A Windows 10 personal computer, which runs an INTEL VxWorks embedded real-time operating system, is an SDN controller to realize the management and control plane functions. The main function of the management plane is to monitor dynamic parameters, while the network topology changes or data services are modified. The controller is responsible for executing all control plane protocols and algorithms (e.g., OSPF, CSPF, link manager protocol (LMP), and CR-LDP). The label-based AOS frame switching and data forwarding are simulated using the hardware module. Figure 4 depicts the framework of the semi-physical simulation platform.

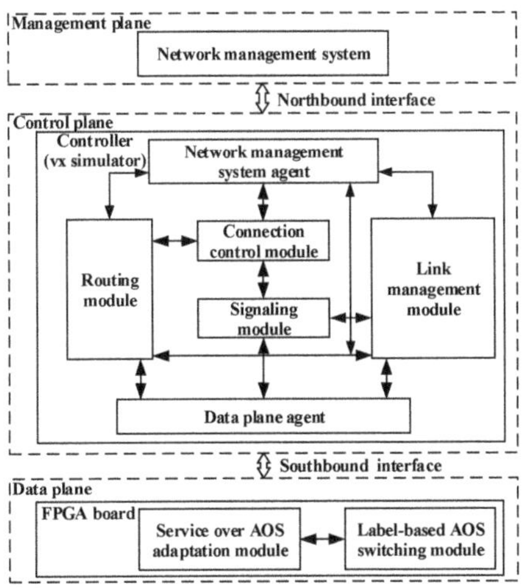

Figure 4. The framework of SDN-based ISTNs simulation platform.

The proposed SDN-based ISTNs are built the same way as a traditional SDN with the management plane (MP), the control plane (CP), and the data plane (DP). The network management system (NMS) in the terrestrial network is capable of managing and controlling the satellite optical network as well as monitoring network status through the northbound interface. As a result, each node, i.e., label edge routers (LERs) and label switching routers, has a network management agent (NMA) that receives, processes, and forwards NMS information. The management plane is logically a star structure since all NMAs are directly governed by NMS. Second, the distributed control plane module in each node includes sub-modules such as the network management system agent, routing module, connection control module, signaling module, link management module, and data plane agent. The data plane, on the other hand, is made up of an FPGA board and communication terminals, e.g., laser and microwave, as well as physical spatial links. The FPGA board uses a XILINX KINTEX-7 (XC7K325) as the primary control chip to provide data forwarding based on label switching, and adaption between data services (e.g., IP data, bitstream data, virtual channel access data, and packet data) and the CCSDS AOS data link protocol.

The routing method, i.e., the (constraint) Dijkstra algorithm, performs routing computation, which is an essential function of the routing module. The routing module computes two types of routes based on the needs of networking protocols:

1. Routing of managing and controlling messages. This sort of routing path is used for the managing and controlling messages from the management plane to the control plane. Since the **data communication network** field is the overhead for managing and controlling communications, each node will construct a routing table for the shortest

way to all other nodes based on the shortest path first (SPF) method (i.e., Dijkstra algorithm) simply in terms of link priority and cost, ignoring the QoS requirements. When a node's link-state database is updated, the routing module will compute a new routing path for managing and controlling messages immediately;

2. Routing of service data. This sort of routing path is an explicit end-to-end routing path based on the CSPF-TE algorithm for traffic flows under service QoS requirements, such as source–destination, forward and backward bandwidth, link duration time, latency, etc.

Due to the restricted bit length of the **data communication network** field, a fixed bits length managing and controlling message is divided into segments and sent in ten continuous AOS frames with the sequence number ranging from one to ten before transmission, as illustrated in Figure 5. After a cyclic redundancy check, the split managing and controlling message segments are reunited in sequence and delivered to the control plane and management plane on the receiving side. An asking-acknowledge re-transmission mechanism is designed to ensure the transmission of managing and controlling messages between nodes, taking into account link turbulence caused by cosmic radiation, particle flow, and other factors. A managing and controlling message from the source node will be re-transmitted if the acknowledgment message from the destination node is not received in the source node. Despite the fact that the routing strategies for managing and controlling messages and service data are diametrically opposed, managing and controlling messages and service data can be transmitted in the same AOS frame if the out-port to destination is the same, i.e., managing and controlling messages using the **data communication network** field while not taking up the **transfer frame data field** for the service data.

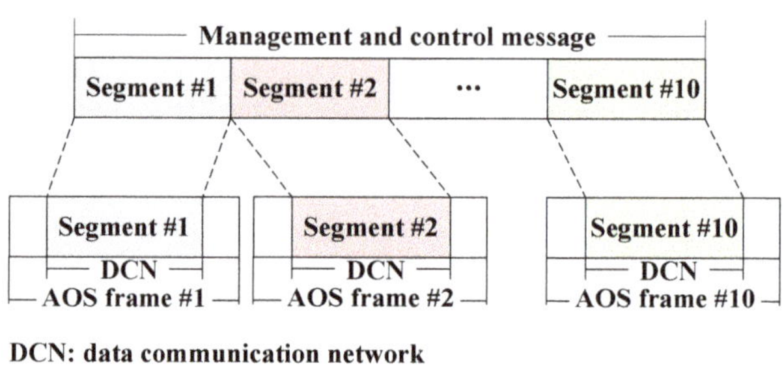

Figure 5. The transmission of a fixed bit length managing and controlling message in the **data communication network** field of the AOS frame.

2.3.2. The Working Principle of the Data Plane in the Semi-Physical Platform

The adaptation of IP and/or non-IP data to AOS frames using a combination of TCP/IP and AOS protocol is one of the latest research topics, specifically the implementation of seamless communication and connection of ISTNs. Both terrestrial networks and on-satellite local area networks (LANs) receive and send data using the Ethernet protocol at the data link layer, which is customized to the CCSDS AOS data link layer for inter-satellite and satellite–terrestrial links, as illustrated in Figure 6.

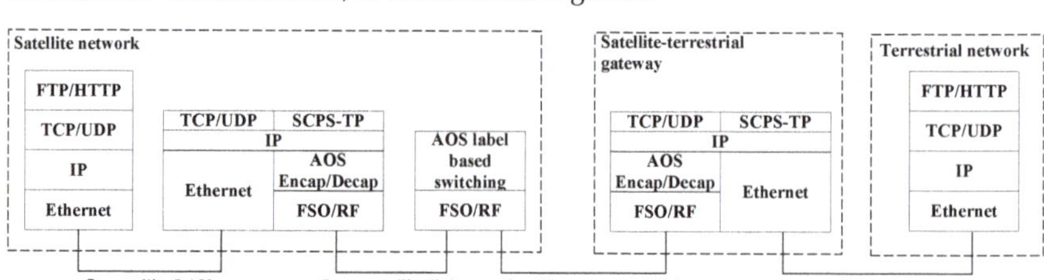

Figure 6. The protocol framework of ISTNs.

In this paper, we propose the label-based AOS frame switching mechanism for the Ethernet IP data service. However, as stated in [12], four sorts of data services may be transmitted by AOS frames, i.e., IP service, bitstream service, virtual channel access service, and packet service. Figure 7 shows the functional module diagram of customizable data services over the AOS frame.

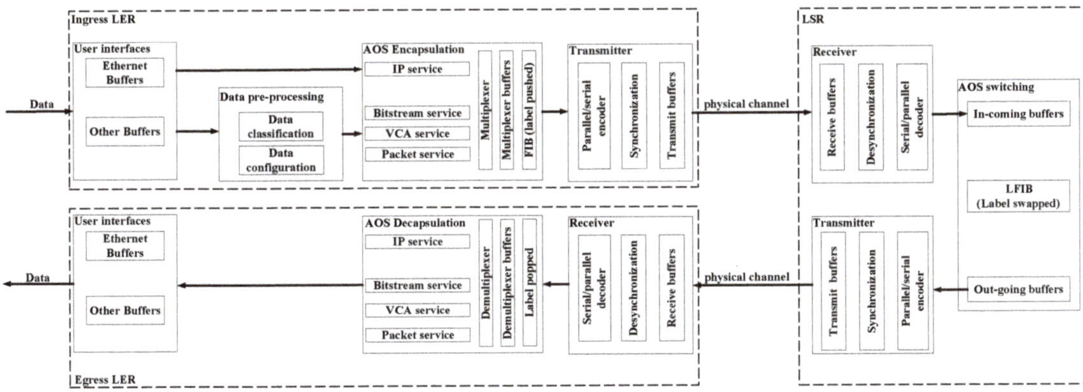

Figure 7. The functional block diagram of four types of service over AOS frame.

- The user interfaces in the ingress label edge router support various types of data access and provide data caching buffers for parallel data pre-processing, where data is classified and transmitted to the corresponding service buffers before being encapsulated as payload in the **transfer frame data field** of the AOS frame. According to the forwarding information base (FIB), the out-label is pushed to the **label** field of the AOS frame before being sent to the transmitter module, and then those parallel data are serialized to 8b/10b encoding for the transmission in the physical channel, e.g., radio and laser;
- Data are recovered to parallel encoding at the label switching router before being transmitted to the AOS switching module, where the value of the in-label of the AOS frame is swapped to the value of the out-label and the AOS frame is switched to an out-port according to the label forwarding information base. Then, the parallel-serial encoding for the AOS frame is processed;
- If the value of in-label is **0xFFFF**, it indicates that the egress label edge router is the last node in the label switching path without further label swapping processing, and the payload in the **transfer frame data field** of the AOS frame should be de-encapsulated to the corresponding data service buffers, and then transmitted to the corresponding user interfaces, and vice versa for the backward label switching path.

The FPGA board designs are as follows:

- Matching data rates. The FPGA board is designed to better match the varied data rates for services and throughput of QoS requirements:

$$\text{Throughput} = \text{clock frequency} \times \text{width of FIFO} \times \text{depth of FIFO}. \quad (1)$$

For example, the clock frequencies for gigabit Ethernet (GE) and 10-gigabit Ethernet (10GE) are 125 MHz and 156.25 MHz, respectively, with parallel data widths of 8 bits and 64 bits and 1 depth of first in first out (FIFO) buffer. Furthermore, the FIFO's configurable depth is designed to fulfill line rates and the QoS requirement (i.e., throughput) for label edge routers and label switching routers along the label switching path;
- Synchronization of data. Only two oscillators with frequencies of 50 MHz and 100 MHz are used in the FPGA board for the simplicity of hardware design, i.e., the oscillator with the frequency of 100 MHz serves as the main working clock source

for the FPGA chip (i.e., XILINX KINTEX-7 (XC7K325)), and the oscillator with the frequency of 50 MHz serves as the input clock source for the programmable clock chip (i.e., Silicon Labs SI5338). So, FPGA asynchronous FIFOs are used for synchronizing data across multiple clock zones. Asynchronous FIFOs are also used for the encoding process from parallel data (i.e., IP data) to serial data (i.e., 8 b/10 b data) in Figure 7, where the input clock source is 125 MHz or 156.25 MHz for writing parallel 8 bit or 64 bit data to the asynchronous FIFOs, and the output clock source is 100 MHz for reading 16 bit data from the asynchronous FIFOs before sending to the physical channel in the transmitter module, and vice versa for the receiver module. The length of the AOS frame and the various fields of the AOS frame are set as recommended in [12] and shown in Figure 2b.

3. Simulation Analysis for the Designed ISTN Protocol Stack

In this section, we will analyze the performance of the designed ISTN protocol stack described in Section 2.

3.1. Simulation Platform Setup

The designed ISTN protocol stack should be evaluated in a simulation platform to ensure its performance. Currently, the majority of those simulations are often demonstrated in OMNeT++ [18], which serves as a global simulated platform that runs on C++ and Python. However, the OMNeT++ programming codes must be designed in the C programming language for the satellite hardware. Instead of OMNeT++, the simulated software is designed and operated in INTEL VxWorks, which is utilized as the on-satellite real-time embedded operating system. The INTEL VxWorks Simulator is a simulated hardware target for use as a prototyping and test-bed environment for VxWorks. The VxWorks Simulator allows users to develop, run, and test VxWorks applications on the host systems (i.e., personal computers), reducing the need for target hardware during development. For external applications needing to interact with a VxWorks target, the capabilities of a VxWorks Simulator instance are identical to those of a VxWorks system running on real-world target hardware. It is worth noting that the versions of the VxWorks Simulator's workbench and image kernel are 3.3 and 6.9 respectively, while the real-world target hardware is a 64-bit MIPS-based microprocessor with VxWorks version 6.8.

3.2. Test of the ISTN Protocol Stack in the Software Platform

For the verification and validation of network protocols and algorithms of the control plane, we firstly conduct the test in a software platform under the circumstances of large-scale network topology with multiple nodes. All managing and controlling messages are transmitted through user datagram protocol (UDP) frames in the software platform. For managing and controlling messages and service data, the multiple routing algorithms are initially evaluated in VxWorks Simulators, which are generated based on the shortest path and constraint path, respectively. Consider the calculation of a routing path from node vx_simulator_01 to node vx_simulator_03. In Figure 8, each node (i.e., vx_simulator_01 to vx_simulator_16, with spacecraft identification (SCID) from SCID 1 to SCID 16) has nine ports (i.e., 1 to 9) bi-directionally connected to neighbor node ports, while the priorities of the ports are increased from 1 to 9 and the costs of the ports are the same value equaling to 1, i.e., the priority and cost for the port 1 are 1 and 1, respectively, the priority and cost for the port 9 are 9 and 1, respectively, etc. According to the routing strategy (i.e., the Dijkstra algorithm) for managing and controlling messages, the computation of the routing path is based on the priority and cost of the port, namely: the path with high priority will be selected under the circumstances of equivalent costs. Link-state databases of all nodes are shown in Figure 9. So, the routing path for managing and controlling messages from node vx_simulator_01 (SCID 1) to node vx_simulator_03 (SCID 3) is shown in Table 1, Figure 10a, and Figure 10b, respectively, under the bi-directional link between vx_simulator_01 (SCID 1) and node vx_simulator_04 (SCID 4). In Figure 10a, the function *manage_control_message_routing_path_show_node_to_node*(1, 3) shows the routing path from the source node (i.e., the first parameter 1) to the destination node (i.e., the second parameter 2). On the other hand, given the constraint of QoS requirements of a link duration time, forward and/or backward bandwidth, etc. the explicit bi-direction routing path of data from node vx_simulator_01 to node vx_simulator_03 is computed based on the CSPF-TE algorithm, which is shown in Table 2. In Figure 10a, the function *cspf_bi_direction_calculation*(1, 2, 500, 200, 300, 1000, 1000) is used for the computation of the bi-direction routing path from the source node (i.e., the first parameter 1) to the destination node (i.e., the second parameter 2) under multiple QoS requirements, e.g., link duration time (i.e., the third parameter 500 s), forward bandwidth (i.e., the fourth parameter 200 megabits per second (Mbps)), backward bandwidth (i.e., the fifth parameter 300 Mbps), forward maximum latency (i.e., the sixth parameter 1000 milliseconds (ms)), and backward maximum latency (i.e., the seventh parameter 1000 ms). The function *print_cspf_bi_direction_route* is for the display of bi-direction explicit routing path.

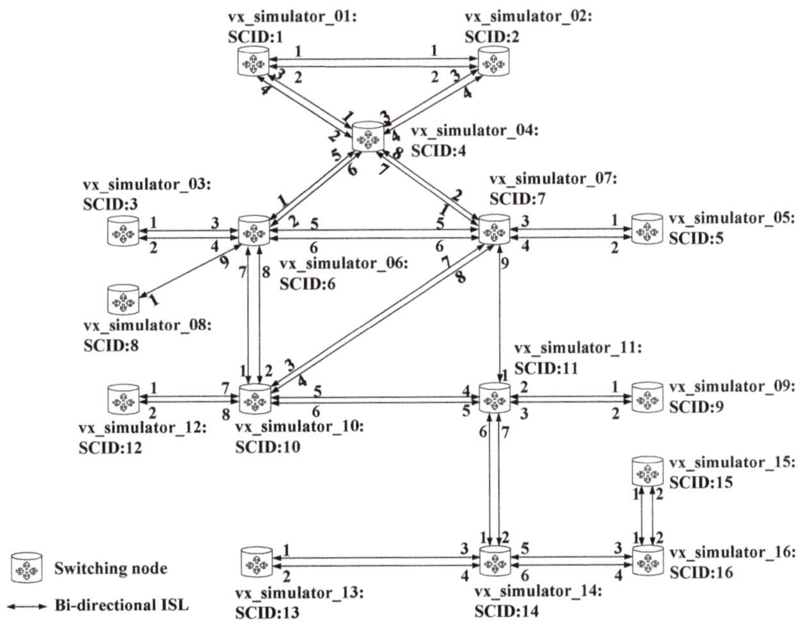

Figure 8. The topology of VxWorks simulated network.

Table 1. The routing path of managing and controlling messages between node vx_simulator_01 (SCID 1) and node vx_simulator_03 (SCID 3).

	vx_simulator_01	vx_simulator_04	vx_simulator_06	vx_simulator_03
Forward path	Out-port: 4	In-port: 2, Out-port: 6	In-port: 2, Out-port: 4	In-port: 2,
Backward path	In-port: 4	Out-port: 2, In-port: 6	Out-port: 2, In-port: 4	Out-port: 2

Figure 9. Link-state database of optimized OSPF in node vx_simulator_01 (SCID 1).

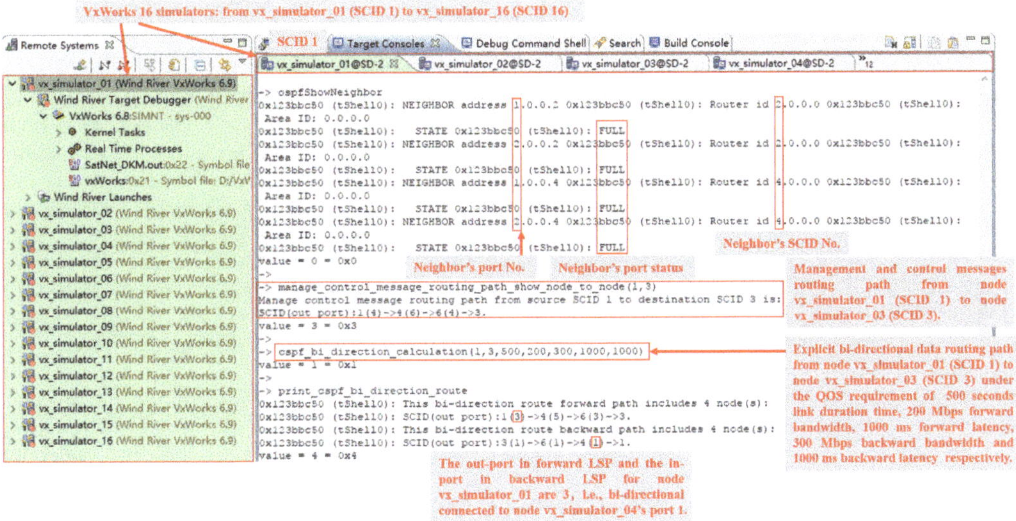

(a)

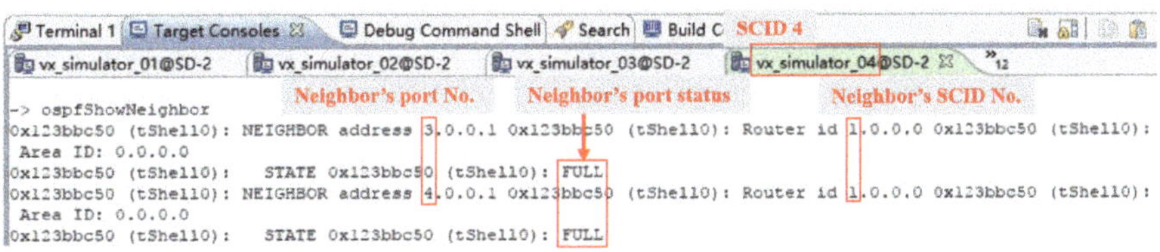

(b)

Figure 10. Simulation in software platform with 16 simulated nodes under the circumstances of bi-directional link. (**a**) The status of a neighbor's ports, routing path for managing and controlling messages, and explicit constraint routing path for data under QoS requirements in node vx_simulator_01 (SCID 1). (**b**) The status of a neighbor's ports in node vx_simulator_04 (SCID 4).

Table 2. A bi-directional label switching path under the circumstances of bi-directional link between node vx_simulator_01 (SCID 1) and node vx_simulator_03 (SCID 3) in the software platform.

	vx_simulator_01	vx_simulator_04	vx_simulator_06	vx_simulator_03
Forward LSP	Out-port: **3** [1]	In-port: **1**, Out-port: 5	In-port: 1, Out-port: 3	In-port: 1
Backward LSP	In-port: **3** [1]	Out-port: **1**, In-port: 5	Out-port: 1, In-port: 3	Out-port: 1

[1] The out/in-ports of forward/backward label switching path for node vx_simulator_01 are the same, i.e., port 3.

We make full use of the uni-directional connection to transmit data while the opposite uni-directional link is down to improve the use of resource-limited satellite networks. The original OSPF protocol, on the other hand, does not allow for the calculation of a uni-directional routing path. We enhance the OSPF protocol in the following way to compute uni-directional routing paths:

1. Add UNI_DIRECTION_SEND and UNI_DIRECTION_RECEIVE to OSPF neighbor's states in addition to DOWN and FULL;
2. The uni-directional receiving node sends hello messages to the uni-directional sending node through other routing paths instead of the down link.

Figure 11 is used as an example to illustrate the working principle of the OSPF uni-directional link as revised in Algorithm 1. If node #2 does not find its own node ID in the

OSPF HELLO packet sent from the peer node #1, then node #1 is considered to be in uni-directional sending status (i.e., UNI_DIRECTION_SEND). Because the former direct link from node #2 to node #1 is down, node #1 cannot receive the uni-directional receiving hello message from node #2 (i.e., UNI_DIRECTION_RECEIVE). In this case, node #2 sends this hello message to node #1 through other routing paths, i.e., from node #2 to node #3 and finally to node #1. When node #1 receives this message from node #3, the neighbor status of node #2 in node #1 is checked as uni-directional receiving, and thus the uni-directional link is established between node #1 and node #2. Note that if there is no other path between node #2 and node #1, the OSPF uni-directional link cannot be established. Figure 12a,b demonstrate the uni-directional link between node vx_simulator_01 (SCID 1) and node vx_simulator_04 (SCID 4) when the uni-directional link from vx_simulator_04's port 1 to vx_simulator_01's port 3 and is down accidentally, leading to the status of vx_simulator_04's port 1 changing from **FULL** to **UNI_DIRECTION_RECEIVE**, and the status of vx_simulator_01's port 3 changing from **FULL** to **UNI_DIRECTION_SEND**.

In addition, we improve the OSPF link-state database with the complement of the satellite link plan of connections (SLPC). SLPC is intended to hold scheduled/planned link information to update satellite network link state periodically, and thus control OSPF flood messages and reduce satellite link bandwidth usage. In [19], we propose a solution called the optimized OSPF with link plan (OOWLP) to overcome the shortcomings in the original rule of OSPF, e.g., the frequent route convergence induced by frequent satellite link validation, and long periods of satellite network instability. SLPC is meant to retain the link connection plan information of the satellite network for a period of time. The terrestrial network management system will upload the new SLPC regularly to the GEO satellites which are directly accessed by the network management system. Then, the GEO satellites will flood the SLPC information to all other satellites for synchronization, e.g., GEO, MEO, LEO satellites. Moreover, the judgment of the flooding procedure of the original OSPF is revised in Algorithm 1 lines 18–24. In the original OSPF, if the on–off action occurs in one link, the two nodes of the link would flood the updated message to all other nodes in the network. The process of route convergence is costly in terms of time and energy, and consumes the limited link resources. In general, the satellite link variation is induced by two scenarios:

1. Most of the time, the link variation is induced by the predicted satellite orbital movement, and there us no need to flood those variations;
2. Rarely, the link variation is unexpectedly caused by nodes out of work, and the variations must be flooded between nodes to update the network topology.

As mentioned before, we periodically update the OSPF LSDB via the SLPC. Thus, the predicted OSPF flooding is restrained, while the unexpected link variation is also under link monitoring, as depicted in Figure 13.

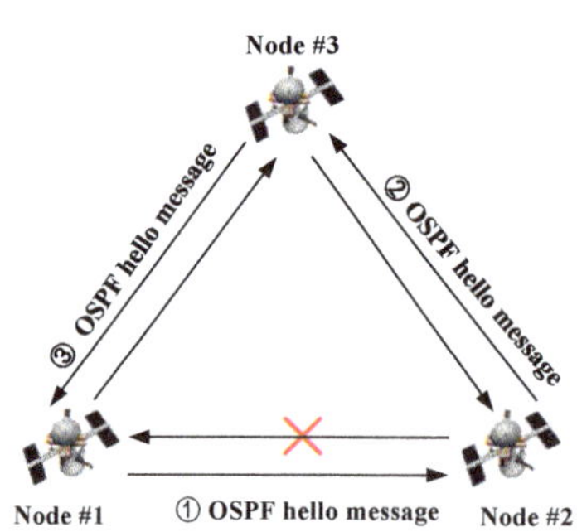

Figure 11. The principle of uni-directional OSPF protocol.

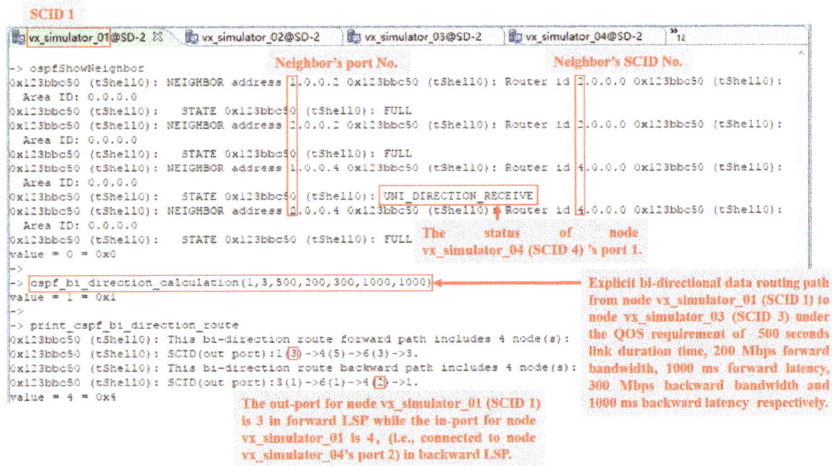

(a)

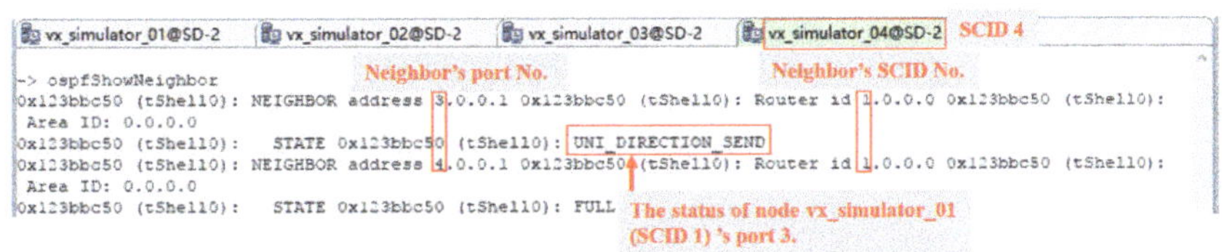

(b)

Figure 12. Simulation in the software platform with 16 simulated nodes under the circumstances of uni-directional link. (**a**) The status of a neighbor's ports and explicit constraint routing path for data under QoS requirements in node vx_simulator_01 (SCID 1). (**b**) The status of a neighbor's ports in node vx_simulator_04 (SCID 4).

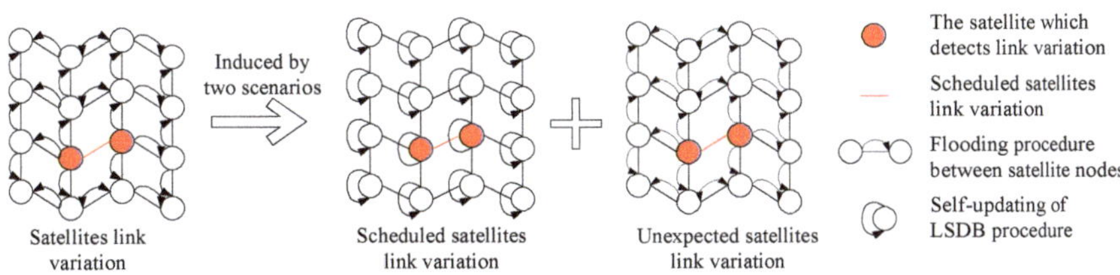

Figure 13. Flooding restraint and flooding procedure for the two scenarios of satellites link variation.

Given the uni-directional link between nodes, the computation of the explicit routing path for data service is clearly different from that of a bi-directional link. Figure 12a,b and Table 3 show the bi-directional routing path for data between node vx_simulator_01 (SCID 1) and node vx_simulator_03 (SCID 3) under the circumstances of the uni-directional link down from node vx_simulator_04 (SCID 4)'s port 1 to node vx_simulator_01's port 3.

Table 3. A Bi-directional label switching path between node vx_simulator_01 (SCID 1) and node vx_simulator_03 (SCID 3) under the circumstances of a uni-directional link in the software platform.

	vx_simulator_01	vx_simulator_04	vx_simulator_06	vx_simulator_03
Forward LSP	Out-port: **3** [1]	In-port: **1**, Out-port: 5	In-port: 1, Out-port: 3	In-port: 1
Backward LSP	In-port: **4** [1]	Out-port: **2**, In-port: 5	Out-port: 1, In-port: 3	Out-port: 1

[1] The out/in-ports of the forward/backward label switching path for node vx_simulator_01 are different, i.e., out-port 3 and in-port 4, respectively.

4. Semi-Physical Platform Setup and Evaluation for SDN-Based ISTNs

In this section, we will establish a semi-physical platform to evaluate the feasibility of label-based AOS frame switching in SDN-based ISTNs.

4.1. Introduction of the Semi-Physical Platform

In Section 3.2, the control plane network protocol and algorithms are successfully tested in the software platform. However, owing to the performance limitations of the VxWorks simulated network card, WindRiver WRTAP has only ten-megabit throughput, and cannot enable label-based AOS frame switching. Real-world end-to-end gigabit-level data transmission cannot be accomplished in the software platform. Therefore, FPGA boards are designed for testing high-level throughput performance based on label switching of AOS frames. As shown in Figure 14a,b the semi-physical platform is made up of a network management system, VxWorks Simulators, and FPGA boards. In the semi-physical platform, just three nodes are evaluated for simplicity. The terrestrial network gateway, i.e., vx_simulator_01, is controlled directly by the network management system, and FPGA boards are connected by low voltage differential signaling (LVDS) cables, which are simulated as satellite–terrestrial connections and inter-satellite links. It should be noted that the network management system may communicate with remote nodes, such as nodes vx_simulator_02 and vx_simulator_03, through the **data communication network** overhead in the AOS frame. The simulated nodes in the control plane communicate with the FPGA boards in the data plane through the southbound interface, which might be a serial port, Ethernet port, or 1553B in this semi-physical platform. The northbound interface communication protocol in the semi-physical platform is the simple network management protocol (SNMP), as the SNMP manager operating in the network management system and the SNMP agent running in the VxWorks simulated node (i.e., the controller). In addition, as illustrated in Figure 15, we create an exclusive management information base (MIB) for parameter setup, retrieval, and status informing/trapping.

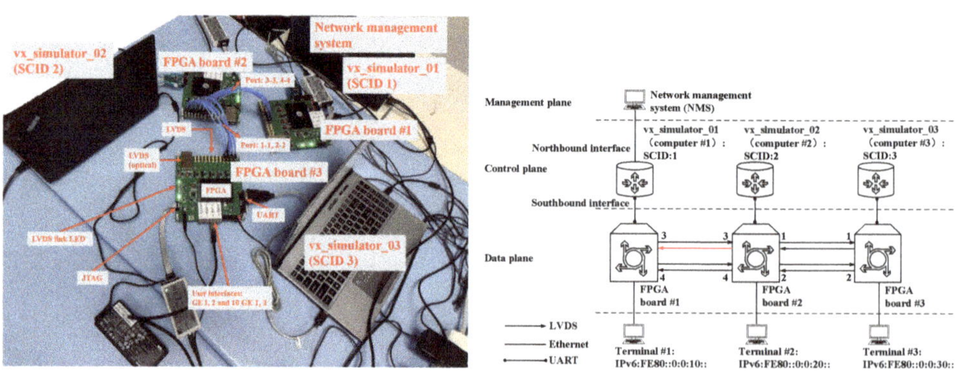

(a) (b)

Figure 14. The semi-physical platform. (**a**) The semi-physical platform with VxWorks simulated nodes and FPGA boards. (**b**) The network topology of the semi-physical platform.

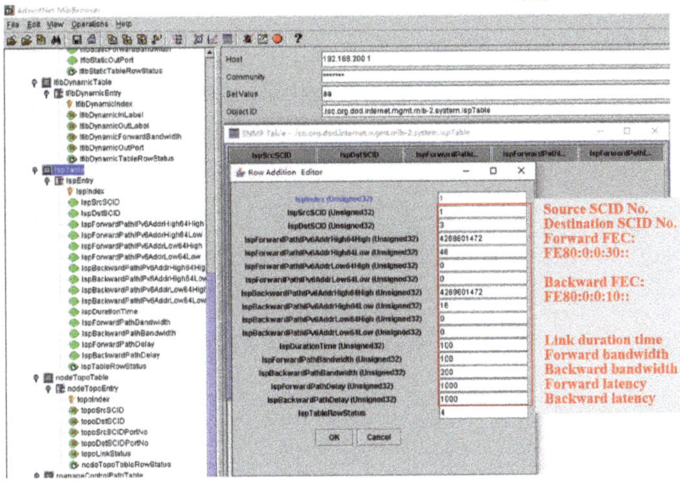

Figure 15. The SNMP MIB of the semi-physical platform.

4.2. Test of Data Communication Network in the Semi-Physical Platform

In Section 3.2, managing and controlling messages (e.g., control plane protocols and network management system messages) are transmitted in UDP frames for the simplicity of validation. As mentioned earlier, managing and controlling messages can be communicated between nodes in ISTNs using the **data communication network** field, the **extended control field**, in the AOS frame, as shown in Figure 5. In Figure 16a,b, there is a data communication network test for the configuration of static forwarding information base in node vx_simulator_02 from the network management system, which uses node vx_simulator_01 as the gateway. After receiving a managing and controlling message from the management plane, e.g., the configuration of static forwarding information base, through the northbound interface, node vx_simulator_01 will verify the destination SCID No. first. That is, if destination SCID No. is not 1, this managing and controlling message will be divided into segments and inserted into the **data communication network** field of ten continuous AOS frames before being transmitted to the destination node along the routing path for managing and controlling messages based on the shortest path algorithm (i.e., Dijkstra algorithm), as shown in Table 4. Figure 17 depicts the ten continuous AOS frames captured by ILA in FPGA board #1's port 4. FPGA board #2, on the other hand, reunites the segments of managing and controlling message by extracting the **data communication network** field from those ten continuous AOS frames. If the destination SCID No. is 2, node vx_simulator_02 then configures a static forwarding information base. Otherwise, node vx_simulator_02 forwards the managing and controlling message to the next node.

Table 4. The routing path for managing and controlling messages from node vx_simulator_01 to node vx_simulator_02 and node vx_simulator_03, respectively, in the semi-physical platform.

	vx_simulator_01	**vx_simulator_02**	**vx_simulator_03**
Routing path	Out-port: 4	In-port: 4, Out-port: 1	In-port: 1

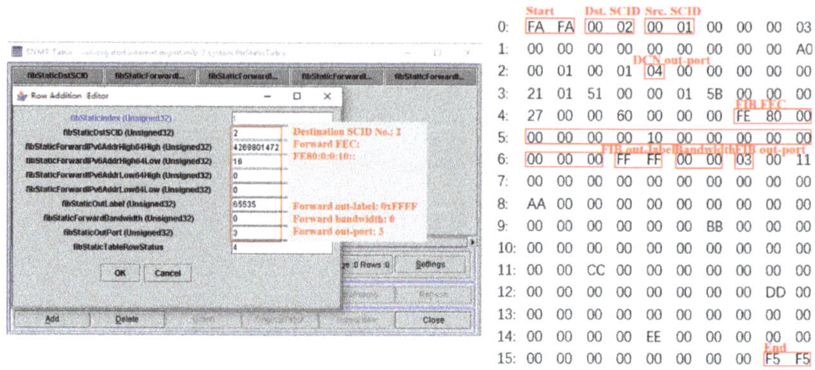

(a) (b)

Figure 16. The configuration from the network management system to the remote node uses the data communication network overhead in the AOS frame. (**a**) The configuration of static forwarding information base from the network management system (SNMP manager). (**b**) The managing and controlling message of static FIB transmitted by node vx_simulator_01 through southbound interface.

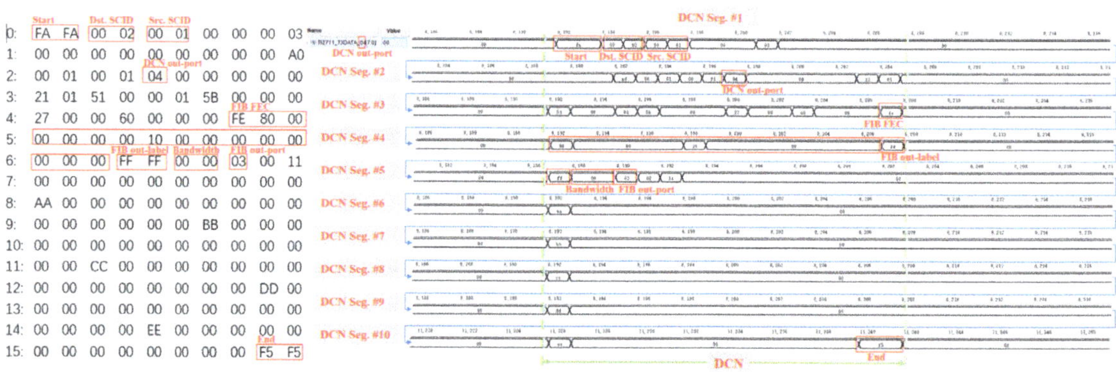

Figure 17. A data communication network sequence diagram for a managing and controlling message, i.e., static FIB, from node vx_simulator_01 (SCID 1) to node vx_simulator_02 (SCID 2).

The satellite constellation constitutes a constantly altering networking architecture owing to the link connection variation caused by satellites' orbital movements. Types of management and controlling flooding messages, i.e., satellite link plan of connections (SLPC) described in Section 3.2, are proposed for a better adaptation to the shifting satellite constellation. These messages make use of the predictability and periodicity of satellite network topology, which repeats itself over time, and also include periodic routing phases with changes in the route computation process for both managing and controlling messages and data. These kinds of managing and controlling messages are flooded to each node in the network for the update of previous OSPF LSDB with new ones according to the current network topology, and the destination SCID No. is **0xFFFF**, as shown in Figure 18.

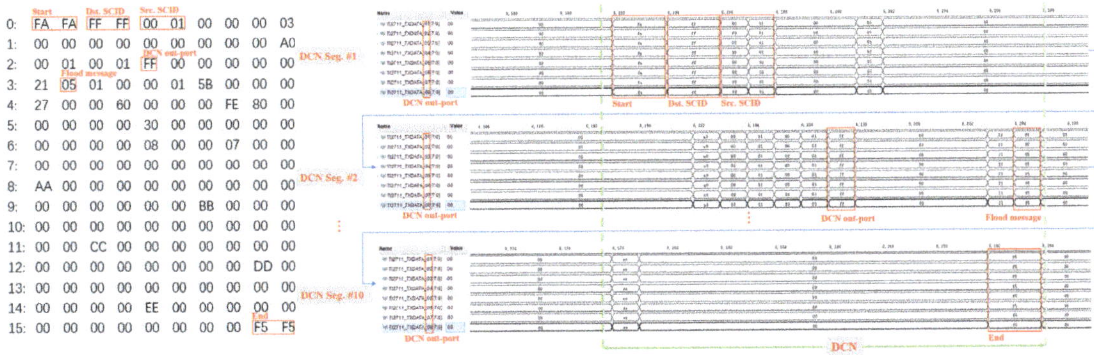

Figure 18. A data communication network sequence diagram for a management and controlling flooding message, i.e., satellite link plan of connections , from node vx_simulator_01 (SCID 1) to other nodes.

4.3. Test of Label-Based AOS Frame Switching in the Semi-Physical Platform

First, a bi-directional label switching path is setup for the label-based AOS frame switching. That is, a bi-directional data path from node vx_simulator_01 to node vx_simulator_03 is built, along which data are sent and received simultaneously via port 3 in node vx_simulator_01 (SCID 1). Before the label switching path is set up, node vx_simulator_01 will compute the explicit constraint routing path under the terminal's QoS requirements, i.e., link duration time, forward and/or backward bandwidth and link latency. Figure 19a–c show the three routers' link-state advertisement of the link-state database in node vx_simulator_01, respectively, which are optimized with metrics of link latency, link duration time, and link bandwidth. In Figure 20a, the function *cspf_bi_direction_calculation* computes the explicit bi-directional data routing path under multiple QoS requirements, i.e., source node SCID 1 (i.e., the first parameter) to destination node SCID 3 (i.e., the second parameter), link duration time 500 s (i.e., the third parameter), forward bandwidth 200 Mpbs (i.e., the fourth parameter), backward bandwidth (i.e., the fifth parameter 300 Mbps), forward maximum latency (i.e., the sixth parameter 1000 ms), and backward maximum latency (i.e., the seventh parameter 1000 ms), as summarized in Table 5. In addition, Figure 20b,c show the explicit bi-directional data routing path under multiple QoS requirements, i.e., source node SCID 3 (i.e., the first parameter) to destination node SCID 1 (i.e., the second parameter) and the neighbors' ports status, respectively. Because of the **FULL** status of the neighbor's port, the out-port of forward label switching path and the in-port of the backward label switching path (i.e., in-port and out-port) in node vx_simulator_01 is port 3 connected by FPGA board #1 and FPGA board #2.

Figure 19. The link-state database of the optimized OSPF in node vx_simulator_01 (SCID 1). (**a**) The router link states of node vx_simulator_01 (SCID 1). (**b**) The router link states of node vx_simulator_02 (SCID 2). (**c**) The router link states of node vx_simulator_03 (SCID 3).

Figure 20. The explicit constraint data routing computation under QoS requirements and the circumstances of a bi-directional link between node vx_simulator_01 (ingress LER, SCID 1), node vx_simulator_02 (LSR, SCID 2), and node vx_simulator_03 (egress LER, SCID 3). (**a**) The status of a neighbor's port and explicit bi-directional data routing path computation in node vx_simulator_01 (SCID 1). (**b**) The explicit uni-directional data routing path computation between node vx_simulator_01 (SCID 1) and node vx_simulator_03 (SCID 3). (**c**) The status of a neighbor's port in node vx_simulator_02 (SCID 2).

The second step is label distribution for each node along the label switching path. SNMP is applied as the northbound interface protocol for the communication between the network management system as the manager in the management plane and the simulated nodes as the agent (i.e., vx_simulator_01) in the control plane. Figure 15 shows the detail configurations to set up the bi-directional label switching path from node vx_simulator_01 (SCID 1) to vx_simulator_03 (SCID 3), i.e., lspSrcSCID equals 1 for the source node SCID, lspDstSCID equals 3 for the destination node SCID, lspForwardPathIPv6AddrHigh64High, lspForwardPathIPv6AddrHigh64Low, lspForwardPathIPv6AddrLow64High, and lspForwardPathIPv6AddrLow64Low equal FE80::0:0:30:: (4269801472 and 48 in decimal format) as forwarding equivalence class of forward label switching path, lspBackwardPathIPv6-AddrHigh64High, lspBackwardPathIPv6AddrHigh64Low, lspBackwardPathIPv6Addr-Low64High, and lspBackwardPathIPv6AddrLow64Low equal FE80::0:0:10:: (4269801472 and 16 in decimal format) as forwarding equivalence class of backward label switching path, lspAliveTime equals 100 s for the link duration time, lspForwardPathBandwidth equals 100 Mbps for the forward bandwidth, lspBackwardPathBandwidth equals 200 Mbps for the backward bandwidth, lspForwardPathDelay equals 1000 ms for the forward latency, and lspBackwardPathDelay equals 1000 ms for the backward latency. The explicit constraint label distribution process is shown in Figure 21 for node vx_simulator_01 (Figure 21a), node vx_simulator_02 (Figure 21b), and node vx_simulator_03 (Figure 21c), respectively, and Table 5 summarizes the forward and backward in/out-labels and in/out-ports of the label switching path. To demonstrate the label-based AOS frame switching, the IP capture tool Wireshark and integrated logic analyzer (ILA) of XILINX VIVADO (version 18.3) are used to probe the IP data, in/out-labels, and in/out-ports of AOS frames, as shown in Figure 22. The IPv6 data transmitted from node vx_simulator_01 to node vx_simulator_03 are captured with Wireshark, as shown in Figure 22a. Figure 22b shows the encapsulation of IPv6 data in the **transfer frame data field** of the AOS frame, which is pushed with the out-label (i.e., 0x00D2) and the forwarding equivalence class (i.e., the destination IPv6 address FE80::0:0:30::) and sent out of node vx_simulator_01's port 3 (ingress LER) according to the forwarding information base. Once the AOS frame with in-label (i.e., 0x00D2) is received, node vx_simulator_02 will immediately swap in-label (i.e., 0x00D2) to out-label (i.e., 0xFFFF, because node vx_simulator_02 is the last but one in the forward label switching path) and send the swapped AOS frame to port 1 according to the label forwarding information base, as shown in Figure 22c. In the end, when node vx_simulator_03 (LER) receives an AOS frame with in-label (i.e., 0xFFFF), it will pop the in-label, de-capsulate the IPv6 data from the **transfer frame data field**, and send the IPv6 data to the corresponding terminal, as shown in Figure 22d, and vice versa for the backward label switching path as shown in Figure 23a–d. Table 5 summarizes the in/out-labels and in/out-ports for the bi-directional label switching path AOS frame switching process.

Table 5. The bi-directional label switching path under QoS requirements and the circumstances of a bi-directional link between node vx_simulator_01, node vx_simulator_02, and node vx_simulator_03 in the semi-physical platform.

	vx_simulator_01	vx_simulator_02	vx_simulator_03
Forward LSP	Out-port: **3** [1] Out-label: 0x00D2	In-port: **3**, Out-port: 1 In-label: 0x00D2, Out-label: 0xFFFF	In-port: 1 In-label: 0xFFFF
Backward LSP	In-port: **3** [1] In-label: 0xFFFF	Out-port: **3**, In-port: 1 Out-label: 0xFFFF, In-label: 0x0054	Out-port: 1 Out-label: 0x0054

[1] The out/in-ports of forward/backward label switching path for node vx_simulator_01 are the same, i.e., port 3.

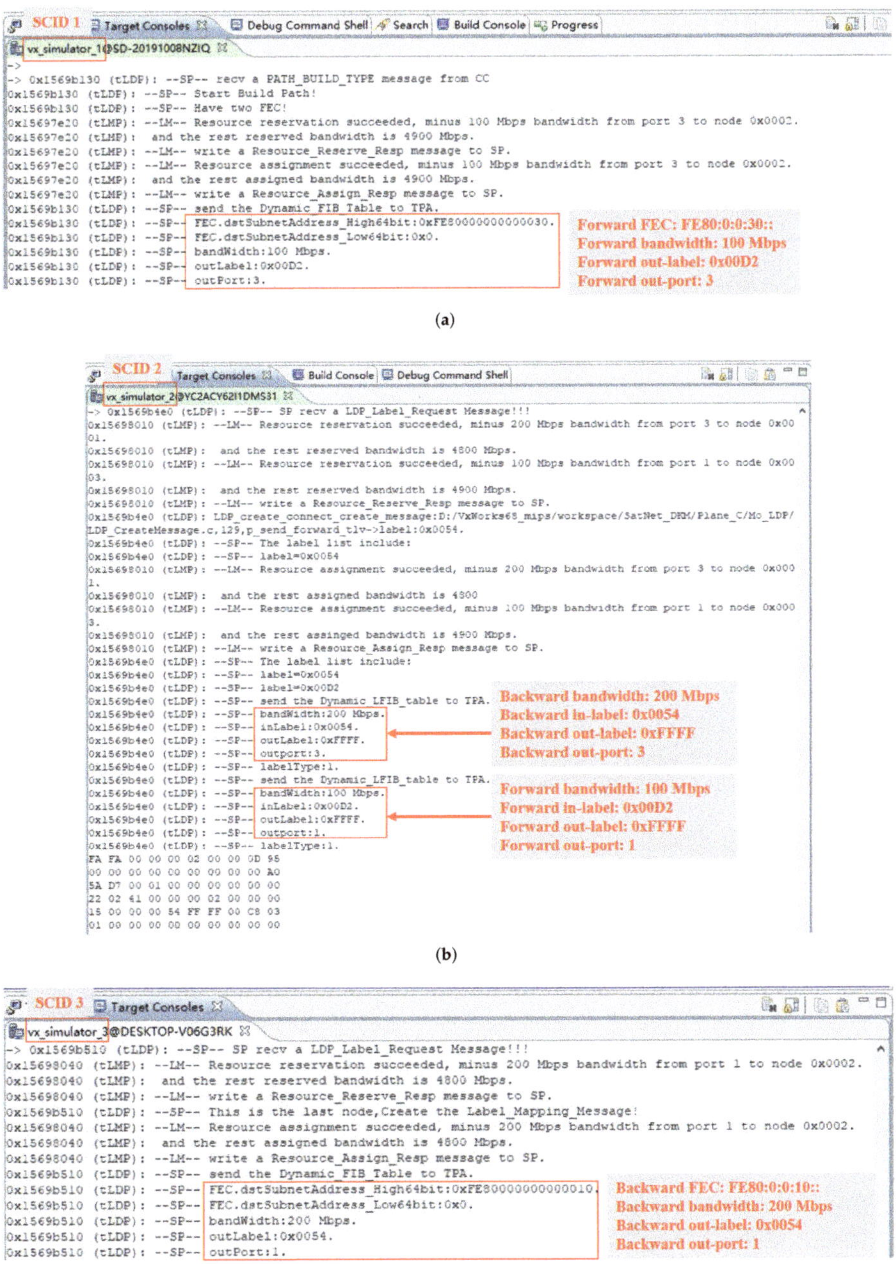

Figure 21. The label distribution process of label switching path setup under QoS requirements and the circumstances of a bi-directional link between node vx_simulator_01 (ingress LER, SCID 1), node vx_simulator_02 (LSR, SCID 2), and node vx_simulator_03 (egress LER, SCID 3) in the semi-physical platform. (**a**) The label distribution process in node vx_simulator_01 (SCID 1). (**b**) The label distribution process in node vx_simulator_02 (SCID 2). (**c**) The label distribution process in node vx_simulator_03 (SCID 3).

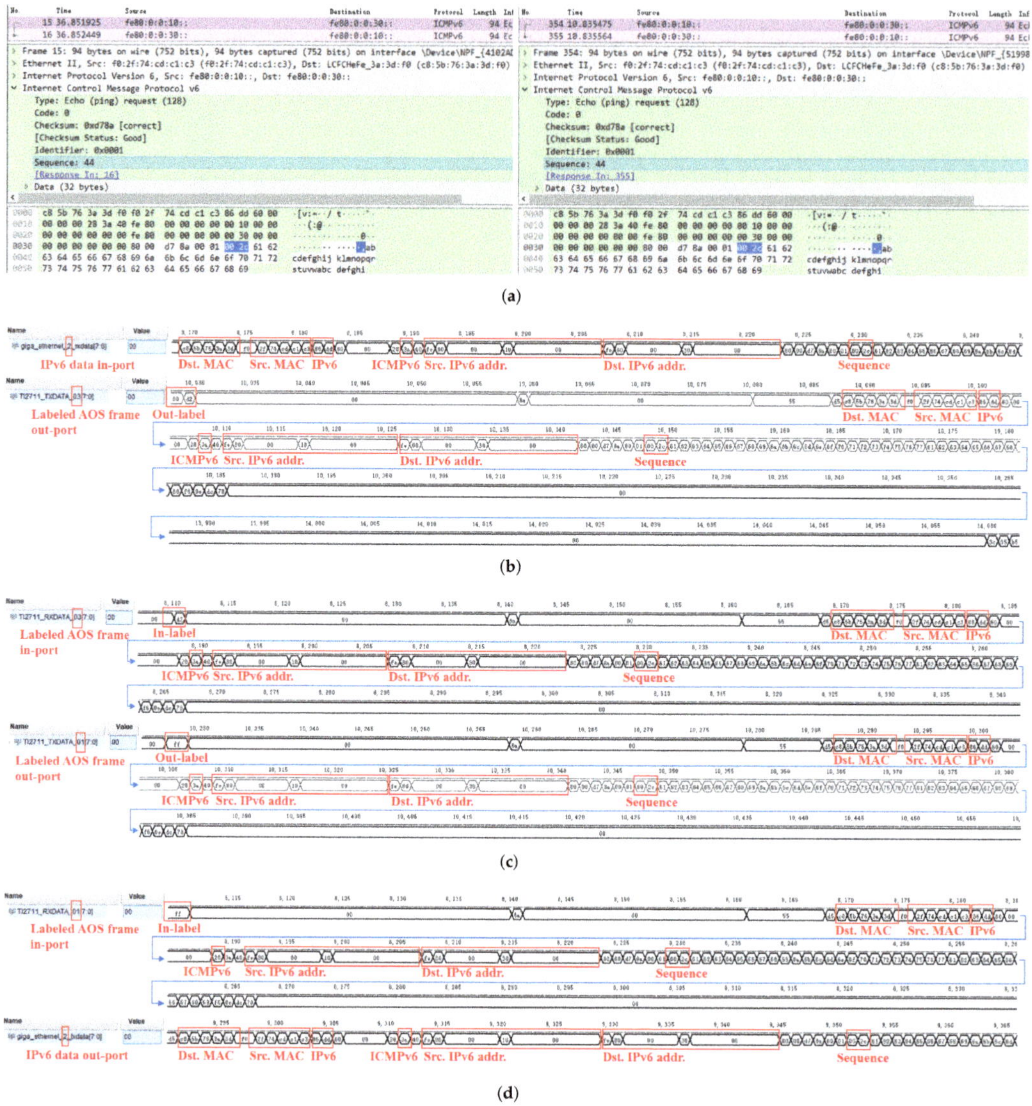

Figure 22. The label-based AOS frame switching in forward label switching path under the circumstances of bi-directional link between node vx_simulator_01 (ingress LER, SCID 1), node vx_simulator_02 (LSR, SCID 2) and node vx_simulator_03 (egress LER, SCID 3) in the semi-physical platform. (**a**) The IPv6 data transmitted from node vx_simulator_01 (SCID 1) to node vx_simulator_03 (SCID 3). (**b**) The forward-FIB-based label pushing of AOS frame in node vx_simulator_01 (SCID 1). (**c**) The forward-LFIB-based label swapping of AOS frame in node vx_simulator_02 (SCID 2). (**d**) The forward-FIB-based label popping of AOS frame in node vx_simulator_03 (SCID 3).

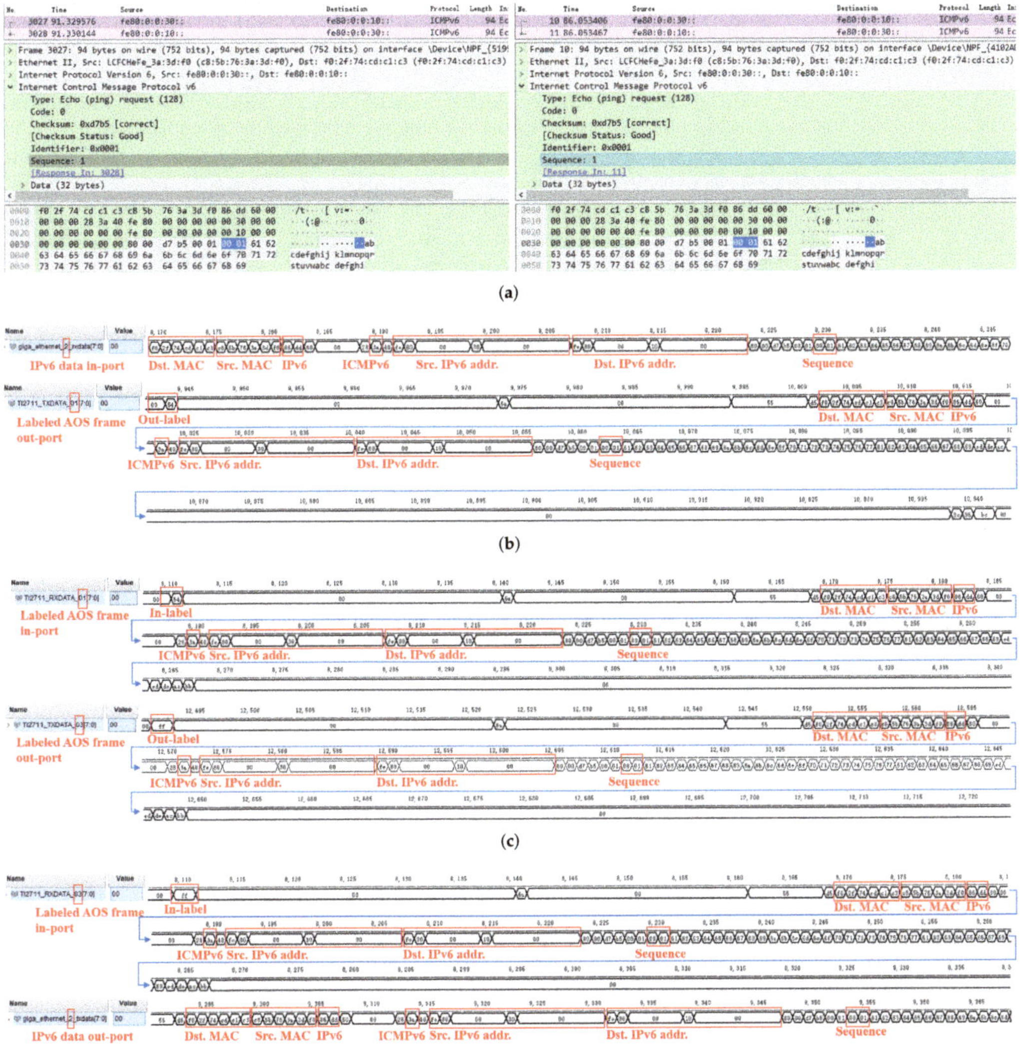

Figure 23. The label-based AOS frame switching in backward label switching path under the circumstances of bi-directional link between node vx_simulator_01 (ingress LER, SCID 1), node vx_simulator_02 (LSR, SCID 2), and node vx_simulator_03 (egress LER, SCID 3) in the semi-physical platform. (**a**) The IPv6 data transmitted from node vx_simulator_03 (SCID 3) to node vx_simulator_01 (SCID 1). (**b**) The backward-FIB-based label pushing of AOS frame in node vx_simulator_03 (SCID 3). (**c**) The backward-LFIB-based label swapping of AOS frame in node vx_simulator_02 (SCID 2). (**d**) The backward-FIB-based label popping of AOS frame in node vx_simulator_01 (SCID 1).

The third step is to test bi-directional label switching path setup under the circumstances of uni-directional link between simulated nodes in the semi-physical platform as tested in the software platform in Section 3.2. Given the link-down occurred from node vx_simulator_02's port 3 to node vx_simulator_01's port 3, as shown in Figure 14a, i.e., the red link down, the status of node vx_simulator_01's port 3 changes from **FULL** to **UNI_DIRECTION_SEND** and the status of node vx_simulator_02's port 3 changes from **FULL** to **UNI_DIRECTION_RECEIVE**, as seen in Figure 24a and Figure 24c , respectively. As a result, the explicit constraint bi-directional data path between node vx_simulator_01 and node vx_simulator_03 is changed, as shown in Figure 24b and summarized in Table 6. Figure 25 shows the process of label switching path setup in node vx_simulator_01 (Figure 25a), vx_simulator_02 (Figure 25b), and vx_simulator_03 (Figure 25c), respectively. Figures 26 and 27 demonstrate the process of IPv6 data (Figure 26a, Figure 27a) encapsulation (Figure 26b, Figure 27b) and de-encapsulation (Figure 26c, Figure 27c) to/from the AOS frame and the process of AOS frame switching (Figure 26d, Figure 27d) in the forward and backward label switching path, respectively, as summarized in Table 6.

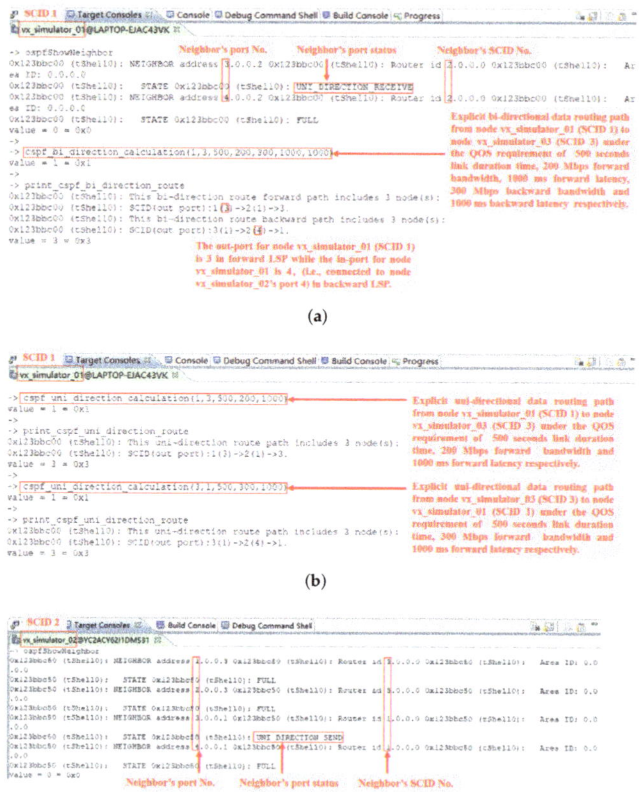

Figure 24. The explicit constraint data routing computation under QoS requirements and the circumstances of uni-directional link between node vx_simulator_01 (ingress LER, SCID 1, uni-directional sending), node vx_simulator_02 (LSR, SCID 2, uni-directional receiving), and node vx_simulator_03 (egress LER, SCID 3). (**a**) The status of a neighbor's port and data routing computation in node vx_simulator_01. (**b**)The explicit uni-directional data routing path computation between node vx_simulator_01 and node vx_simulator_03. (**c**) The status of a neighbor's port in node vx_simulator_02.

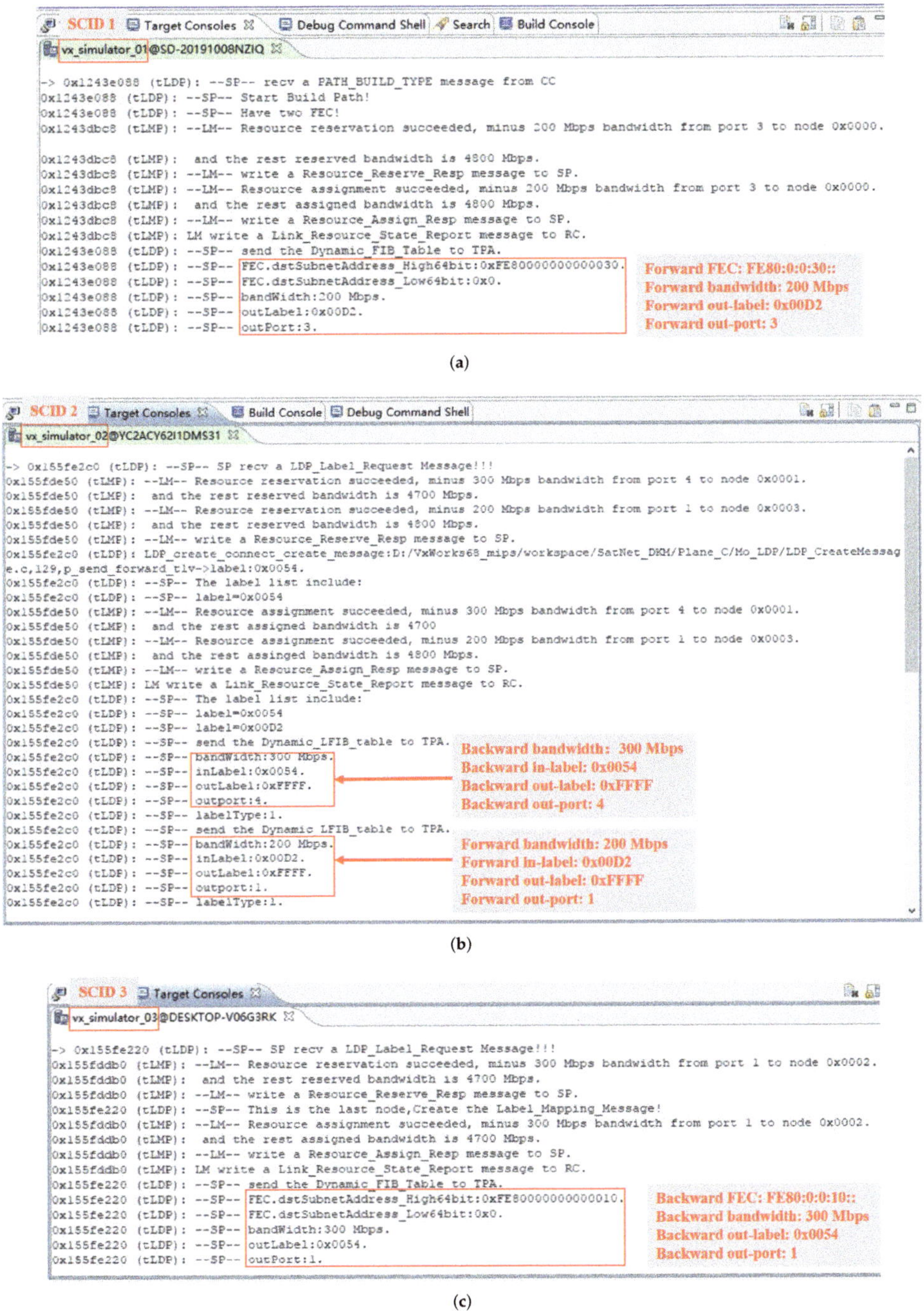

Figure 25. The label distribution process of label switching path setup under QoS requirements and the circumstances of uni-directional link between node vx_simulator_01 (ingress LER, SCID 1), node vx_simulator_02 (LSR, SCID 2), and node vx_simulator_03 (egress LER, SCID 3). (a) The label distribution process in node vx_simulator_01 (SCID 1). (b) The label distribution process in node vx_simulator_02 (SCID 2). (c) The label distribution process in node vx_simulator_03 (SCID 3).

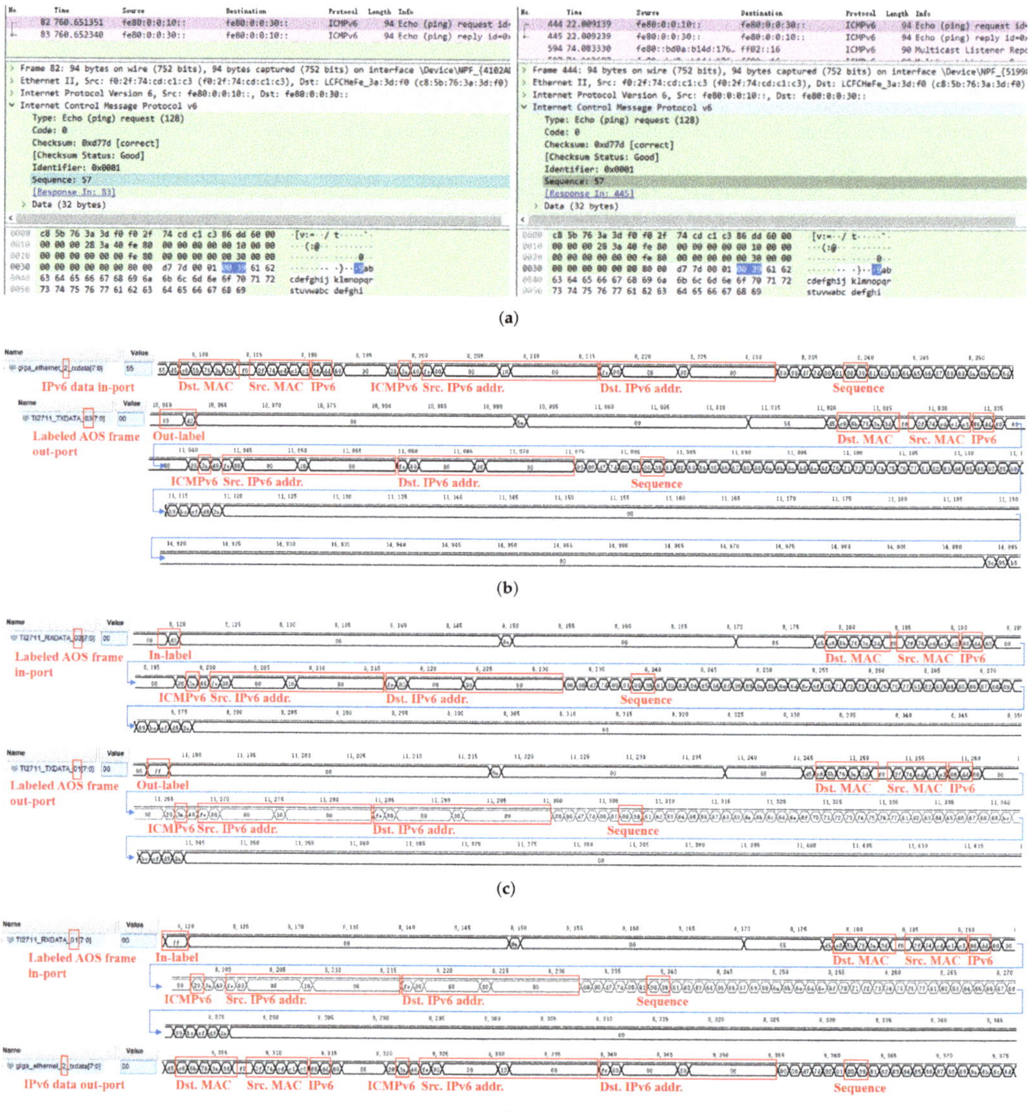

Figure 26. The label-based AOS frame switching in forward label switching path under the circumstances of uni-directional link between node vx_simulator_01 (uni-directional sending) and node vx_simulator_02 (uni-directional receiving) in the semi-physical platform. (**a**) The IPv6 data transmitted from node vx_simulator_01 (ingress LER, SCID 1) to node vx_simulator_03 (egress LER, SCID 3). (**b**) The forward-FIB-based label pushing of AOS frame in node vx_simulator_01 (SCID 1). (**c**) The forward-LFIB-based label swapping of AOS frame in node vx_simulator_02 (SCID 2). (**d**) The forward-FIB-based label popping of AOS frame in node vx_simulator_03 (SCID 3).

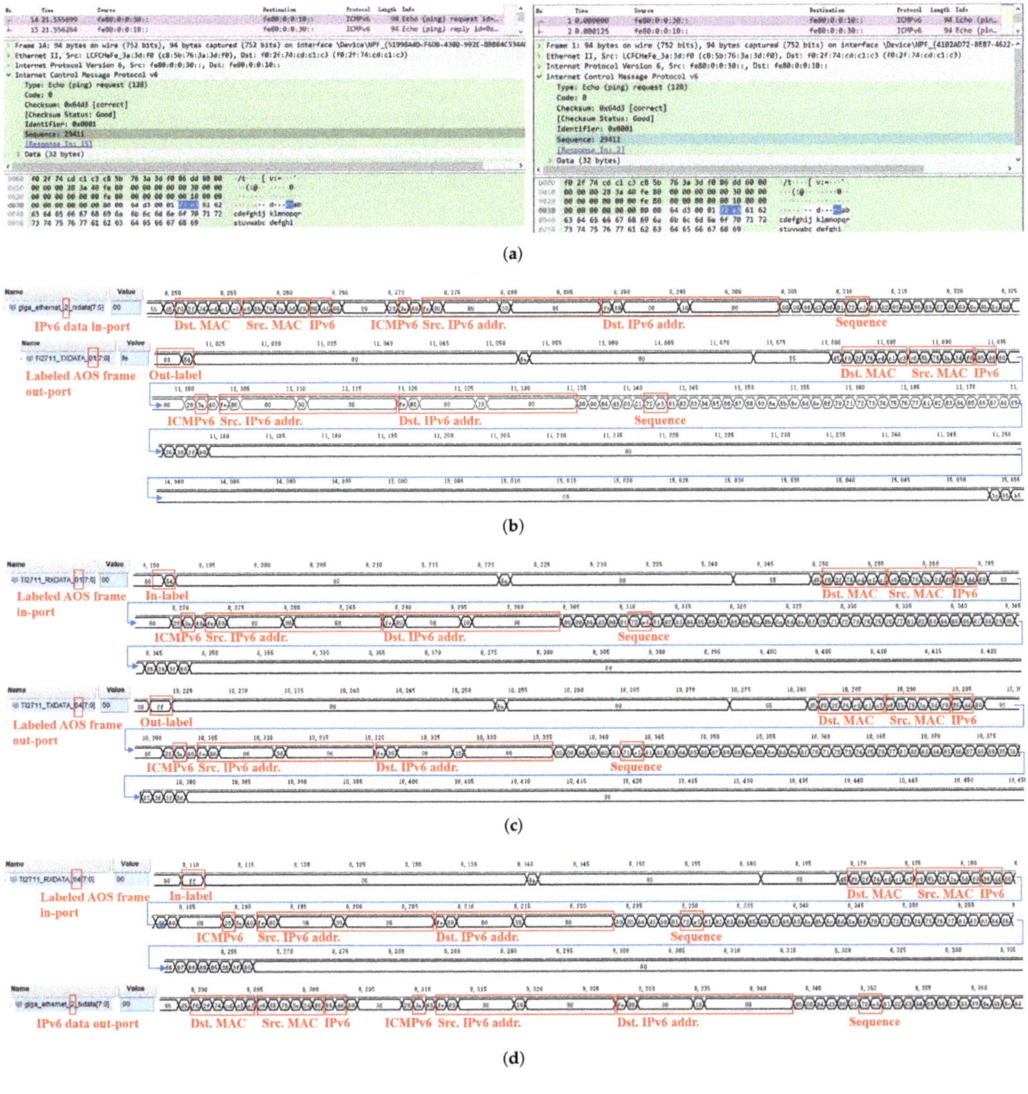

Figure 27. The label-based AOS frame switching in backward label switching path under the circumstances of uni-directional link between node vx_simulator_01 (uni-directional sending) and node vx_simulator_02 (uni-directional receiving) in the semi-physical platform. (**a**) The IPv6 data transmitted from node vx_simulator_03 (egress LER, SCID 3) to node vx_simulator_01 (ingress LER, SCID 1). (**b**) The backward-FIB-based label pushing of AOS frame in node vx_simulator_03 (SCID 3). (**c**) The backward-LFIB-based label swapping of AOS frame in node vx_simulator_02 (LSR, SCID 2). (**d**) The backward-FIB-based label popping of AOS frame in node vx_simulator_01 (SCID 1).

Table 6. The bi-directional label switching path under QoS requirements and the circumstances of uni-directional link between node vx_simulator_01 and node vx_simulator_03 in the semi-physical platform.

	vx_simulator_01	vx_simulator_02	vx_simulator_03
Forward LSP	Out-port: **3** [1] Out-label: 0x00D2	In-port: **3**,Out-port: 1 In-label: 0x00D2, Out-label: 0xFFFF	In-port: 1 In-label: 0xFFFF
Backward LSP	In-port: **4** [1] In-label: 0xFFFF	Out-port: **4**, In-port: 1 Out-label: 0xFFFF, In-label: 0x0054	Out-port: 1 Out-label: 0x0054

[1] The out/in-ports of forward/backward label switching path for node vx_simulator_01 are different, i.e., out-port 3 and in-port 4, respectively.

5. Conclusions

In this paper, we present a novel SDN-based integrated satellite–terrestrial network architecture, in terms of a label-based AOS frame switching mechanism and optimized network protocols as follows:

1. The label-based AOS frame switching mechanism is proposed to replace the unnecessary on-satellite time–energy–cost process of data unpacking–repacking caused by IP forwarding. In addition, a data communication network overhead is used for the managing and controlling message communication between nodes;
2. Distributed routing and label switching path setup is designed to realize terrestrial–terrestrial, satellite–terrestrial, and intra-satellite data transmission using the optimized OSPF protocol (i.e., flood-restraint and routing path computation under uni-directional link) and the constraint CSPF protocol and LDP protocol (i.e., under the multiple QoS requirements of port cost, port priority, link duration time, forward/backward bandwidth, and link latency);
3. A semi-physical platform constitutes a software platform and hardware that is used to validate the proposed mechanisms and optimized protocols.

Still, there will be more work to do in the future:

1. Source–destination SCIDs are still needed for the ingress–egress label edge routers in the label switching path setup. We aim to use the neighbor discovery protocol (NDP) for IP/MAC address learning, and thus fetch terminal IP addresses for the network management system, which qualifies source–destination IP addresses for label switching path setup without the source–destination SCIDs;
2. The **OAM** field in the **extended control field** of the AOS frame will be upgraded for fast link connection detection, enabling 1+1, 1:1, and 1:n working path protection with fast rerouting;
3. The completely optical switching approach will be used for high-speed data flows (e.g., 10/40/100 Gbps) without the necessity of optical-electronic transformation;
4. Since the ISTNs are primarily composed of GEO satellites, MEO satellites, and terrestrial networks, LEO satellites are the primary access network utilized by terrestrial terminals. Thus, the orbital speed and direction of LEO constellations need to be considered. In addition, the locations of multiple ground stations are critical for ISTNs;
5. Recent innovations and research on satellite communication will be studied, such as semantic routing, semantic addressing, and variable-length addressing for space-based infrastructure.

Author Contributions: Conceptualization, W.Z., X.J., Q.L., S.H., B.G. and S.L.; methodology, W.Z. and B.G.; software, W.Z., X.T. and M.M.; hardware, T.F.; validation, W.Z. and X.S.; writing-original draft preparation, W.Z.; writing-review and editing, X.J. and X.S.; supervision, X.J.; project administration, W.Z. All authors have read and agreed to the published version of the manuscript.

Funding: This research is fully supported by GUET Excellent Graduate Thesis Program (Grant No. 19YJPYBS03), Innovation Project of Guangxi Graduate Education (Grant No. YCBZ2022109), and New Technology Research University Cooperation Project of the 34th Research Institute of China Electronics Technology Group Corporation, 2021 (Grant No. SF2126007).

Institutional Review Board Statement: Not applicable.

Informed Consent Statement: Not applicable.

Data Availability Statement: Not applicable

Acknowledgments: We acknowledge the support given by Yu Zhang, Yixiang Wang, Mingjiang Fu, Yiting Liu, Kaiqiang Wang, Xinbin Cui, Huilin Ren, Chengguang Pang, Huada Gong and Liang Meng during the research.

Conflicts of Interest: The authors declare that they have no known competing financial interest or personal relationships that could have appeared to influence the work reported in this paper. The funders have no role in the design of the work; in the collection, analysis, or interpretation of research data; in the writing of the manuscript, or in the decision to publish the results.

Abbreviations

The following abbreviations are used in this manuscript:

CSPF-TE	constraint shortest path first–traffic engineering
DiffServ	differentiated services
DP	data plane
DSCP	differentiated services code point
FEC	forwarding equivalence class
FIB	forwarding information base
FIFO	first in first out
FPGA	field programmable gate array
FSO	free-space optical
GEO	geostationary orbit
HILS	hardware-in-the-loop-simulation
IETF	Internet engineering task force
IGSO	inclined geosynchronous orbit
IntServ	integrated services
ISL	inter-satellite link
ISTNs	integrated satellite–terrestrial networks
LANs	local area networks
LDP	label distribution protocol
LEO	low earth orbit
LER	label edge router
LFIB	label forwarding information base
LMP	link manager protocol
LSDB	link-state database
LSP	label switching path
LSR	label switching router
LVDS	low voltage differential signaling
Mbps	Megabits per second
MEO	middle earth orbit
MIPS	microprocessor without interlocked piped stages
MP	management plane
MPLS	multi-protocol label switching
NMA	network management agent
NMS	network management system
NP-hard	non-deterministic polynomial-hard
OAM	operation administration and maintenance
OSPF	open shortest path first
PowerPC	Performance optimization with enhanced RISC–performance computing
QoS	quality of service
RSVP-TE	resource reservation protocol-traffic engineering
SCID	spacecraft identification
SDNs	software defined networks
SNMP	simple network management protocol

SPF	shortest path first
TE	traffic engineering
TCP	transmission control protocol
TLV	type length value
UART	universal asynchronous receiver/transmitter
UAVs	unmanned aerial vehicles
UDP	user datagram protocol

References

1. King, D.; Farrel, A.; Chen, Z. An Evolution of Optical Network Control: From Earth to Space. In Proceedings of the 2020 2nd International Conference on Transparent Optical Networks (ICTON), Bari, Italy, 19–23 July 2020; pp. 1–4. [CrossRef]
2. Ma, Z.; Zhao, Y.; Wang, W.; Zhang, J. Demonstration of Highly Dynamic Satellite Optical Networks Supporting Rapid Reconfiguration. In Proceedings of the 2021 17th International Conference on the Design of Reliable Communication Networks (DRCN), Milano, Italy, 19–22 April 2021; pp. 1–3. [CrossRef]
3. Zhai, H.; Zhang, Z.; Zhang, H.; Wang, B.; Zhao, Y.; Zhang, J. Design and Implementation of the Hardware Platform of Satellite Optical Switching Node. In Proceedings of the 2021 19th International Conference on Optical Communications and Networks (ICOCN), Qufu, China, 23–27 August 2021; pp. 1–3. [CrossRef]
4. Zhang, Y.; Yuan, Y.; Guo, B.; Luo, Q.; Zhao, B.; Zhou, W.; Jiang, M.; Wang, Y.; Fu, M.; Liu, Y.; et al. A Research Study on Protocol Stack of Space-Based Optical Backbone Network. *Appl. Sci.* **2021**, *11*, 2367. [CrossRef]
5. Zhang, T.; Sun, X.; Wang, C. On Optimizing the Divergence Angle of an FSO-Based Fronthaul Link in Drone-Assisted Mobile Networks. *IEEE Internet Things J.* **2022**, *9*, 6914–6921. [CrossRef]
6. Guo, Q.; Gu, R.; Dong, T.; Yin, J.; Liu, Z.; Bai, L.; Ji, Y. SDN-Based End-to-End Fragment-Aware Routing for Elastic Data Flows in LEO Satellite-Terrestrial Network. *IEEE Access* **2019**, *7*, 396–410. [CrossRef]
7. Guo, X.; Guo, B.; Li, K.; Fan, C.; Yang, H.; Huang, S. A SDN-enabled Integrated Space-Ground Information Network Simulation Platform. In Proceedings of the 2019 18th International Conference on Optical Communications and Networks (ICOCN), Huangshan, China, 5–8 August 2019; pp. 1–3. [CrossRef]
8. Nunes, B.A.A.; Mendonca, M.; Nguyen, X.N.; Obraczka, K.; Turletti, T. A Survey of Software-Defined Networking: Past, Present, and Future of Programmable Networks. *IEEE Commun. Surv. Tutor.* **2014**, *16*, 1617–1634. [CrossRef]
9. Dong, Y.; Zhao, S.; Ran, H.d.; Li, Y.; Zhu, Z. Routing and wavelength assignment in a satellite optical network based on ant colony optimization with the small window strategy. *J. Opt. Commun. Netw.* **2015**, *7*, 995–1000. [CrossRef]
10. Cao, S.; Zhang, T. Congestion Control Based on OSPF in LEO Satellite Constellation. In Proceedings of the 2019 IEEE 19th International Conference on Communication Technology (ICCT), Xi'an, China, 16–19 October 2019; pp. 1111–1115. [CrossRef]
11. Wang, Y.; Chen, T.; Zhou, N. Space-based Optical Burst Switching Assembly Algorithm Based on QoS Adaption. In Proceedings of the 2019 IEEE 11th International Conference on Communication Software and Networks (ICCSN), Chongqing, China, 12–15 June 2019; pp. 101–105. [CrossRef]
12. Aos Space Data Link Protocol. *CCSDS 732.0-B-4, Blue Book*; CCSDS: Washington, DC, USA, 2021.
13. IP Over CCSDS Space Links. *CCSDS 702.1-B-1, Blue Book*; CCSDS: Washington, DC, USA, 2012.
14. Andersson, L.; Doolan, P.; Feldman, N.; Fredette, A.; Thomas, B. *LDP Specification*; RFC3036; IETF: Wilmington, DE, USA, 2001. [CrossRef]
15. Awduche, D.; Berger, L.; Gan, D.; Li, T.; Srinivasan, V.; Swallow, G. *RSVP-TE: Extensions to RSVP for LSP Tunnels; RFC 3209*; IETF: Wilmington, DE, USA, 2001. [CrossRef]
16. Braden, R.; Zhang, L.; Berson, S.; Herzog, S.; Jamin, S. *Resource ReSerVation Protocol (RSVP)—Version 1 Functional Specification*; RFC 2205; IETF: Wilmington, DE, USA, 1997. [CrossRef]
17. Jamoussi, B.; Andersson, L.; Callon, R.; Dantu, R.; Wu, L.; Doolan, P.; Worster, T.; Feldman, N.; Fredette, A.; Girish, M.; et al. *Constraint-Based LSP Setup Using LDP*; RFC 3212; IETF: Wilmington, DE, USA, 2002. [CrossRef]
18. Varga, A.; Hornig, R. An overview of the OMNeT++ simulation environment. In Proceedings of the 1st International Conference on Simulation Tools and Techniques for Communications, Networks and Systems & Workshops, SimuTools 2008, Marseille, France, 3–7 March 2008. [CrossRef]
19. Fu, M.; Guo, B.; Yang, H.; Pang, C.; Huang, S. *Routing Optimization Based on OSPF in Multi-Layer Satellite Network*; IOS Press: Amsterdam, The Netherlands, 2022; Volume 345, pp. 597–602. [CrossRef]

applied sciences

MDPI

Review

A Review of Optical Neural Networks

Danni Zhang and Zhongwei Tan *

Key Lab of All Optical Network & Advanced Telecommunication Network of EMC, Institute of Lightwave Technology, Beijing Jiaotong University, Beijing 100044, China; dannizhang@bjtu.edu.cn
* Correspondence: zhwtan@bjtu.edu.cn

Abstract: With the continuous miniaturization of conventional integrated circuits, obstacles such as excessive cost, increased resistance to electronic motion, and increased energy consumption are gradually slowing down the development of electrical computing and constraining the application of deep learning. Optical neuromorphic computing presents various opportunities and challenges compared with the realm of electronics. Algorithms running on optical hardware have the potential to meet the growing computational demands of deep learning and artificial intelligence. Here, we review the development of optical neural networks and compare various research proposals. We focus on fiber-based neural networks. Finally, we describe some new research directions and challenges.

Keywords: optical neural network; fiber; deep learning

1. Introduction

Deep learning has been integrated into various subject fields in recent years, allowing the rejuvenation of fields that had previously reached their development limits. Deep learning has gained momentum in applications including image processing [1], physics [2], and natural language processing [3]. In addition, it is being used as a foundation to develop custom industrial applications rapidly. Deep learning is the process of learning the intrinsic laws and representation levels of sample data and combining low-level features to form more abstract high-level representations of attribute categories or features by building a neural network. It can mimic the behavior of the human brain in the process of analytical learning to discover distributed feature representations of data, allowing complex learning tasks such as classification to be performed with simple models [4,5]. The tools that support deep learning have existed for decades and have evolved continuously over the past few years due to the availability of large datasets and hardware developments. High-quality algorithms, such as neural networks for deep learning, have become an essential driver regarding the technological aspects of artificial intelligence [6–8].

Three major factors are essential for developing deep learning approaches and technologies: data, algorithms, and computational capacity. Data processing and algorithm implementation require the technical support of the chips in the application terminal, with current deep learning chip technologies primarily including graphics processing units, field-programmable gate arrays, and application-specific integrated circuits. With the dramatic increase in the amount of data to be processed, the bottlenecks of traditional electrical chip technology such as the von Neumann architecture or the available complementary metal oxide semiconductor (CMOS) process and devices are emerging, limiting the application and popularity of deep learning by introducing the problems of chip power consumption and performance enhancement. These limitations make it challenging to adopt deep learning in end equipment [9]. Neuromorphic systems are the systems that attempt to map machine learning and artificial intelligence algorithms into large-scale distributed hardware capable of simulating the human brain at the physical level for computation and simulation purposes. In the past, neuromorphic system applications have been comprehensively studied within the field of electronics, which has ultimately resulted in the rendering and

Citation: Zhang, D.; Tan, Z. A Review of Optical Neural Networks. *Appl. Sci.* **2022**, *12*, 5338. https://doi.org/10.3390/app12115338

Academic Editors: Fabio Cavaliere and Luca Potì

Received: 6 April 2022
Accepted: 23 May 2022
Published: 25 May 2022

refinement of innovations such as the memristor, the field-effect transistor, as well as a host of other neuromorphic-based chip devices. Recently, owing to the prominence of the von Neumann bottleneck, the inherent difficulties of electronic wiring, and the rapid growth of chip power consumption, people have started to focus more upon optical neuromorphic systems as alternatives.

Optical neural networks and photonic circuits can provide a novel dedicated neural network accelerator scheme to exploit the parametric changes of optics for computing. Optical computing systems can be massively parallel or combined with small form factor devices. Photonics offers the advantages of high speed, high bandwidth, and low power consumption compared with electronics. Photonic solutions can significantly reduce the power consumption of logic and data operations [10–12]. It is well known that an ordinary lens can perform a Fourier transform without consuming any energy and that certain matrix operations can be performed with no energy consumption. Many inherent optical nonlinearities can be directly used to implement nonlinear operations with photons. Once a neural network has been trained, the architecture can be passive, and computations on the optical signal can be performed without additional energy input [13]. Optical interconnects can allow hybrid optoelectronic deep neural networks, where low-energy, highly parallel integration techniques can be used as part of an analog optical processor.

Despite half a century of development, optical computing, which shows great potential, has not yet become a mature general-purpose practical technology [14,15]. However, deep learning is well-suited to the implementation of all-optical or hybrid optoelectronic systems, especially for visual computing applications [16]. Here, we review the development of optical neural networks and compare various research proposals. We focus on fiber-based optical neural networks. We then highlight some new research directions and challenges.

2. History

During the 1980s, the field of nonlinear optics and neural computing fields both experienced rapid improvements [17–19]. Because of their inherent parallelism, large neural network models can be implemented using the speed and capability of light to process two-dimensional data arrays without conversion bottlenecks. Nonlinear optical applications, such as associative memory, Hopfield networks, and self-organizing networks, can be implemented in an all-optical manner using nonlinear optical processing elements [20–22]. Nevertheless, research enthusiasm for dedicated optical hardware diminished in the 1990s, owing to the technical immaturity of optoelectronic implementations of nonlinear activation functions and the difficulties associated with controlling analog computations.

The concept of deep learning was introduced by Hinton et al. in 2006 [23]. The unsupervised greedy layer-by-layer training algorithm based on a deep belief network was later proposed to help solve optimization challenges related to deep structures. However, the convolutional neural network proposed by Lecun et al. is the first true multilayer structure learning algorithm. This algorithm uses spatial relativity to reduce the number of parameters and improve the training performance [24]. After deep convolutional neural networks were proposed, many improved structural schemes emerged, and deep learning developed rapidly [25–29]. As deep neural networks have become one of the main algorithmic approaches for many applications, the development path of optical neural networks has changed accordingly. In addition, significant improvements in optoelectronics and silicon photonics have led many researchers to revisit the idea of implementing neural networks using optical technologies [30–32].

3. Development Processes and Categories

3.1. Silicon-Based Optical Neural Networks

More recently, the rapid development of silicon optical integration technology has provided technical support for research into photonic integrated neural network chips. Photonic integrated neural networks are used to perform matrix multiplication and addition calculations (MACs) in neural networks, using photon integration technology to run deep learning algorithms and implement machine learning-related applications. Conceptual exploratory research work has already been conducted on the use of photonic integrated neural network chips to implement neural network operations, and there is the hope of using photonic integrated neural network chips to accomplish tasks such as image classification and object detection for applications in servers, autonomous driving, and other scenarios. Representative research schemes include the Mach–Zehnder modulator scheme [33–38], micro-loop modulator scheme [39,40], and the three-dimensional integration scheme [41]. Such research has focused on simulation modeling and experimental applications of photonic neural network chips. In 2019, Lightelligence released the world's first prototype board for photonic integrated neural network chips, which were then used to implement the MNIST handwritten digital dataset recognition application with a recognition accuracy of over 90% and a speed approximately two orders of magnitude faster than that of traditional electric chips, encouraging photonic integrated neural network chip applications and development. The number of Mach–Zehnder interferometers (MZIs) is proportional to the square number of elements N in a vector and is a necessary side effect of solving arbitrary matrices [33,36]. Loss, noise, and defects are serious problems that affect a system as the size of a photonic circuit increases [42].

Contemporary deep neural network architectures consist of linear cascades of layers followed by multiple iterations of nonlinear activation functions. Nonetheless, meshes based on devices such as MZIs, while capable of arbitrary matrix multiplication, cannot perform important nonlinear operations. As an alternative to MZI-based MACs, SiN-conjugated waveguides can be used to construct a fully optical neurosynaptic network that is based on phase change materials (PCMs) [43–47]. By varying the number of optical pulses to set the weights randomly, the time delay can modulate the weights to achieve a nonlinear activation function similar to a rectified linear unit with a synaptic plasticity consistent with the well-known Hebbian learning or spike-dependent plasticity rule [48]. Optical neural networks are based on silicon integration as shown in Figure 1.

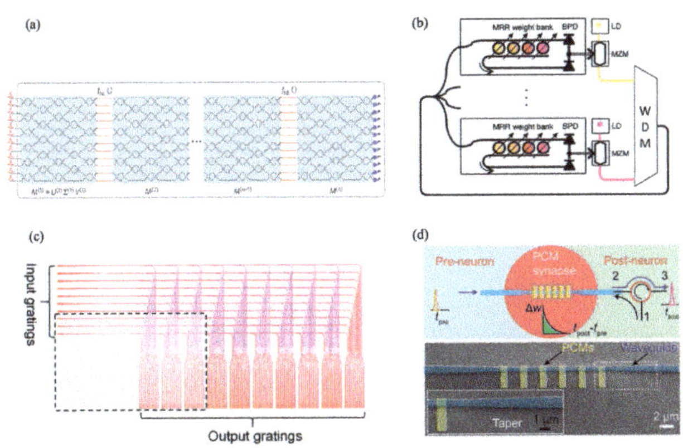

Figure 1. Implementing a neural network based on silicon integration. (**a**) A silicon photonic integrated circuit comprising programmable MZIs realizes the optical interference unit. This architecture enables the operation of the weight matrix using the MZIs through singular value decomposition [33]. (**b**) Using the micro-ring resonator as the neuron of the broadcast and weight network, the wavelength division multiplexed signals are weighted using reconfigurable continuous value filters and then summed for total power detection [40]. (**c**) A biplane signal distribution network is realized through the stacking of waveguides, which can complete the all-around connection of the feedforward neural network [41]. (**d**) Photonic synapses are fabricated using PCM in combination with an integrated silicon nitride waveguide, and the synaptic weights are set randomly by varying the number of optical pulses sent by the waveguide [43].

Photonic neural networks offer a promising alternative to microelectronic and hybrid optoelectronic implementations, where classical neural networks rely heavily on fixed matrix multiplication. However, the need for phase stabilization and many neurons with large optical components, such as fibers and lenses, has been a major obstacle to achieving this conversion. Integrated photonics solves this problem by providing a scalable solution. Photon particles are too large to be integrated at the same high level as electrons [49]. In addition, present optical neural network solutions inevitably rely on electronics. Submicron-scale etching techniques are still not available, and height differences and spacing between optics and electronics and process incompatibilities make optoelectronic integration challenging. Furthermore, nonlinear limitations and photonics integration constraints make it impossible to scale chips such as electronic neuromorphic computing chips, which are mere arrays. In addition, nonvolatile photonic storage and weighting, as well as low power consumption, are necessary to ensure neurosynaptic functions [50,51]. Even though photonic nonlinear neural operations have been demonstrated in materials such as PCMs, there is still a large gap to electronic nonlinear neuromorphic operations, and there is still room for development in the energy efficiency and fast switching of new cumulable materials. As of yet, the goal of a large-scale, rapidly reprogrammable photonic neural network chip remains unrealized.

3.2. Deep Diffraction Neural Networks

Optical neural networks have been studied in Fourier optics for decades, and the forward physical structure of multilayer coherent neural networks is thought to be promising [18,52,53]. Deep diffraction neural networks (D^2NNs) have been proposed for various classification tasks and can approach high-dimensional information capacity via optical processing, with millions of neurons and hundreds of billions of connections [54–61]. D^2NNs also enable image saliency detection in Fourier space [53,62–65]. Diffraction is a prevalent physical phenomenon of incoherent light propagation, as seen in Figure 2. Any point in the plane perpendicular to the direction of propagation can be interpreted as a coherent superposition of complex amplitudes of integral points in the reference plane with a certain diffraction distance [66]. The superposition of diffraction meets the basic requirements of deep complex neural networks. Therefore, as an optical mechanism, coherent diffraction can provide an alternative method to fully connect multi-valued neurons. The training input of the underlying D^2NN model is operated via spatial features. The input domain is filtered in the object or Fourier space by introducing a series of ordered passive filters prior to the diffraction network. This allows for the parallel processing of optical information.

D^2NNs do not have nonlinear capabilities. In a D^2NN made of linear materials, nonlinear optical processes, including surface nonlinearity, are negligible. The only form of nonlinearity in the forward optical model occurs in the photoelectron detector plane [53,67]. Deep neural networks using this scheme are capable of inference only for fixed tasks. Nonlinear operations are an indispensable and vital part of neuromorphic computing. Training, one of the key steps in the vast majority of neural network algorithms, has still not been implemented for such networks [68]. Optical neural networks are a promising energy-efficient method for implementing matrix multiplication. However, in practice, the neural network needs to be trained to be applied and cannot be implemented on hardware in which information can only flow in the forward direction.

3.3. Fiber-Based Optical Neural Networks

The human brain is complex, with 10^{11} neurons and 10^{15} synapses occupying a minimal space and consuming very little power. The average power consumption of a human brain is only 15 W [69]. In addition, the human brain is thought to be the fastest and most intelligent processing system in existence. The human brain relies on an interconnected network of organic biological microfibers, also known as neurons, to propagate information through the body. Using electrical action potentials, these signals are processed via different spatiotemporal principles [70].

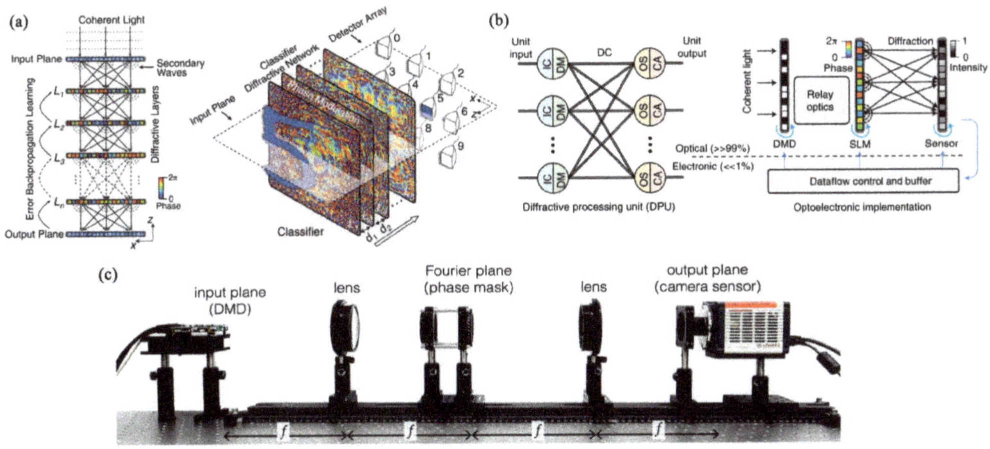

Figure 2. Implementing D²NNs. (**a**) Several transmission or reflection layers are 3D printed, and each point on the layer is an artificial neuron with a given weight connected to other neurons in the next layer via optical diffraction [52]. (**b**) Diffractive process units (DPUs) are designed as a large-scale perceptron class optoelectronic computing building block that can be programmed to build different deep neural networks implemented with programmable optoelectronic devices, such as digital micromirror devices (DMDs), phased spatial light modulators (SLMs), and CMOS sensors [61]. (**c**) An optical convolutional layer with an optimizable phase mask is conceived to implement a convolutional neural network using the intrinsic convolution performed using a linear spatial non-turning image system [60].

Optical fibers have been widely used in communication, sensors, and illumination since the 1970s. Optical fibers are used as a transmission medium to transmit information via light. Optical fibers work similarly to biological neurons, where the neurons are activated by the input information and recognize it. The following section focuses on applying and developing optical fibers in neural networks.

3.3.1. Microfibers

Throughout the years, the field of neuromorphic engineering has been dedicated to the development of practical artificial neurocomputing devices that mimic the functions of the biological brain. To date, neuromorphic systems have been demonstrated with software running on conventional computers or complex electronic circuit configurations [43,71,72]. However, the efficiency of such systems is low compared with that of human neuromorphic systems. Accordingly, the search for alternative materials and structures to achieve efficient neuromorphic components is an area of intense research.

Optical fibers have been implemented in functional materials such as semiconductor compounds [73–75]. The physical properties of sulfur-based alloys are of interest because they are altered by light and exhibit brain-like functions in terms of plasticity and their form of transmission [76–78]. The signals of this type of microfiber display characteristics of biological signals transmitted in the brain via light transmission as seen in Figure 3 [79–81]. In addition, microfibers can monitor neural activity on the cortical surface and in deep brain regions [82,83]. Accordingly, microfibers appear to be a competitive category of artificial neuromorphic components and will likely be widely used in neuroscience research and medical applications in the future.

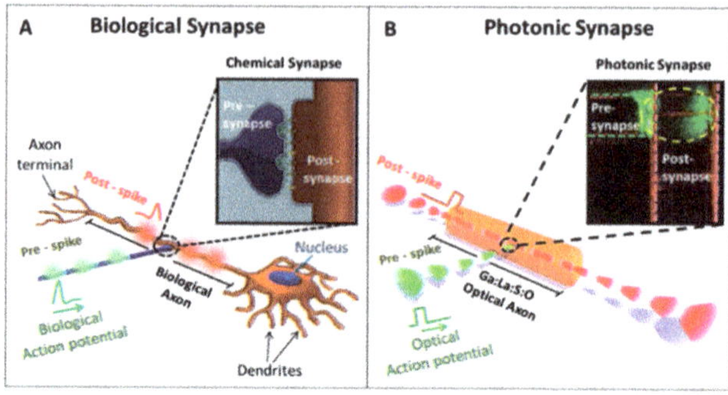

Figure 3. Comparison of photons with biological neurons and synapses [80]. (**A**,**B**) is biological synapse and photonic synapse, respectively.

3.3.2. Multimode Fibers

Since the 1980s, researchers have been studying the effect of scattering media on imaging systems and have found it possible to conduct imaging using the scattering media itself. The results indicate that imaging via scattering media not only improves the resolution of the imaging system but also eliminates image noise. Single-fiber imaging can be achieved if an optical fiber is treated as a scattering medium [84]. A multimode fiber (MMF) is a highly scattering medium that disrupts the light propagating through it and outputs speckle patterns that the human eye cannot decode. These patterns are called scattered spots. The behavior of a system consisting of input patterns has deterministic propagation through an MMF and a detector [85]. Multiple studies have shown that image sending or imaging through an MMF can be performed by simulating phase conjugation. With the development of digital holography and the increasing maturity of spatial light modulation devices and digital micromirror devices, beam wavefront and phase adjustments are becoming increasingly flexible and precise [86]. The proposed systems suffer from overly complex optical components and image distortion, resulting from variations in the path length experienced by the various modes of light propagation in optical fibers. Such problems can be effectively solved using neural networks. The idea of combining MMF imaging with neural networks for image classification has been around for more than three decades [87,88].

Currently, deep learning applications can solve more complex problems, and MMFs are becoming one of the choices available for optically implemented neural networks, as seen in Figure 4. The concept of combining a complex fix mapping with a simple programmable processor to achieve a powerful overall system has been applied to a variety of deep learning algorithms. Optical neural networks based on MMFs learn the nonlinear relationship between the amplitude of the scatter pattern obtained at the fiber output and the phase or amplitude at the fiber input [89–91]. In addition, optical reservoir computing (RC) systems have been shown to significantly improve the computational performance [92]. The reservoir layer of an all-optical reservoir computing system consists of an optical fiber and other optical process devices, and nonlinearity is reflected in the system. No electro-optical conversions are required in the reservoir layer, which dramatically reduces the problem of time consumption [93–99]. What is more, a picosecond-pulsed fiber laser is used to illuminate the subject, and the spatial information of the detected image is expanded in the time domain by using the inter-mode dispersion property of the multimode fiber. Time domain information is accordingly detected using a single-pixel detector, and the resulting two-dimensional image information can then ultimately be recovered from the one-dimensional time domain information with a neural network training method [100].

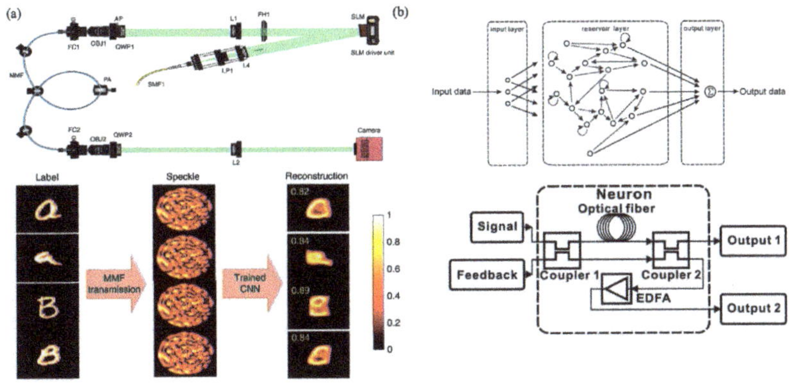

Figure 4. (**a**) CNNs project arbitrary patterns and perform transfer learning via MMF [89]. (**b**) MMF-based free-space all-optical RC system using spatial light modulators and diffractive optical elements [98].

The key challenge when designing a viable optical neuromorphic computer is to combine the linear part of the optical system with the nonlinear elements and input–output interfaces while maintaining the speed and power efficiency of the optical interconnect. It is possible to combine the linear and nonlinear portions of an optical system in the shared body of an MMF. At the same time, a large number of spatial modes are densely supported in the MMF, maintaining the traditional high parallelism of optics while retaining the compact form factor. Therefore, it appears that MMF-based analog optical computers can be highly energy-efficient and versatile while having performance comparable to digital computers.

3.3.3. Time Stretch

Photonic time stretch (PTS) has been developing for more than two decades since its introduction in the 1990s [101]. It has become a mature optical technology that can slow down, amplify, and capture ultrafast events by slowing down the optical signal in real-time to bridge the bandwidth gap between electronics and photonics. PTS consists of a dispersive optical link that uses group velocity dispersion to convert the spectrum of a broadband optical pulse into a time-stretched temporal waveform. This method is used to perform Fourier transforms of the optical signal at a high frame rate on a single transmission basis for real-time analyses of fast dynamic processes. Via dispersion, the information residing in the spectrum is stretched over time. Group delay dispersion for time stretching can be experimentally implemented using various different device designs. The main designs of the proposed devices are internal Raman amplified fibers, chirped fiber Bragg gratings, array optical waveguide gratings, multimode waveguides (fiber or a pair of planar reflectors), and curved spatially mapped chromatic dispersion devices [102,103].

Deep learning extracts patterns and information from rich multidimensional datasets and is widely used for image recognition. Its earliest combination with PTS was for recognizing and classifying large numbers of cells. The Optical Fluid Time-Stretch Quantitative Phase Imaging (OTS-QPI) system reconstructs bright-field and quantitative phase images of flowing cells based on spectral interferograms of time-stretch light pulses, captures images of flowing cells with minimal motion distortion at rates greater than 10,000 cells/s, and extracts multiple biophysical features of individual cells, making it a powerful instrument for biomedical applications [104–108]. OTS-QPI was first implemented in combination with fully connected neural networks for image classification as seen in Figure 5A [104]. In addition, with the rapid development of deep learning and convolutional neural networks, combinations with generative adversarial networks and other image enhancement algorithms have been proposed to solve additional deep learning problems, combined with PTS [109,110].

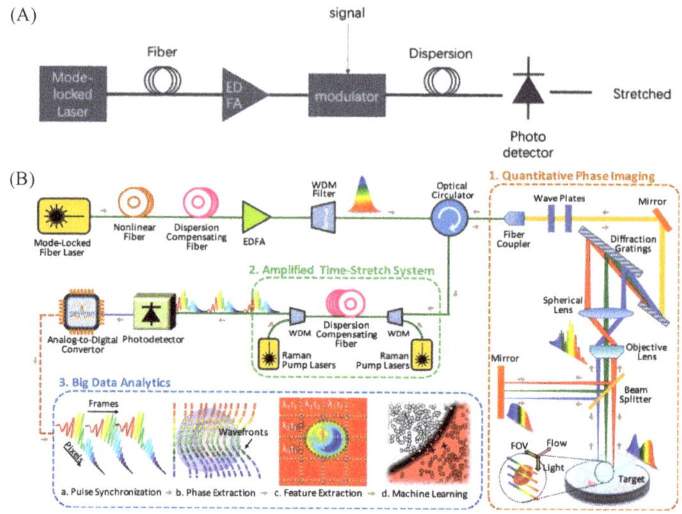

Figure 5. (**A**) PTS schematic diagram of principle. (**B**) OTS-QPI system operational process [104].

Deep learning has evolved to the point where there is an urgent need for computationally small, low-latency datasets to reduce the computational effort associated with computational power limits and large datasets. When inputting data, nonlinear optical dynamics enables linear learning algorithms to learn nonlinear functions or determine boundaries to classify the data into correct classes. The nonlinear Schrödinger kernel, also called the Lambda kernel, is the core of this system [111,112]. The Lambda kernel, as seen in Figure 5B, is capable of nonlinearly casting data that have been modulated onto the spectrum of femtosecond pulses into a space where data that are non-dividable in linear terms are converted to be linearly divisible in the new space. In terms of functionality, there are similarities between the Lambda kernel and the concept of "kernel projection" in the machine learning literature. Using this approach, the spectrum of the data is mapped onto femtosecond light pulses and projected into an implicit high-dimensional space using nonlinear optical dynamics, thereby improving the accuracy and reducing the latency in the data classification by several orders of magnitude. The nonlinear dynamics are introduced into the data before processing the output data with an optical classifier.

Fiber optic imaging and nonlinear phenomena have been popular research topics in the field of optics. Control of the nonlinear propagation under spatiotemporal conditions can be achieved via mechanical perturbation of the fiber or by shaping the pumped light. Such an approach enables optically controllable computation within nonlinear optical fibers. Meanwhile, optical neural networks of optical fibers have good energy efficiency and scalability, providing a favorable approach for photonic neuromorphic computations.

Photonic integration renders ultrafast artificial neural networks possible, and photonic neuromorphic computing results in challenges different from those of electronic computing. Algorithms running on such hardware can meet the growing demand for deep learning and artificial intelligence in areas such as medical diagnostics, telecommunications, and high-performance and scientific computing. All calculations will be performed at the speed of light, which is much faster than the analog speed of digital computers, therefore processing exponentially more data. For real-time observations, time-stretch processors are ideal output devices. Even though photonic neuromorphic computing is still in its infancy, this research area has great potential to expand the frontiers of deep learning and information processing.

In this article, we highlight the diversity of optical neural networks. Optical neural networks are a rapidly growing field, with many schemes currently under investigation, all of which have their own advantages and disadvantages, as shown in Table 1.

Table 1. Comparison of the different approaches to optical neural networks.

Technology	Categories	Advantages	Disadvantages	Chip Capabilities
Silicon-based optical neural networks	Mach–Zehnder modulator scheme	Matrix multiplication has high speed and low power consumption	O/E conversion	Inference
	Micro-loop modulator scheme			
	3D integration scheme			
D^{2}NN	Optical diffraction	Handles large amounts of data	Not conducive to reuse	Inference
Fiber-based optical neural networks	Microfiber	Wavelength division multiplexing, biological-like neurons	Large size	No chip
	MMF	It has both linear and nonlinear functions	Not easy to control	
	PTS	Photon DAC	Massive system	

Photonic neuromorphic computing is a system comprising active and passive devices, light sources, and transistors. However, no fabrication platform is yet capable of implementing all of these functions upon a single mold based on current fabrication levels. Present optical neural network solutions still exclusively rely on electronics. Optoelectronic integration processes remain challenging because submicron-scale etching techniques are still unavailable, and issues such as height disparities, spacing restrictions, and process incompatibilities are still prevalent. Currently, on-chip light sources use a concatenation of III–V families of materials or a direct epitaxy of quantum dot lasers on silicon. However, the complexity of the preparation process and associated reliability cannot meet the commercial mass production standards. Several approaches for integrating photonic systems and CMOS sensors for ultimately achieving a successful on-chip photonic neural network design are presently being explored.

Difficulties associated with on-chip optical delays are also a challenge being evaluated currently. Nonlinear operations are an indispensable and important part of neuromorphic computing. Moreover, nonvolatile photonic storage and weighting as well as low power consumption are necessary to ensure neurosynaptic function. Although photonic nonlinear neural operations have been demonstrated in materials such as PCM, a large gap still remains regarding electronic nonlinear neuromorphic operations. Accordingly, there is still room for development in energy efficiency and fast switching of new accumulable materials. In addition, training is one of the key steps for the vast majority of neural network algorithms, which to date still requires implementation.

Optical neural networks have been shown to be a promising energy-efficient method for implementing matrix multiplication [33,34]. However, these neural networks must be trained to be applied in practice, and they cannot be applied in hardware where information can flow in only the forward direction. Finally, current photonic platforms lack the common and important storage in electronic computers. In adaptive learning and training, weights need to be updated frequently, a situation that requires fast memories such as Dynamic Random Access Memory (DRAM), Resistive Random Access Memory (RRAM), and Ferroelectric Random Access Memory (FRAM). In summary, neuromorphic photonic systems can follow the same architecture, and the fabrication of stand-alone all-optical computers is therefore unlikely, owing to the lack of photonic memory devices.

In addition to being popular candidates for neuromorphic systems, optical neural networks exhibit several advantages. Neuromorphic processors are massively distributed hardware. Therefore, they heavily rely on parallel interconnections such as those which exist between components. Optical computing for neural network computation has the advantages of low latency, low power consumption, and high bandwidth, while the use of optical interconnections can offset the disadvantages of requiring broadband in electronic devices in exchange for interconnectivity. In addition, photonic synaptic devices modulated via photonic signals can simulate retinal neurons in the real eye, and physical device computing is conceptually close to the visual system and brain computing to simulate complex neural functions through optical modulation. Furthermore, the use of optical fiber can solve the problem of on-chip optical delays. In the future, co-packaging with electronic memory can be used to solve the existing challenges.

4. Discussion

The demand for increased data processing volumes and operation speeds is constantly growing, and it is urgent to circumvent the structural von Neumann bottleneck and design new structures to accommodate efficient neural network training and testing. Optical processors, with their advantages of high speeds and low power consumption, have been gaining attention as a result of the rapid development of hardware dedicated to the inference and training of optical neural networks. Optical neural networks have shown great promise in terms of low-energy consumption and high-speed parallel computing, and the development trend of optical neuromorphic computing appears to be unstoppable. According to research findings, photonic neuromorphic computing is currently in its preliminary developmental stage, and various optimization solutions have emerged, even though photonic neural networks still have many challenges to overcome.

In this paper, we reviewed the development of optical neural networks, emphasized optical neural networks with fiber dispersion, and provided some perspectives for the future development of optical neuromorphic computing (Table 1). Compared with other optical neural networks, optical neural networks with fiber dispersion have better nonlinear performance and neuron-like properties, thus exhibiting greater potential value. As optical technology matures, deep learning will further evolve with enhanced neuromorphic computing and optical neural networks as well as the application of increasingly high-performance integrated optics. In the future, hyper-scale and programmable optical neural networks will likely be implemented and applied to speech processing, image recognition, and target tracking, which will provide a bright future for neuromorphic photonics.

Author Contributions: Resources, D.Z.; data curation, D.Z.; writing—original draft preparation, D.Z.; writing—review and editing, D.Z. and Z.T.; supervision, Z.T.; project administration, Z.T.; funding acquisition, Z.T. All authors have read and agreed to the published version of the manuscript.

Funding: This research was funded by the National Natural Science Foundation of China (61875008).

Conflicts of Interest: The authors declare no conflict of interest.

References

1. Hemanth, D.J.; Estrela, V.V. *Deep Learning for Image Processing Applications*; IOS Press: Amsterdam, The Netherlands, 2017; Volume 31.
2. Huggins, W.J.; McClean, J.R.; Rubin, N.C.; Jiang, Z.; Wiebe, N.; Whaley, K.B.; Babbush, R. Efficient and noise resilient measurements for quantum chemistry on near-term quantum computers. *NPJ. Quantum Inf.* **2021**, *7*, 23. [CrossRef]
3. Young, T.; Hazarika, D.; Poria, S.; Cambria, E. Recent trends in deep learning based natural language processing. *IEEE Comput. Intell. Mag.* **2018**, *13*, 55–75. [CrossRef]
4. LeCun, Y.; Bengio, Y.; Hinton, G. Deep learning. *Nature* **2015**, *521*, 436–444. [CrossRef] [PubMed]
5. Goodfellow, I.; Bengio, Y.; Courville, A. *Deep Learning*; MIT Press: Cambridge, MA, USA, 2016.
6. Simonyan, K.; Zisserman, A. Very deep convolutional networks for large-scale image recognition. *arXiv* **2014**, arXiv:1409.1556.
7. Jouppi, N.P.; Young, C.; Patil, N.; Patterson, D.; Agrawal, G.; Bajwa, R.; Bates, S.; Bhatia, S.; Boden, N.; Borchers, A.; et al. In-datacenter performance analysis of a tensor processing unit. In Proceedings of the 44th Annual International Symposium on Computer Architecture, Toronto, ON, Canada, 24–28 June 2017; pp. 1–12.
8. Basu, J.K.; Bhattacharyya, D.; Kim, T.H. Use of artificial neural network in pattern recognition. *Int. J. Softw. Eng. Appl.* **2010**, *4*.
9. De Vries, A. Bitcoin's growing energy problem. *Joule* **2018**, *2*, 801–805. [CrossRef]
10. Prucnal, P.R.; Shastri, B.J.; Teich, M.C. *Neuromorphic Photonics*; CRC Press: Boca Raton, FL, USA, 2017.
11. Padovani, A.; Woo, J.; Hwang, H.; Larcher, L. Understanding and optimization of pulsed SET operation in HfO x-based RRAM devices for neuromorphic computing applications. *IEEE Electron Device Lett.* **2018**, *39*, 672–675. [CrossRef]

12. Eltes, F.; Villarreal-Garcia, G.E.; Caimi, D.; Siegwart, H.; Gentile, A.A.; Hart, A.; Stark, P.; Marshall, G.D.; Thompson, M.G.; Barreto, J.; et al. An integrated optical modulator operating at cryogenic temperatures. *Nat. Mater.* **2020**, *19*, 1164–1168. [CrossRef]
13. Ying, Z.; Wang, Z.; Zhao, Z.; Dhar, S.; Pan, D.Z.; Soref, R.; Chen, R.T. Silicon microdisk-based full adders for optical computing. *Opt. Lett.* **2018**, *43*, 983–986. [CrossRef]
14. Solli, D.R.; Jalali, B. Analog optical computing. *Nat. Photonics* **2015**, *9*, 704–706. [CrossRef]
15. Sawchuk, A.A.; Strand, T.C. Digital optical computing. *Proc. IEEE* **1984**, *72*, 758–779. [CrossRef]
16. Mennel, L.; Symonowicz, J.; Wachter, S.; Polyushkin, D.K.; Molina-Mendoza, A.J.; Mueller, T. Ultrafast machine vision with 2D material neural network image sensors. *Nature* **2020**, *579*, 62–66. [CrossRef] [PubMed]
17. Psaltis, D.; Farhat, N. Optical information processing based on an associative-memory model of neural nets with thresholding and feedback. *Opt. Lett.* **1985**, *10*, 98–100. [CrossRef] [PubMed]
18. Caulfield, H.J.; Kinser, J.; Rogers, S.K. Optical neural networks. *Proc. IEEE* **1989**, *77*, 1573–1583. [CrossRef]
19. Denz, C. *Optical Neural Networks*; Springer Science & Business Media: Berlin/Heidelberg, Germany, 2013.
20. Lee, L.S.; Stoll, H.; Tackitt, M. Continuous-time optical neural network associative memory. *Opt. Lett.* **1989**, *14*, 162–164. [CrossRef] [PubMed]
21. Farhat, N.H.; Psaltis, D.; Prata, A.; Paek, E. Optical implementation of the Hopfield model. *Appl. Opt.* **1985**, *24*, 1469–1475. [CrossRef] [PubMed]
22. Lu, T.T.; Francis, T.; Gregory, D.A. Self-organizing optical neural network for unsupervised learning. *Opt. Eng.* **1990**, *29*, 1107–1113. [CrossRef]
23. Hinton, G.E. Deep belief networks. *Scholarpedia* **2009**, *4*, 5947. [CrossRef]
24. LeCun, Y.; Bottou, L.; Bengio, Y.; Haffner, P. Gradient-based learning applied to document recognition. *Proc. IEEE* **1998**, *86*, 2278–2324. [CrossRef]
25. Krizhevsky, A.; Sutskever, I.; Hinton, G.E. Imagenet classification with deep convolutional neural networks. *Adv. Neural Inf. Process. Syst.* **2012**, *60*, 84–90. [CrossRef]
26. Szegedy, C.; Liu, W.; Jia, Y.; Sermanet, P.; Reed, S.; Anguelov, D.; Erhan, D.; Vanhoucke, V.; Rabinovich, A. Going deeper with convolutions. In Proceedings of the IEEE Conference on Computer Vision and Pattern Recognition, Boston, MA, USA, 7–12 June 2015; pp. 1–9.
27. Szegedy, C.; Ioffe, S.; Vanhoucke, V.; Alemi, A.A. Inception-v4, inception-resnet and the impact of residual connections on learning. In Proceedings of the Thirty-First AAAI Conference on Artificial Intelligence, San Francisco, CA, USA, 4–9 February 2017.
28. Mateen, M.; Wen, J.; Song, S.; Huang, Z. Fundus image classification using VGG-19 architecture with PCA and SVD. *Symmetry* **2018**, *11*, 1. [CrossRef]
29. Anand, R.; Shanthi, T.; Nithish, M.; Lakshman, S. Face recognition and classification using GoogleNET architecture. In *Soft Computing for Problem Solving*; Springer: Berlin/Heidelberg, Germany, 2020; pp. 261–269.
30. Thomson, D.; Zilkie, A.; Bowers, J.E.; Komljenovic, T.; Reed, G.T.; Vivien, L.; Marris-Morini, D.; Cassan, E.; Virot, L.; Fédéli, J.M.; et al. Roadmap on silicon photonics. *J. Opt.* **2016**, *18*, 073003. [CrossRef]
31. Majumder, A.; Shen, B.; Polson, R.; Menon, R. Ultra-compact polarization rotation in integrated silicon photonics using digital metamaterials. *Opt. Express* **2017**, *25*, 19721–19731. [CrossRef]
32. Li, J.; Huang, X.; Gong, J. Deep neural network for remote-sensing image interpretation: Status and perspectives. *Natl. Sci. Rev.* **2019**, *6*, 1082–1086. [CrossRef]
33. Shen, Y.; Harris, N.C.; Skirlo, S.; Prabhu, M.; Baehr-Jones, T.; Hochberg, M.; Sun, X.; Zhao, S.; Larochelle, H.; Englund, D.; et al. Deep learning with coherent nanophotonic circuits. *Nat. Photonics* **2017**, *11*, 441–446. [CrossRef]
34. Hamerly, R.; Bernstein, L.; Sludds, A.; Soljačić, M.; Englund, D. Large-scale optical neural networks based on photoelectric 3multiplication. *Phys. Rev. X* **2019**, *9*, 021032.
35. Fang, M.Y.S.; Manipatruni, S.; Wierzynski, C.; Khosrowshahi, A.; DeWeese, M.R. Design of optical neural networks with component imprecisions. *Opt. Express* **2019**, *27*, 14009–14029. [CrossRef]
36. Pai, S.; Bartlett, B.; Solgaard, O.; Miller, D.A. Matrix optimization on universal unitary photonic devices. *Phys. Rev. Appl.* **2019**, *11*, 064044. [CrossRef]
37. Bangari, V.; Marquez, B.A.; Miller, H.; Tait, A.N.; Nahmias, M.A.; De Lima, T.F.; Peng, H.T.; Prucnal, P.R.; Shastri, B.J. Digital electronics and analog photonics for convolutional neural networks (DEAP-CNNs). *IEEE J. Sel. Top. Quantum Electron.* **2019**, *26*, 1–13. [CrossRef]
38. Tait, A.N.; De Lima, T.F.; Nahmias, M.A.; Miller, H.B.; Peng, H.T.; Shastri, B.J.; Prucnal, P.R. Silicon photonic modulator neuron. *Phys. Rev. Appl.* **2019**, *11*, 064043. [CrossRef]
39. Huang, C.; Bilodeau, S.; Ferreira de Lima, T.; Tait, A.N.; Ma, P.Y.; Blow, E.C.; Jha, A.; Peng, H.T.; Shastri, B.J.; Prucnal, P.R. Demonstration of scalable microring weight bank control for large-scale photonic integrated circuits. *APL Photonics* **2020**, *5*, 040803. [CrossRef]
40. Tait, A.N.; De Lima, T.F.; Zhou, E.; Wu, A.X.; Nahmias, M.A.; Shastri, B.J.; Prucnal, P.R. Neuromorphic photonic networks using silicon photonic weight banks. *Sci. Rep.* **2017**, *7*, 7430. [CrossRef] [PubMed]
41. Chiles, J.; Buckley, S.M.; Nam, S.W.; Mirin, R.P.; Shainline, J.M. Design, fabrication, and metrology of 10 × 100 multi-planar integrated photonic routing manifolds for neural networks. *APL Photonics* **2018**, *3*, 106101. [CrossRef]

42. Martens, D.; Bienstman, P. Study on the limit of detection in MZI-based biosensor systems. *Sci. Rep.* **2019**, *9*, 5767. [CrossRef]
43. Cheng, Z.; Ríos, C.; Pernice, W.H.; Wright, C.D.; Bhaskaran, H. On-chip photonic synapse. *Sci. Adv.* **2017**, *3*, e1700160. [CrossRef]
44. Feldmann, J.; Youngblood, N.; Wright, C.D.; Bhaskaran, H.; Pernice, W.H. All-optical spiking neurosynaptic networks with self-learning capabilities. *Nature* **2019**, *569*, 208–214. [CrossRef]
45. Joshi, V.; Le Gallo, M.; Haefeli, S.; Boybat, I.; Nandakumar, S.R.; Piveteau, C.; Dazzi, M.; Rajendran, B.; Sebastian, A.; Eleftheriou, E. Accurate deep neural network inference using computational phase-change memory. *Nat. Commun.* **2020**, *11*, 2473. [CrossRef]
46. Miscuglio, M.; Sorger, V.J. Photonic tensor cores for machine learning. *Appl. Phys. Rev.* **2020**, *7*, 031404. [CrossRef]
47. Wu, C.; Yu, H.; Lee, S.; Peng, R.; Takeuchi, I.; Li, M. Programmable phase-change metasurfaces on waveguides for multimode photonic convolutional neural network. *Nat. Commun.* **2021**, *12*, 96. [CrossRef]
48. Caporale, N.; Dan, Y. Spike timing–dependent plasticity: A Hebbian learning rule. *Annu. Rev. Neurosci.* **2008**, *31*, 25–46. [CrossRef]
49. de Valicourt, G.; Chang, C.M.; Eggleston, M.S.; Melikyan, A.; Zhu, C.; Lee, J.; Simsarian, J.E.; Chandrasekhar, S.; Sinsky, J.H.; Kim, K.W.; et al. Photonic integrated circuit based on hybrid III–V/silicon integration. *J. Lightwave Technol.* **2017**, *36*, 265–273. [CrossRef]
50. Guo, X.; He, A.; Su, Y. Recent advances of heterogeneously integrated III–V laser on Si. *J. Semicond.* **2019**, *40*, 101304. [CrossRef]
51. Zhai, Y.; Yang, J.Q.; Zhou, Y.; Mao, J.Y.; Ren, Y.; Roy, V.A.; Han, S.T. Toward non-volatile photonic memory: Concept, material and design. *Mater. Horiz.* **2018**, *5*, 641–654. [CrossRef]
52. Lin, X.; Rivenson, Y.; Yardimci, N.T.; Veli, M.; Luo, Y.; Jarrahi, M.; Ozcan, A. All-optical machine learning using diffractive deep neural networks. *Science* **2018**, *361*, 1004–1008. [CrossRef]
53. Mengu, D.; Luo, Y.; Rivenson, Y.; Ozcan, A. Analysis of diffractive optical neural networks and their integration with electronic neural networks. *IEEE J. Sel. Top. Quantum Electron.* **2019**, *26*, 1–14. [CrossRef]
54. Maktoobi, S.; Froehly, L.; Andreoli, L.; Porte, X.; Jacquot, M.; Larger, L.; Brunner, D. Diffractive coupling for photonic networks: How big can we go? *IEEE J. Sel. Top. Quantum Electron.* **2019**, *26*, 1–8. [CrossRef]
55. Xiao, Y.L.; Li, S.; Situ, G.; You, Z. Unitary learning for diffractive deep neural network. *Opt. Lasers Eng.* **2021**, *139*, 106499. [CrossRef]
56. Xiao, Y.L.; Liang, R.; Zhong, J.; Su, X.; You, Z. Compatible Learning for Deep Photonic Neural Network. *arXiv* **2020**, arXiv:2003.08360.
57. Zhao, Q.; Hao, S.; Wang, Y.; Wang, L.; Xu, C. Orbital angular momentum detection based on diffractive deep neural network. *Opt. Commun.* **2019**, *443*, 245–249. [CrossRef]
58. Fu, T.; Zang, Y.; Huang, H.; Du, Z.; Hu, C.; Chen, M.; Yang, S.; Chen, H. On-chip photonic diffractive optical neural network based on a spatial domain electromagnetic propagation model. *Opt. Express* **2021**, *29*, 31924–31940. [CrossRef]
59. Lu, L.; Zhu, L.; Zhang, Q.; Zhu, B.; Yao, Q.; Yu, M.; Niu, H.; Dong, M.; Zhong, G.; Zeng, Z. Miniaturized diffraction grating design and processing for deep neural network. *IEEE Photonics Technol. Lett.* **2019**, *31*, 1952–1955. [CrossRef]
60. Bernstein, L.; Sludds, A.; Hamerly, R.; Sze, V.; Emer, J.; Englund, D. Freely scalable and reconfigurable optical hardware for deep learning. *Sci. Rep.* **2021**, *11*, 3144. [CrossRef] [PubMed]
61. Zhou, T.; Lin, X.; Wu, J.; Chen, Y.; Xie, H.; Li, Y.; Fan, J.; Wu, H.; Fang, L.; Dai, Q. Large-scale neuromorphic optoelectronic computing with a reconfigurable diffractive processing unit. *Nat. Photonics* **2021**, *15*, 367–373. [CrossRef]
62. Li, J.; Mengu, D.; Luo, Y.; Rivenson, Y.; Ozcan, A. Class-specific differential detection in diffractive optical neural networks improves inference accuracy. *Adv. Photonics* **2019**, *1*, 046001. [CrossRef]
63. Yan, T.; Wu, J.; Zhou, T.; Xie, H.; Xu, F.; Fan, J.; Fang, L.; Lin, X.; Dai, Q. Fourier-space diffractive deep neural network. *Phys. Rev. Lett.* **2019**, *123*, 023901. [CrossRef]
64. Rahman, M.S.S.; Li, J.; Mengu, D.; Rivenson, Y.; Ozcan, A. Ensemble learning of diffractive optical networks. *Light Sci. Appl.* **2021**, *10*, 14. [CrossRef]
65. Chang, J.; Sitzmann, V.; Dun, X.; Heidrich, W.; Wetzstein, G. Hybrid optical-electronic convolutional neural networks with optimized diffractive optics for image classification. *Sci. Rep.* **2018**, *8*, 12324. [CrossRef]
66. Matsushima, K.; Schimmel, H.; Wyrowski, F. Fast calculation method for optical diffraction on tilted planes by use of the angular spectrum of plane waves. *J. Opt. Soc. Am. A* **2003**, *20*, 1755–1762. [CrossRef]
67. Xiao, Y.; Qian, H.; Liu, Z. Nonlinear metasurface based on giant optical kerr response of gold quantum wells. *ACS Photonics* **2018**, *5*, 1654–1659. [CrossRef]
68. Silva, I.N.D.; Hernane Spatti, D.; Andrade Flauzino, R.; Liboni, L.H.B.; Reis Alves, S.F.D. Artificial neural network architectures and training processes. In *Artificial Neural Networks*; Springer: Berlin/Heidelberg, Germany, 2017; pp. 21–28.
69. Marković, D.; Mizrahi, A.; Querlioz, D.; Grollier, J. Physics for neuromorphic computing. *Nat. Rev. Phys.* **2020**, *2*, 499–510. [CrossRef]
70. Radivojevic, M.; Jäckel, D.; Altermatt, M.; Müller, J.; Viswam, V.; Hierlemann, A.; Bakkum, D.J. Electrical identification and selective microstimulation of neuronal compartments based on features of extracellular action potentials. *Sci. Rep.* **2016**, *6*, 31332. [CrossRef] [PubMed]
71. Sokolov, A.S.; Abbas, H.; Abbas, Y.; Choi, C. Towards engineering in memristors for emerging memory and neuromorphic computing: A review. *J. Semicond.* **2021**, *42*, 013101. [CrossRef]
72. Kim, S.G.; Han, J.S.; Kim, H.; Kim, S.Y.; Jang, H.W. Recent advances in memristive materials for artificial synapses. *Adv. Mater. Technol.* **2018**, *3*, 1800457. [CrossRef]

73. Ballato, J.; Hawkins, T.; Foy, P.; Stolen, R.; Kokuoz, B.; Ellison, M.; McMillen, C.; Reppert, J.; Rao, A.; Daw, M.; et al. Silicon optical fiber. *Opt. Express* **2008**, *16*, 18675–18683. [CrossRef] [PubMed]
74. Gambling, W.A. The rise and rise of optical fibers. *IEEE J. Sel. Top. Quantum Electron.* **2000**, *6*, 1084–1093. [CrossRef]
75. Lu, P.; Lalam, N.; Badar, M.; Liu, B.; Chorpening, B.T.; Buric, M.P.; Ohodnicki, P.R. Distributed optical fiber sensing: Review and perspective. *Appl. Phys. Rev.* **2019**, *6*, 041302. [CrossRef]
76. Pickett, M.D.; Medeiros-Ribeiro, G.; Williams, R.S. A scalable neuristor built with Mott memristors. *Nat. Mater.* **2013**, *12*, 114–117. [CrossRef]
77. Shi, J.; Ha, S.D.; Zhou, Y.; Schoofs, F.; Ramanathan, S. A correlated nickelate synaptic transistor. *Nat. Commun.* **2013**, *4*, 2676. [CrossRef]
78. Wright, C.D.; Au, Y.Y.; Aziz, M.M.; Bhaskaran, H.; Cobley, R.; Rodriguez-Hernandez, G.; Hosseini, P.; Pernice, W.H.; Wang, L. Novel Applications Possibilities for Phase-Change Materials and Devices. 2013. Available online: http://hdl.handle.net/10871/20347 (accessed on 5 April 2022).
79. Gholipour, B.; Zhang, J.; MacDonald, K.F.; Hewak, D.W.; Zheludev, N.I. An all-optical, non-volatile, bidirectional, phase-change meta-switch. *Adv. Mater.* **2013**, *25*, 3050–3054. [CrossRef]
80. Gholipour, B.; Bastock, P.; Craig, C.; Khan, K.; Hewak, D.; Soci, C. Amorphous metal-sulphide microfibers enable photonic synapses for brain-like computing. *Adv. Opt. Mater.* **2015**, *3*, 635–641. [CrossRef]
81. Ramos, M.; Bharadwaj, V.; Sotillo, B.; Gholipour, B.; Giakoumaki, A.N.; Ramponi, R.; Eaton, S.M.; Soci, C. Photonic implementation of artificial synapses in ultrafast laser inscribed waveguides in chalcogenide glass. *Appl. Phys. Lett.* **2021**, *119*, 031104. [CrossRef]
82. Miyamoto, D.; Murayama, M. The fiber-optic imaging and manipulation of neural activity during animal behavior. *Neurosci. Res.* **2016**, *103*, 1–9. [CrossRef] [PubMed]
83. Schlegel, F.; Sych, Y.; Schroeter, A.; Stobart, J.; Weber, B.; Helmchen, F.; Rudin, M. Fiber-optic implant for simultaneous fluorescence-based calcium recordings and BOLD fMRI in mice. *Nat. Protoc.* **2018**, *13*, 840–855. [CrossRef]
84. Fischer, B.; Sternklar, S. Image transmission and interferometry with multimode fibers using self-pumped phase conjugation. *Appl. Phys. Lett.* **1985**, *46*, 113–114. [CrossRef]
85. Psaltis, D.; Moser, C. Imaging with multimode fibers. *Opt. Photonics News* **2016**, *27*, 24–31. [CrossRef]
86. Vasquez-Lopez, S.A.; Turcotte, R.; Koren, V.; Plöschner, M.; Padamsey, Z.; Booth, M.J.; Cižmár, T.; Emptage, N.J. Subcellular spatial resolution achieved for deep-brain imaging in vivo using a minimally invasive multimode fiber. *Light Sci. Appl.* **2018**, *7*, 110. [CrossRef]
87. Aisawa, S.; Noguchi, K.; Matsumoto, T. Remote image classification through multimode optical fiber using a neural network. *Opt. Lett.* **1991**, *16*, 645–647. [CrossRef]
88. Marusarz, R.K.; Sayeh, M.R. Neural network-based multimode fiber-optic information transmission. *Appl. Opt.* **2001**, *40*, 219–227. [CrossRef]
89. Rahmani, B.; Loterie, D.; Konstantinou, G.; Psaltis, D.; Moser, C. Multimode optical fiber transmission with a deep learning network. *Light Sci. Appl.* **2018**, *7*, 69.
90. Caramazza, P.; Moran, O.; Murray-Smith, R.; Faccio, D. Transmission of natural scene images through a multimode fibre. *Nat. Commun.* **2019**, *10*, 2029. [CrossRef]
91. Teğin, U.; Yıldırım, M.; Oğuz, İ.; Moser, C.; Psaltis, D. Scalable optical learning operator. *Nat. Comput. Sci.* **2021**, *1*, 542–549. [CrossRef]
92. Tanaka, G.; Yamane, T.; Héroux, J.B.; Nakane, R.; Kanazawa, N.; Takeda, S.; Numata, H.; Nakano, D.; Hirose, A. Recent advances in physical reservoir computing: A review. *Neural Netw.* **2019**, *115*, 100–123. [CrossRef] [PubMed]
93. Vandoorne, K.; Dierckx, W.; Schrauwen, B.; Verstraeten, D.; Baets, R.; Bienstman, P.; Van Campenhout, J. Toward optical signal processing using photonic reservoir computing. *Opt. Express* **2008**, *16*, 11182–11192. [CrossRef]
94. Fiers, M.A.A.; Van Vaerenbergh, T.; Wyffels, F.; Verstraeten, D.; Schrauwen, B.; Dambre, J.; Bienstman, P. Nanophotonic reservoir computing with photonic crystal cavities to generate periodic patterns. *IEEE Trans. Neural Netw. Learn. Syst.* **2013**, *25*, 344–355. [CrossRef]
95. Vinckier, Q.; Duport, F.; Smerieri, A.; Vandoorne, K.; Bienstman, P.; Haelterman, M.; Massar, S. High-performance photonic reservoir computer based on a coherently driven passive cavity. *Optica* **2015**, *2*, 438–446. [CrossRef]
96. Mesaritakis, C.; Syvridis, D. Reservoir computing based on transverse modes in a single optical waveguide. *Opt. Lett.* **2019**, *44*, 1218–1221. [CrossRef]
97. Scofield, A.C.; Sefler, G.A.; Shaw, T.J.; Valley, G.C. Recent results using laser speckle in multimode waveguides for random projections. *Opt. Data Sci.* **2019**, *10937*, 17–24.
98. Cheng, T.Y.; Chou, D.Y.; Liu, C.C.; Chang, Y.J.; Chen, C.C. Optical neural networks based on optical fiber-communication system. *Neurocomputing* **2019**, *364*, 239–244. [CrossRef]
99. Sunada, S.; Kanno, K.; Uchida, A. Using multidimensional speckle dynamics for high-speed, large-scale, parallel photonic computing. *Opt. Express* **2020**, *28*, 30349–30361. [CrossRef]
100. Liu, Z.; Wang, L.; Meng, Y.; He, T.; He, S.; Yang, Y.; Wang, L.; Tian, J.; Li, D.; Yan, P.; et al. All-fiber high-speed image detection enabled by deep learning. *Nat. Commun.* **2022**, *13*, 1433. [CrossRef]
101. Caputi, W.J. Stretch: A time-transformation technique. *IEEE Trans. Aerosp. Electron. Syst.* **1971**, *AES-7*, 269–278. [CrossRef]

102. Lei, C.; Guo, B.; Cheng, Z.; Goda, K. Optical time-stretch imaging: Principles and applications. *Appl. Phys. Rev.* **2016**, *3*, 011102. [CrossRef]
103. Mahjoubfar, A.; Churkin, D.V.; Barland, S.; Broderick, N.; Turitsyn, S.K.; Jalali, B. Time stretch and its applications. *Nat. Photonics* **2017**, *11*, 341–351. [CrossRef]
104. Chen, C.L.; Mahjoubfar, A.; Tai, L.C.; Blaby, I.K.; Huang, A.; Niazi, K.R.; Jalali, B. Deep learning in label-free cell classification. *Sci. Rep.* **2016**, *6*, 21471. [CrossRef]
105. Wu, Y.; Zhou, Y.; Huang, C.J.; Kobayashi, H.; Yan, S.; Ozeki, Y.; Wu, Y.; Sun, C.W.; Yasumoto, A.; Yatomi, Y.; et al. Intelligent frequency-shifted optofluidic time-stretch quantitative phase imaging. *Opt. Express* **2020**, *28*, 519–532. [CrossRef]
106. Mahjoubfar, A.; Chen, C.L.; Lin, J.; Jalali, B. AI-augmented time stretch microscopy. In Proceedings of the High-Speed Biomedical Imaging and Spectroscopy: Toward Big Data Instrumentation and Management II, San Francisco, CA, USA, 28 January–2 February 2017; Volume 10076, p. 10076J.
107. Guo, B.; Lei, C.; Kobayashi, H.; Ito, T.; Yalikun, Y.; Jiang, Y.; Tanaka, Y.; Ozeki, Y.; Goda, K. High-throughput, label-free, single-cell, microalgal lipid screening by machine-learning-equipped optofluidic time-stretch quantitative phase microscopy. *Cytom. Part A* **2017**, *91*, 494–502. [CrossRef]
108. Guo, B.; Lei, C.; Wu, Y.; Kobayashi, H.; Ito, T.; Yalikun, Y.; Lee, S.; Isozaki, A.; Li, M.; Jiang, Y.; et al. Optofluidic time-stretch23 quantitative phase microscopy. *Methods* **2018**, *136*, 116–125. [CrossRef]
109. Lo, M.C.; Lee, K.C.; Siu, D.M.; Lam, E.Y.; Tsia, K.K. Augmented multiplexed asymmetric-detection time-stretch optical microscopy by generative deep learning. In Proceedings of the High-Speed Biomedical Imaging and Spectroscopy VI, Online, 6–12 March 2021; Volume 11654, p. 1165410.
110. Suthar, M.; Jalali, B. Natural algorithms for image and video enhancement. In Proceedings of the AI and Optical Data Sciences II, Online, 6–12 March 2021; Volume 11703, p. 1170315.
111. Zhou, T.; Scalzo, F.; Jalali, B. Nonlinear Schrodinger Kernel for hardware acceleration of machine learning. *J. Lightwave Technol.* **2022**, *40*, 1308–1319. [CrossRef]
112. Jalali, B.; Zhou, T.; Scalzo, F. Time Stretch Computing for Ultrafast Single-shot Data Acquisition and Inference. In Proceedings of the 2021 Optical Fiber Communications Conference and Exhibition (OFC), Washington, DC, USA, 6–11 June 2021; pp. 1–3.

applied sciences

MDPI

Article

Evolution of System Embedded Optical Interconnect in Sub-Top-of-Rack Data Center Systems

Richard Pitwon [1,2,3,*], Anil Reddy [2], Aditya Jain [2], Kevin Gomez [2], Sebastian A. Schulz [1], Liam O'Faolain [1,4], Kai Wang [5], Allen Miller [6,†] and Vivienne Davies [6,‡]

1. School of Physics and Astronomy, University of St Andrews, St Andrews KY16 9AJ, UK; sas35@st-andrews.ac.uk (S.A.S.); William.Whelan-Curtin@mtu.ie (L.O.)
2. Seagate Research Group, Seagate Technology, Shakopee, MN 55379, USA; anil.j.reddy@seagate.com (A.R.); aditya.jain@seagate.com (A.J.); kevin.gomez@seagate.com (K.G.)
3. Resolute Photonics Ltd., Eastleigh SO53 3AL, UK
4. Centre for Advanced Photonics and Process Analysis (CAPPA), Munster Technology University, T12 P928 Cork, Ireland
5. WaveOptics Ltd., Abingdon OX14 4SR, UK; k.wang@enhancedworld.com
6. Seagate Systems UK Ltd., Havant PO9 1SA, UK; millermountain2@hotmail.com

* Correspondence: rcap@st-andrews.ac.uk

† Retired.

‡ Deceased.

Abstract: In this paper we review key technological milestones in system embedded optical interconnects in data centers that have been achieved between 2014 and 2020 on major European Union research and development projects. This includes the development of proprietary optically enabled data storage and switch systems and optically enabled data storage and compute subsystems. We report on four optically enabled data center system demonstrators: LightningValley, ThunderValley2, Pegasus and Aurora, which include advanced optical circuits based on polymer waveguides and fibers and proprietary electro-optical connectors. We also report on optically enabled subsystems including Ethernet-connected hard disk drives and microservers. Both are designed in the same pluggable carrier form factor and with embedded optical transceiver and connector interfaces, thus allowing, for the first time, both compute and storage nodes to be optically interchangeable and directly interconnectable over long distances. Finally, we present the Nexus platform, which allows different optically enabled data center test systems and subsystems to be interconnected and comparatively characterized within a data center test environment.

Keywords: data centers; integrated photonics; silicon photonics; fiber optics; polymer waveguides; copackaged optics; high-performance computers; optical interconnects; optical communications

Citation: Pitwon, R.; Reddy, A.; Jain, A.; Gomez, K.; Schulz, S.A.; O'Faolain, L.; Wang, K.; Miller, A.; Davies, V. Evolution of System Embedded Optical Interconnect in Sub-Top-of-Rack Data Center Systems. *Appl. Sci.* **2022**, *12*, 1565. https://doi.org/10.3390/app12031565

Academic Editors: Fabio Cavaliere and Luca Potì

Received: 18 October 2021
Accepted: 14 December 2021
Published: 31 January 2022

Publisher's Note: MDPI stays neutral with regard to jurisdictional claims in published maps and institutional affiliations.

1. Introduction

The past decade is exemplified by the surge in digital information being captured, processed, stored and moved from one location to another. This data explosion has been precipitated for the most part by the widespread adoption of mobile data devices—predominantly smartphones and tablets—and is pushing modern information and communications systems beyond their design limits and towards a crippling "data cliff".

A major consequence of the adoption of smaller portable mobile data devices over larger static computer terminals (PCs) is that a dramatic shift is now occurring in where customers need to store their information. While it was sufficient to store data locally (such as on the user's local laptop or desktop computer hard drive), the average size of data objects generated, such as high-definition pictures or short videos, has grown to the extent that the storage available on mobile devices is rapidly becoming insufficient for long-term accumulation and retention of data. This has given rise to the emergence of "cloud" services where customers can outsource their data storage and increased computing requirements to very large and secure data centers typically comprising at least 100,000 servers and associated data storage and network switching. These "hyperscale" data centers are run by internet content providers (ICPs) such as Amazon, Google and Microsoft and provide the dedicated compute, storage and server equipment required to meet the remote and diverse data processing and storage requirements of cloud environments. However, in order to cope with rapidly changing customer demand, the architectures underlying the data centers themselves need to evolve, and a critical part of that evolution is the deployment of optical connections at all levels of the data center environment.

Printed circuit boards (PCBs) are at the heart of all modern information and communication technology (ICT) systems. The increase in data communication speeds incurs a toll on ICT systems, such as servers and switches, when higher frequency electronic signals are conveyed along the metal channels used in conventional PCBs. As frequencies of these electronic signals increase, dielectric absorption, skin effect and other resistive loss mechanisms attenuate them more strongly, while signal reflections, signal skew and interference from other electronic channels distort their integrity. Furthermore, the environmental effects of system operation, such as temperature and humidity, cause changes in the circuit board substrate, thus altering the carefully balanced characteristics of the electronic channels. Many of these constraints can be mitigated to some degree, however, at an ever-mounting cost to the overall system design and with an increasing power penalty.

Embedded optical interconnect technologies, whether deployed at the cable level, circuit board level or chip level, offer significant performance and power advantages over conventional electronic interconnect. Performance gains include higher data rates, reduced electromagnetic interference, reduced power consumption, higher channel density and a corresponding reduction in the amount of cable or PCB materials used. Therefore, in order to cope with the exponential increase in capacity, processing power and bandwidth density inside information communication systems, there has been a trend over the past decade to migrate optical channels down from the higher communication tier optical fiber networks into the data communication system enclosure itself. One area in which this is particularly apparent is in modern data centers, where the migration of optical interconnect into top-of-rack (TOR) or other network switch enclosures can substantially mitigate the communications bottlenecks resulting from the increase in both data rate and internal interconnect link lengths. As is discussed toward the end of the paper, a great deal of research is currently underway into "copackaged optics" (CPO) for data center switch ASICs, which have aggregated bandwidths expected to exceed 100 Tb/s by 2026. Intel has been leading the effort to develop CPO based on silicon photonics microtransceivers or "chiplets" coassembled with switch ASICs onto common carriers [1–3] to accommodate these astronomical bandwidths in data center switch ASICs. However, the vast majority of systems in data centers, what we refer to as sub-TOR systems, are servers and data storage arrays, which in hyperscale data centers number in the hundreds of thousands.

In order to assess the viability of embedding optical links within modern sub-TOR data center architectures, three generations of data center systems were developed over the past 5 years based on current storage switch enclosure form factors. In this paper, we review the evolution of system embedded optical interconnect technologies in three key data center demonstration prototype platforms produced by Seagate Systems in the UK: LightningValley, ThunderValley2 and Pegasus, which were adapted from different data storage switch enclosures to allow selected internal high-speed electronic transmission lines to be converted to optical links.

LightningValley is a partially optically interconnected data storage system. It was developed based on a modified 4U24 OneStor enclosure from Seagate Technology, in which 12 Gb/s Serial Attached SCSI (SAS) traffic was conveyed optically between the SAS protocol switches on two internal controller cards along 24 PCB embedded polymer optical waveguide channels, thereby showing, for the first time, how in-system optical channels could be successfully deployed within a 12G SAS architecture [4].

ThunderValley2 is a fully optically enabled data storage array that was developed based on a 2U24 OneStor enclosure from Seagate Technology, in which all internal high-speed links were implemented optically. This required the deployment of commercial midboard optical transceivers, an electro-optical midplane and proprietary pluggable optical connectors for hard disk drives [5].

Pegasus is a fully optically enabled 24 drive Ethernet data storage, switch and compute platform, for which interchangeable optical transceiver mezzanine cards and a proprietary electro-optical drive connector were developed.

Aurora is a test and measurement platform developed to allow comparative characterization of different types of more advanced optical interconnect technologies including embedded and discrete polymer and glass waveguide circuits, optical PCB connectors and transceiver and switch technologies including advanced silicon photonics devices [6].

A converged data center test rack, Nexus, was developed to allow different optically enabled platforms to be interconnected and validated with respect to each other. In particular, this allows for more advanced, less mature technologies such as glass waveguide circuits and silicon photonic transceivers to be quickly validated in fully operational, optically enabled data center systems. Table 1 lists the key parameters of the three optically enabled data center system demonstration platforms showing both the communication protocol used in the reported demonstrators and the next generation of communication protocols, for which the need for system-level optical interconnect would be greater.

Table 1. Optically enabled data center platforms.

System Designation	Communication Protocol for System/ Current (Reported) Protocol Generation/ Next Protocol Generation	System Cards/Modules	Bisection Optical Bandwidth for Current Generation/Next Generation
LightningValley2	Serial Attached SCSI (SAS)/ SAS3 = 12 Gbps/ SAS4 = 24 Gbps	SAS switch controller (2) 3.5″ hard disk drives (24) Midplane (1)	144 Gbps/288 Gbps
ThunderValley	Serial Attached SCSI (SAS)/ SAS3 = 12 Gbps/ SAS4 = 24 Gbps	SAS switch controller (2) 2.5″ hard disk drives (24) Midplane (1)	288 Gbps/572 Gbps
Pegasus	Ethernet/ 10 GbE = 10 Gbps/ 25 GbE = 25 Gbps	SAS switch controller (2) 3.5″ hard disk drives or 3.5″ microservers (24) Midplane (1)	240 Gbps/600 Gbps

Given the low bandwidth requirements of these sub-TOR systems compared to switch enclosures, there has been very limited effort beyond the research and development described in this paper to address optically enabling these systems, though the work described herein sets out a technology ecosystem, which laid the foundation for CPO.

Figure 1 provides a schematic overview of all platforms reported in this paper.

The outlook for the next generation of polymer-based optical interconnects and its combination with silicon photonics and nanophotonics is then reviewed. The LightningValley, ThunderValley2, and Pegasus demonstrations indicate how data center infrastructure architects may be able to configure rack systems in order to maximize the data bandwidth through top-of-rack (TOR) switches. The enablers for this are storage systems that utilize a combination of optical interface cards and burst mode transmission of data. If adopted, this strategy would provide another building block in the shift to disaggregated infrastructure: racks consisting of storage alone could be accessed by a remote server while fully utilizing the bandwidth of the TOR switch.

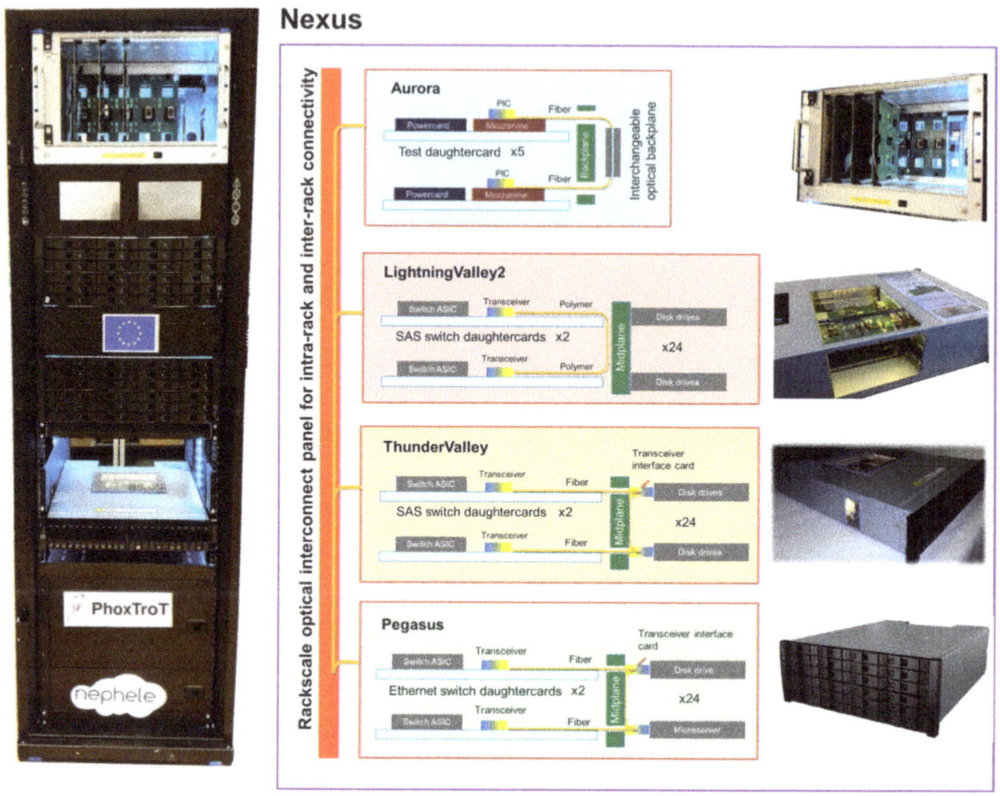

Figure 1. Schematic view of three optically enabled data center platforms (LightningValley2, ThunderValley and Pegasus) and the Aurora test and measurement platform contained within the Nexus rack [7,8], which allows intra-rack and inter-rack connectivity between different platforms.

2. Midboard Optical Transceivers

The main advantage of midboard optical transceivers over more conventional card edge pluggable transceivers, such as the Quad Small Form-Factor Pluggable (QSFP) format [9], is that they can be placed close to the electronic signal source, such as a CPU, a Serial Attached SCSI expander switch chip or an Ethernet switch chip. This substantially reduces the electronic transmission line lengths between signal source and optical transceiver compared to edge pluggable transceivers. The resulting reduced signal integrity degradation enables energy savings at the port of the signal source by limiting or eliminating altogether the need for signal conditioning functions such as pre-emphasis or signal equalization. Furthermore, the reduced signal attenuation will allow the amplitude of the differential signal generated at the signal source to be reduced accordingly, and thus the power consumed by the signal port is also reduced. Depending on the type of signal source and the number of high-speed signal ports, this can result in a reduction in power consumption by as much as half in the signal source chip when the optical module is placed within 5 cm of the signal source chip [10].

2.1. Midboard Optical Transceivers

The choice of midboard optical transceiver on the controller boards for LightningValley, ThunderValley2 and Pegasus was the board-mounted optical assembly (BOA) from Finisar, supporting 12 bidirectional channels [11]. The BOAs used are protocol-independent and support SAS or Ethernet traffic at any line-rate up to 12 Gb/s, thus yielding a maximum bidirectional aggregate bandwidth of 144 Gb/s (12 × 12 Gb/s). Figure 2a shows the composition of the BOA module, which comprises an electrical interposer card, the optical transceiver multichip module, an optical receptacle supporting a 2 × 12 way MT ferrule connection and a heat sink. Figure 2b shows a photo of a BOA module mounted on a LightningValley controller card with a double layer polymer waveguide flexible ribbon attached. Figure 2c shows a photo of a BOA module mounted on an interchangeable mezzanine card in the Pegasus system.

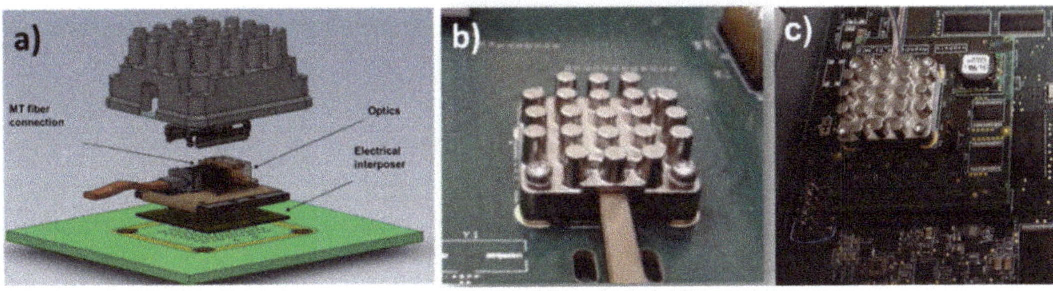

Figure 2. Finisar midboard optical transceiver module package: (**a**) schematic view of Finisar "BOA" module [3], (**b**) photo of Finisar BOA module assembled directly onto a PCB, (**c**) photo of Finisar BOA module on a pluggable small mezzanine card [12].

2.2. *Low Port-Count Optical Transceivers*

In distributed array systems where optical interconnects are required between arrays of end node devices such as hard disk drives or microservers, low port-count transceivers will be required, as shown in Figure 3.

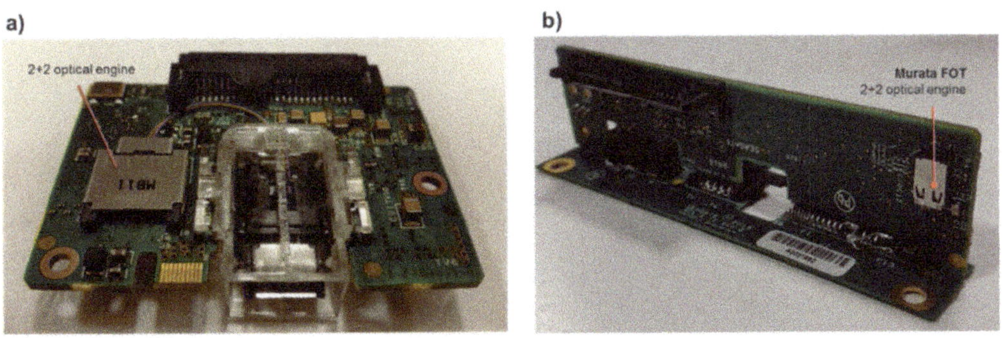

Figure 3. Low port-count transceivers on end node interface cards: (**a**) Avago McLight 2 + 2 transceiver on ThunderValley2 disk drive interface card, (**b**) Murata FOT 2 + 2 transceiver on Pegasus end node interface card.

3. LightningValley—First-Generation Optically Enabled Data Center System

The LightningValley system (Figure 4b) was adapted from a 4U24 OneStor data storage array platform (Figure 4a). The system enclosure is 4U (177.8 mm) high and 19″ (482.6 mm) wide and supports an array of 24 3.5″ hard disk drives, which gives rise to the prefix 4U24. The LightningValley platform was modified to allow 12 bidirectional 12 Gb/s SAS channels to be conveyed optically from the expander chip of one controller daughtercard across an optical bridge on the midplane to the expander chip of the other controller daughtercard in the system (Figure 4c).

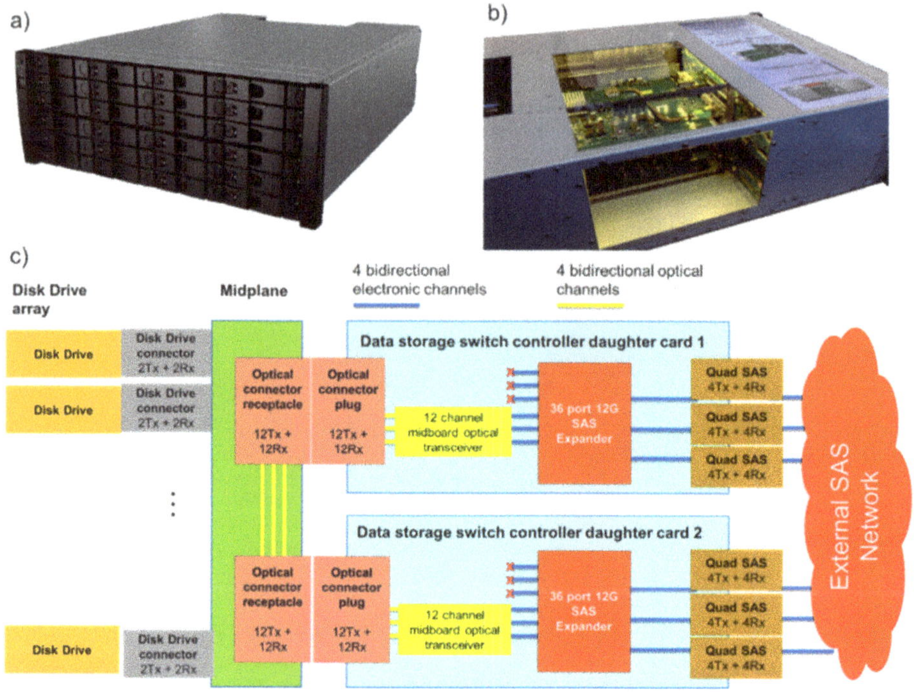

Figure 4. Partially optically enabled data storage system (LightningValley): (**a**) standard 4U24 OneStor storage array system by Seagate Technology, (**b**) photo of LightningValley system, (**c**) schematic view of optically enabled storage system [12].

The LightningValley platform has been described in previous work [12–14]. In this section we focus on the optical interconnect technology and performance.

3.1. Optically Enabled Data Storage Switch Controller Daughtercards with Polymer Waveguide Interconnect

The controller daughtercards were modified such that 12 high-speed bidirectional electronic links from the SAS switch (expander) on the controller daughtercard to the midplane were converted to optical signals through the use of a Finisar BOA midboard optical transceiver module. The optical link on the controller daughtercards between the midboard optical transceiver modules and the optical midplane connector was conveyed across discrete, interchangeable dual-layer polymer optical ribbons. Figure 5 shows the optically modified LightningValley controller card with one BOA midboard optical transceiver mounted and a freestanding polymer waveguide flexible ribbon assembled providing the optical link between the midboard transceiver and the optical midplane connector. Each ribbon comprised two separate layers of polymer waveguides fabricated on an S-curved Kapton polyimide strip. Each layer contained 12 multimode polymer waveguides with a square cross-sectional profile of size 50 µm × 50 µm and separated by a center-to-center pitch of 250 µm, thus matching the horizontal interchannel separation used in MT ferrules. At each end of the ribbon cable, both layers of 12 waveguides were terminated into a bespoke ferrule, the interface of which was fully compliant with a single 2 × 12 MT interface.

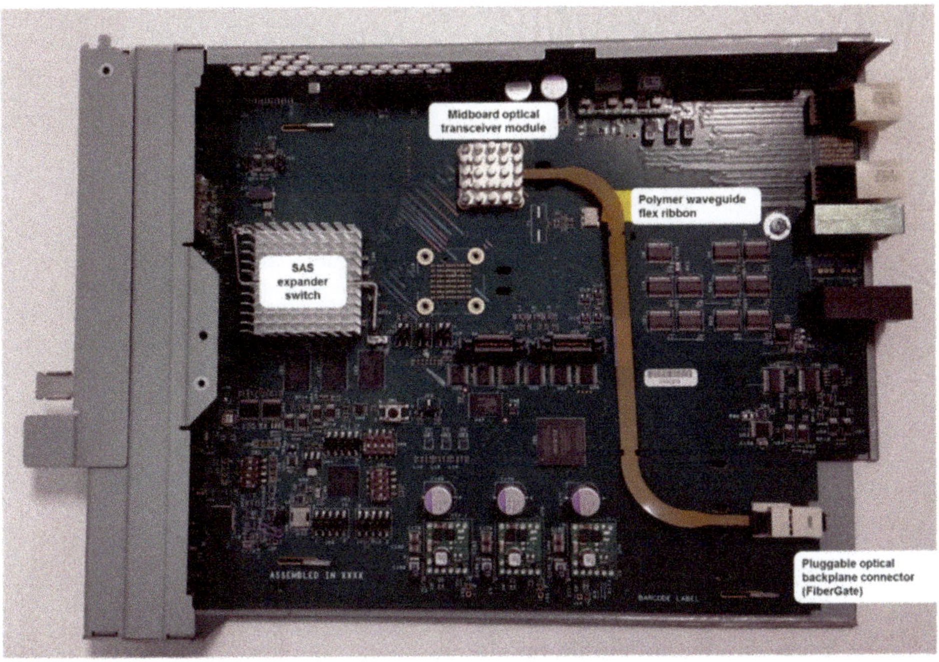

Figure 5. LightningValley optically enabled data storage switch controller daughtercards with polymer waveguide flexible ribbon and Huber + Suhner FiberGate optical backplane connector.

The Swiss company Vario-optics manufactured the two types of dual-layer polymer optical waveguide flexible ribbon. The proprietary polymer acrylate material sets used to fabricate the waveguides were compatible with PCB manufacturing processes such as lamination, through-hole plating and reflow soldering and exhibited acceptable loss changes for long-term (2000 h) tests at elevated temperature (85 °C) and relative humidity (85%) as part of standard reliability tests, e.g., Telcordia GR-1221 [15].

3.2. Polymer Interconnect Measurement Set-Up and Results

Insertion loss measurements were carried out on the polymer waveguide flex using an 850 nm VCSEL test light source. The reason for this is that all commercial midboard transceivers deployed in the reported systems use 850 nm VCSELs, which are high-volume commodity items. The 850 nm VCSEL test source was connected to a 5 m multimode graded-index optical fiber of category OM3 with a core size of 50 µm. The fiber was wound 10 times across a 50 mm diameter mandrel to produce a normalized modal launch condition. The full optical link path shown in Figure 5 is described as follows:

(a) An optical signal generated in the BOA module on the first controller daughtercard couples into the 2 × 12 MT compliant ferrule of the first S-curved polymer waveguide flexible cable.
(b) The signal propagates along the first S-curved flexible polymer waveguide cable.
(c) At the card edge, the optical signal couples into the 2 × 12 MT compliant ferrule of the straight midplane polymer waveguide flex through the Huber + Suhner FiberGate connector.
(d) The signal propagates along the midplane polymer waveguide flex from the first daughtercard location to the second controller daughtercard location.

(e) The signal couples into the 2 × 12 MT compliant ferrule of the second S-curved flexible polymer waveguide cable through the FiberGate connector on the second controlled daughtercard.
(f) The signal propagates along the second S-curved flexible polymer waveguide cable.
(g) The signal couples into the optical interface of the BOA module on the second controller daughtercard.

The total transmitted power was measured across each of the 24 channels and compared to reference measurements in order to evaluate the insertion loss. The results are shown in Figure 6. The result for channel 24 was discarded as the channel was damaged. An average insertion loss of 6.76 dB was measured with a large standard deviation of 3.15 dB with 87% of waveguide links falling below the estimated BOA transceiver link budget of 8 dB. The size and variation in polymer waveguide insertion loss were too large to be considered for future variants of on-board interconnect, and so optical fiber flexplanes were adopted in the subsequent ThunderValley2 and Pegasus systems. It should be noted that since deployment in this early prototype, optical polymer formulations have continued to improve with polysiloxanes from Dow Corning [16] and Ormocer formulations such as "Sunconnect" by Nissan Chemicals [17].

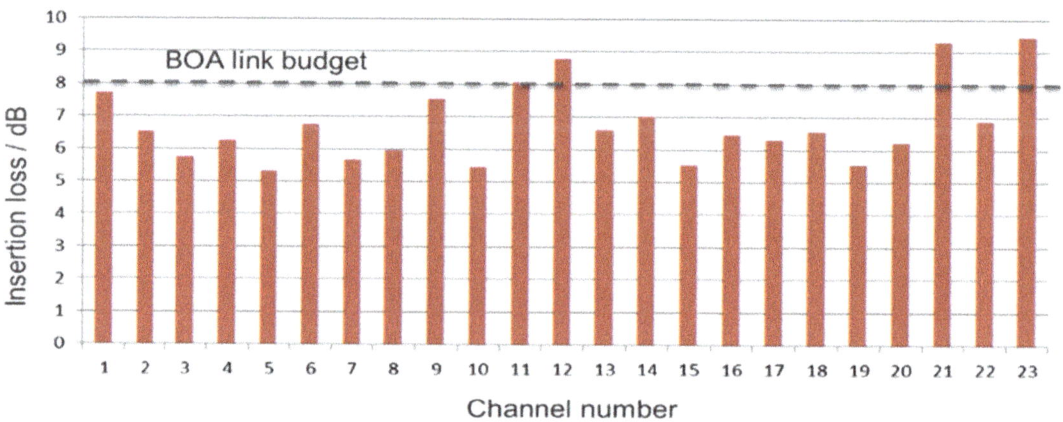

Figure 6. Insertion loss measurements of complete end-to-end link from BOA1 to BOA2.

4. ThunderValley2—Second-Generation Optically Enabled Data Center System

The ThunderValley2 platform was adapted from an existing 2U (89 mm) high, 19″ (482.6 mm) wide OneStor system enclosure (Figure 7a) and included two optically enabled 12G SAS switch controller modules and an electro-optical midplane with a full aggregate bandwidth capacity of 2.3 Tb/s. Additionally, it provides the option to optically plug 24 conventional 2.5″ disk drives to the midplane (Figure 7b). The system architecture is shown in Figure 7c. As with the LightningValley system, the ThunderValley2 system supports two controller daughtercards, but it also allows 24 bidirectional 6 Gb/s SAS channels to be conveyed optically from the expander chip of each controller daughtercard to each of the 24 2.5″ disk drives across a fiber-optic flexplane attached to the midplane. An optical interface card is present on each disk drive to provide the opto-electronic signal conversion.

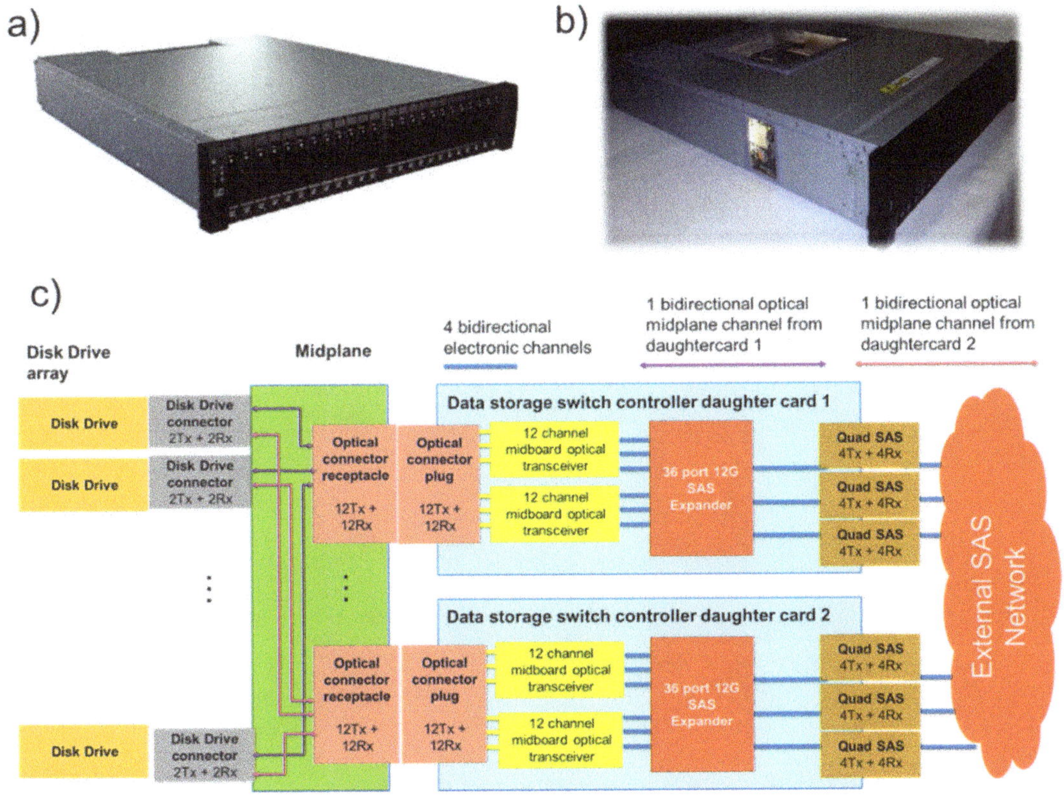

Figure 7. Fully optically enabled SAS data storage system (ThunderValley2): (**a**) standard 2U24 OneStor storage array system, (**b**) photo of ThunderValley2 system [12], (**c**) schematic view of fully optically enabled SAS storage system.

The ThunderValley2 platform was validated, with 6 Gb/s SAS data traffic successfully transmitted along the internal optical links, and has been described in previous work [5]. In this section we focus on the optical interconnect and interface technology and performance.

4.1. Optically Enabled Data Storage Switch Controller Daughtercards with Fiber Interconnect

The controller daughtercards were modified such that all 24 high-speed bidirectional electronic links from the SAS switch (expander) to the midplane were converted to optical signals through the use of two Finisar BOA midboard optical transceiver modules.

The optical links between the two midboard optical transceiver modules and two of the ferrule ports on the commercial Molex HBMT optical midplane connector were conveyed across discrete, interchangeable dual-layer multimode OM3 optical fiber ribbons (Figure 8).

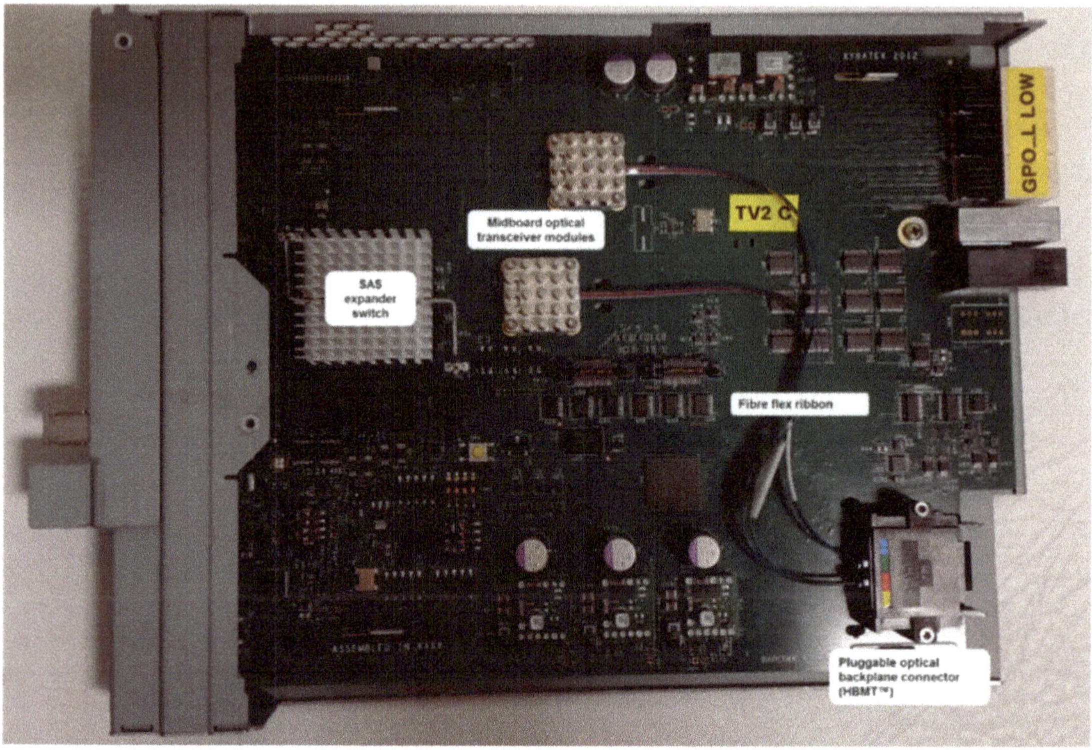

Figure 8. Optically enabled data storage controller cards.

4.2. Electro-Optical Midplane

The electro-optical midplane design was a modified version of the standard 2U midplane, in which all high-speed electrical transmission lines had been removed and provision made for a separate fiber flexible circuit laminate to be attached to the reduced electronic midplane PCB (Figure 9a). Consequently, the number of electronic layers in the midplane PCB was reduced by 55% and the open area available for airflow increased by 20%. The fiber flexplane comprised separate right-hand and left-hand sections (Figure 9b), each providing connectivity to 12 of the 24 slots.

The high-availability interconnect topology, in a passive dual-star configuration, required that each disk drive support two duplex data links situated on the midplane, one to each controller module. As a result, the midplane of the 24 SAS drive enclosure needed to support at least 48 duplex links (96 multimode OM3 fibers). To fully exploit the density advantages of optical interconnect, the midplane was designed for up to 96 duplex links (192 multimode OM3 fibers), thus providing for the future possibility of quad-lane drive interfaces (four independent duplex links per drive), for example through the deployment of PCIe drives, MultiLink SAS or Quad Ethernet interfaces. The midplane provided each drive with a separate, small, electrical connector for the purpose of supplying power and low-speed control signals with an optical midplane receptacle for high-speed SAS signals. The 192-fiber flexplane was terminated with four 2 × 12 MT ferrules on the controller side and twenty-four 1 × 12 MT ferrules on the drive side (of which only eight fibers per ferrule were populated). The flexplane was produced by TE Connectivity as part of the PhoxTroT project, and the optical link performance on all 192 fibers was measured using the same launch and measurement set-up described in Section 3.2 (Figure 9c). The optical link performance showed an average insertion loss of 0.32 dB with six fibers exhibiting abnormally high losses, due to partial damage during assembly. This shows that fiber, due to its maturity, is still far more suited for on-board interconnect than polymer waveguides and will be so for the foreseeable future.

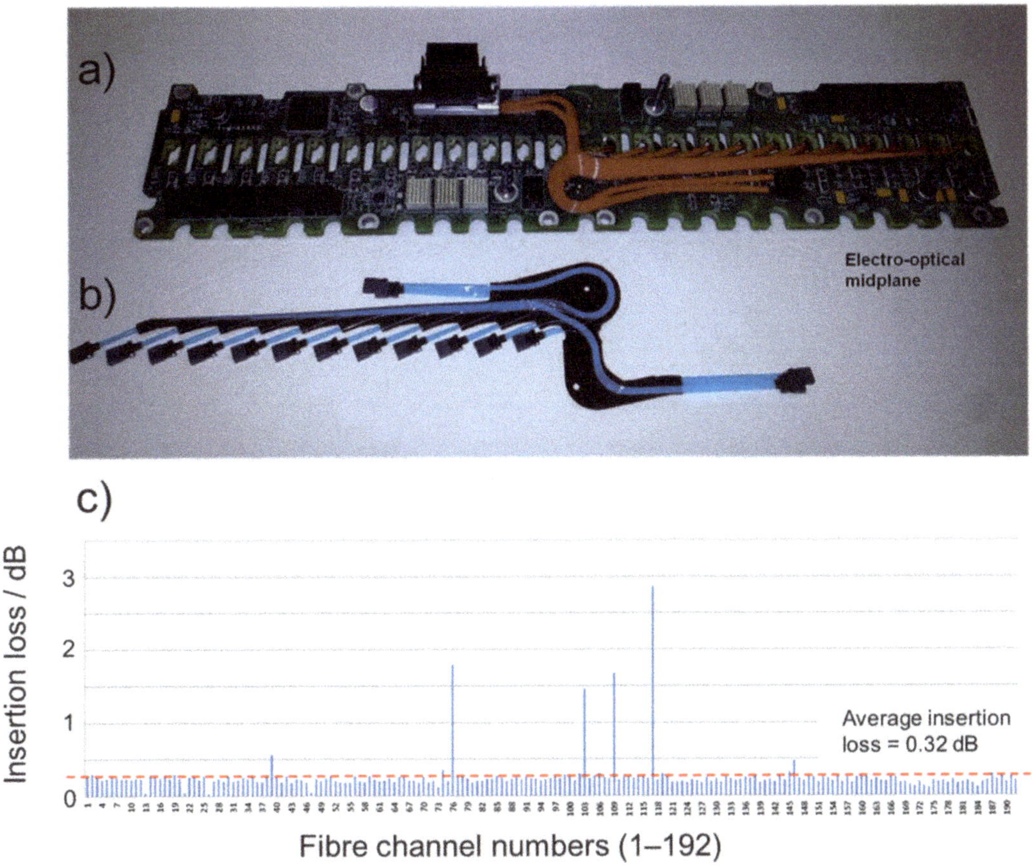

Figure 9. Electro-optical midplane for data storage array: (**a**) electro-optical midplane with only right-hand fiber flexplane populated [5,12], (**b**) left-hand fiber flexplane with 92 fibers [5,12], (**c**) optical link performance of fiber flexplane.

4.3. Optical Disk Drive Interface Card

A special interface card was designed to fit in the disk drive carrier and allow a standard 2.5″ hard disk drive to communicate and connect optically to the electro-optical midplane. Figure 10a shows the link topology of each optical interface card, which contained a proprietary "MorningStar" pluggable optical connector [18], two Avago McLink dual-channel board-mounted optical transceivers, a six-port SAS expander and an electrical connector compliant with conventional SAS/SATA disk drive interfaces. The hard disk drive supported 6 Gb/s SAS and maintained two bidirectional links to allow high availability connectivity. The SAS protocol allows optical communication between SAS devices only when a special mode, "optical OOB", is supported.

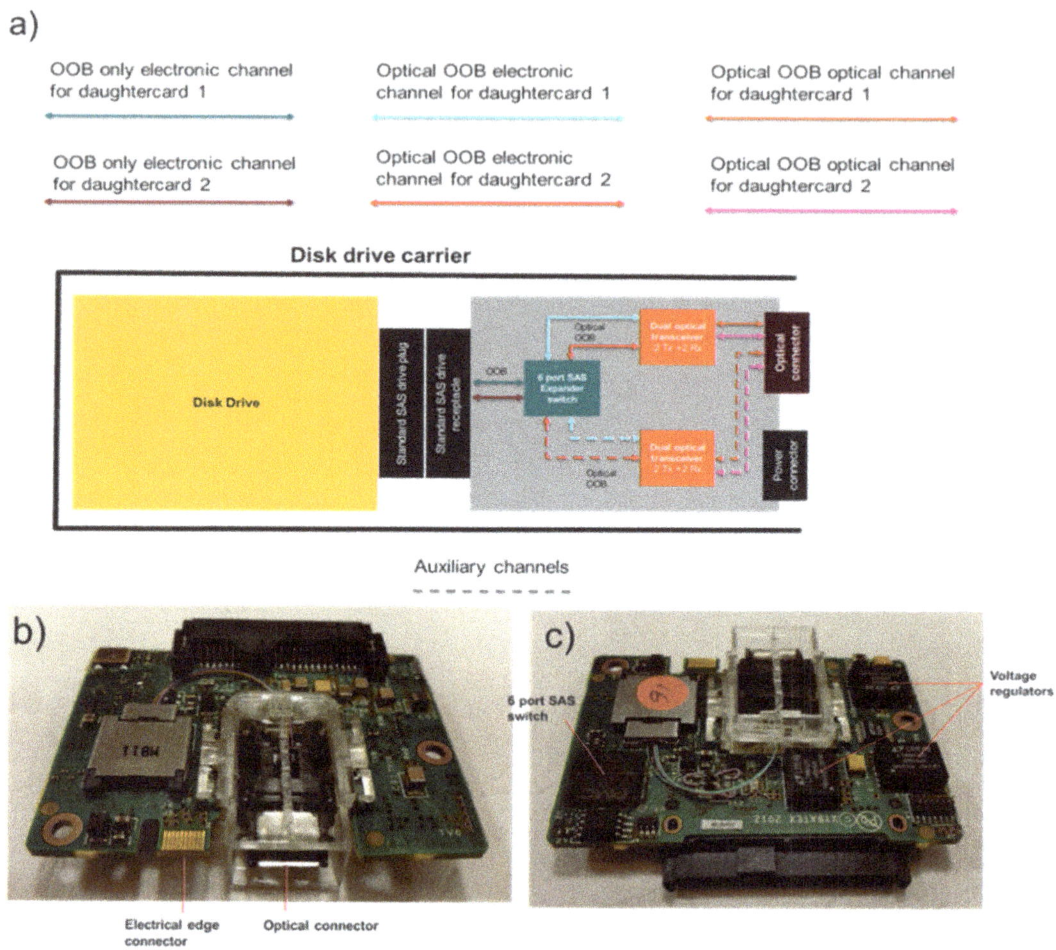

Figure 10. ThunderValley2 optical interface card: (**a**) schematic architecture of optical interface card, (**b**) top side of interface card showing MorningStar connector and first McLink transceiver module, (**c**) bottom side showing SAS switch and second McLink transceiver.

The hard disk drive itself, however, did not support the optical OOB mode, and was restricted to electrical OOB, which inserts electrical idles of defined periods onto the high-speed link, making it unsuitable for direct connection to an optical transceiver. For this purpose, a six-port SAS expander was provided to serve as a bridge between the classical OOB of the disk drive and the optical OOB required to convey the signal through an optical transceiver. Although one dual-channel optical transceiver was sufficient to fully convey the two bidirectional links between the disk drive and the midplane, a second dual optical transceiver was added to serve as an auxiliary link, as shown with dashed lines in Figure 9a. In order to use this transceiver, the six-port SAS switch would simply need to be configured to convey the auxiliary links to the disk drive. The top and bottom sides of the optical interface card are shown in Figure 10b,c respectively.

4.4. Optical Disk Drive "MorningStar" Connector

The MT ferrule in turn was held within a Seagate proprietary "MorningStar" connector plug, which was attached to the edge of the interface card as shown in Figure 11. This parallel optical connector system comprises a plug that resides on the edge of the interface card (Figure 11a–c) and a receptacle that resides on the electro-optical midplane (Figure 11d,e). Guiding features included on the plug allow a receiving MT ferrule from the midplane receptacle to be connected precisely with a compliant MT ferrule in the plug section. The plug and receptacle enable a pluggable optical connection between the disk drive and the electro-optical midplane via the interface card.

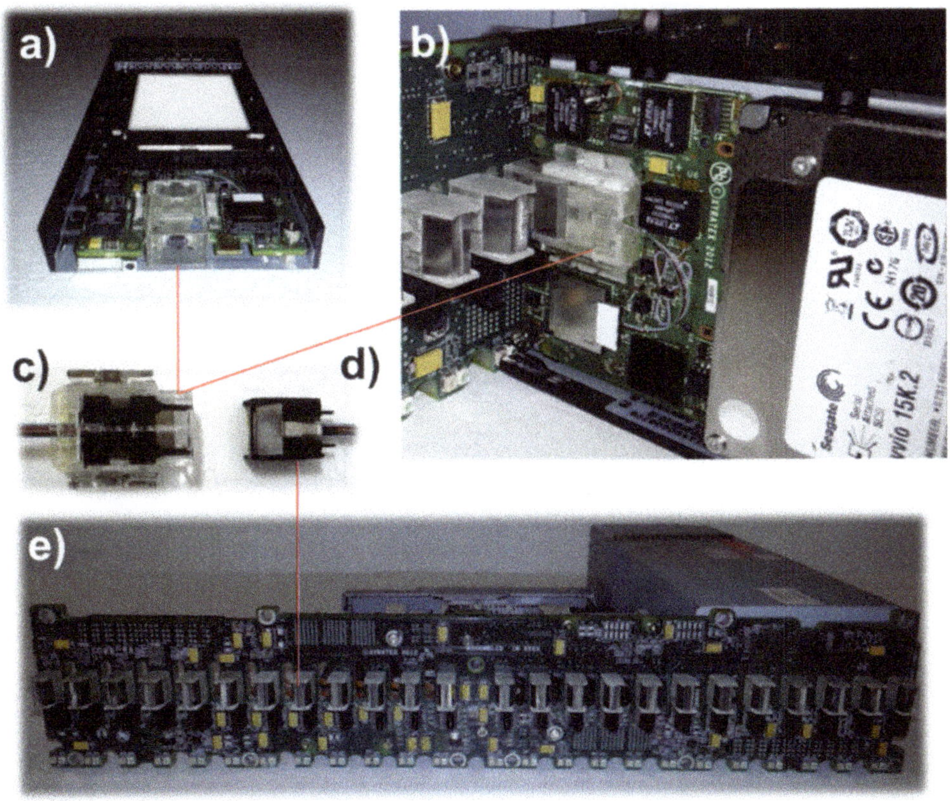

Figure 11. Optical disk drive interface card connectivity: (**a**) optical disk drive interface card assembled into a disk drive carrier connected to a hard disk drive, (**b**) interface card enabled disk drive carrier connecting to an electro-optical midplane, (**c**) MorningStar optical connector plug, (**d**) MorningStar optical connector receptacle, (**e**) electro-optical midplane with MorningStar receptacles highlighted [12].

5. Pegasus—Third-Generation Optically Enabled Data Center System

The final-generation optically enabled data center system, Pegasus, comprised a converged switch, storage and compute platform. The switch controller modules and the compute and storage end nodes were optically interchangeable, independent entities within an object-oriented Ethernet framework, which was demonstrated on the Nephele project [19].

The Pegasus enclosure was based on a Seagate Technology 4U24 OneStor enclosure form factor (Figure 12a). The Pegasus interconnect topology, shown in Figure 12d, is based on a dual star configuration, whereby each end node (optically pluggable disk drive or microserver) supports two bidirectional data links on a proprietary electro-optical midplane (Figure 12c), one to each of two separate prototype Ethernet switch controller modules (Figure 12b).

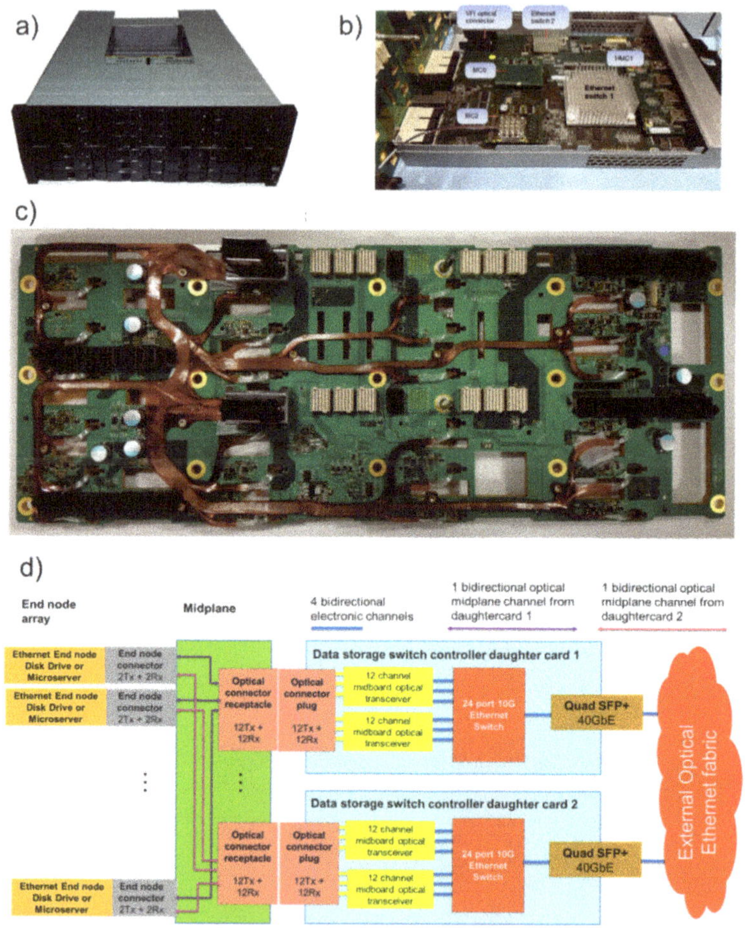

Figure 12. Pegasus—optically enabled, object-oriented converged switch, storage and compute platform: (**a**) photo of Pegasus system [8], (**b**) optically enabled controller card with mezzanine transceiver cards, (**c**) electro-optical midplane with 192-fiber flexplane [8], (**d**) schematic view of interconnect topology.

The Pegasus platform has been described and characterized in previous work [8]. Here we focus on the optical interface technologies.

Optical End Node Connector Interface

To connect the storage or compute end nodes we developed a proprietary electro-optical connector [20] (Figure 13). The connector was based on an SFF-type hard disk drive connector, allowing it to be integrated onto a hard disk drive. This offers the potential for backward-compatible connectivity between current and future generations of disk drives. The connector retains power and high-speed electronic data interfaces; however, one section is reserved for high-density optical communication. This section includes a receptacle for a partially floating PrizmMT ferrule, which can accommodate up to 64 optical channels in a small space.

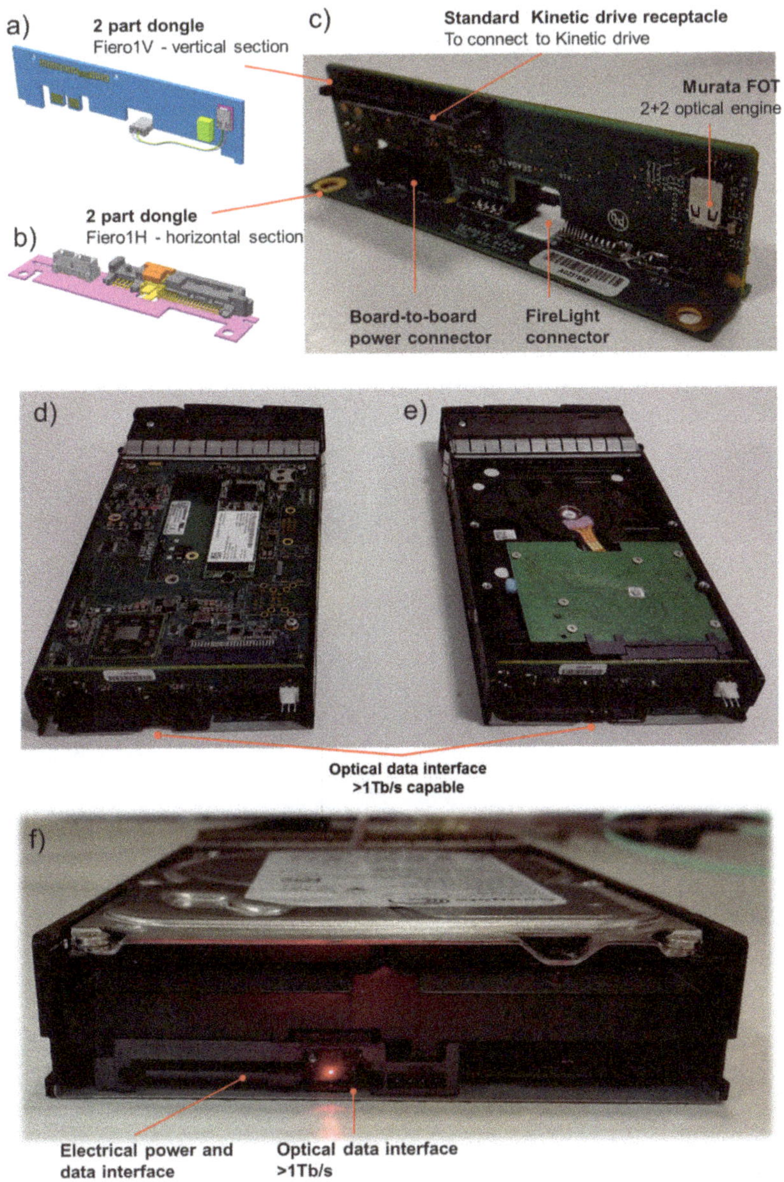

Figure 13. Optical end node connector and interface with proprietary optical end node connector assembled onto a hard disk drive, (**a**) CAD view of vertical interface cards, (**b**) CAD view of horizontal interface card, (**c**) photo of vertical and horizontal interface cards connected, (**d**) microserver end node with optical interface, (**e**) data storage end node with optical interface, (**f**) Ethernet disk drive with proprietary electro-optical connector.

We designed a special two-part interface card (Figure 13c) to fit into a 3.5″ disk drive carrier in order to convert Ethernet data from optical signals from the midplane to electrical signals to the storage or compute end node. These comprised a vertical section holding a 2 + 2 Murata FOT optical transceiver (Figure 13a) and a horizontal section (Figure 13b) holding the electro-optical connector. Two varieties of optically interchangeable end node device were developed: (1) a storage device comprising a 3.5″ hard disk drive with an Ethernet communications interface (Figure 13d) and (2) a compute device, which comprised a microserver platform with an Ethernet communications interface (Figure 13e). Both included the optical interface card and electro-optical connector allowing them to be optically pluggable and interchangeable.

6. Aurora—Hyperscale Integrated Optical and Photonics Ecosystem Demonstrator Platform

The FP7 PhoxTroT, H2020 Nephele and H2020 COSMICC projects developed a test and measurement platform, named "Aurora", that enabled the systematic characterization of optical transceivers, board-to-board optical connectors and both embedded and passive optical circuit boards, as shown in Figure 14 [21,22].

Figure 14. Images taken of EU H2020 "Aurora" hyperscale integrated photonic ecosystem demonstrator during its exhibition at ECOC 2019 in Dublin: (**a**) complete data center rack compliant Aurora platform; (**b**) test daughtercard, which includes two microphotonic test mezzanine cards hosting different PICs; (**c**) silicon PIC with attached vertical grating coupler and fiber array unit; (**d**) silicon PIC with integrated adiabatic coupler and polymer waveguide ribbon; (**e**) microphotonic test mezzanine card; (**f**) glass waveguide backplane provided by Fraunhofer IZM [23].

The cross-project "Aurora" platform is designed to hold different types of advanced optical interconnect technologies including embedded and discrete polymer and glass waveguide circuits, optical PCB connectors and transceiver and switch technologies including advanced silicon photonics devices.

Aurora comprised a test enclosure with interlocking and interchangeable mezzanine electro-optical test boards, electro-optical daughterboards and electro-optical backplanes. This allows the characterization of different optical components with different technical maturity levels either alone or in combination with other technologies. The initial work was carried out during the PhoxTroT project (which concluded in May 2017) [24]. The H2020 Nephele [25] and H2020 COSMICC [26] projects further developed the Aurora universal test platform, building it into an open-source test board design portfolio, which was widely disseminated. Organizations or consortia can use Aurora to test diverse integrated optical and photonic interconnect technologies on a common platform. Researchers only need to focus on the design of small simple test cards using the common design form factors, thus giving significant time and cost savings. The need to design a new evaluation platform from scratch is avoided, allowing researchers to make more technology iterations and accelerate the movement of the target technology through the TRL levels. The test boards can be used on their own or connected to appropriate communication platforms.

An international standard (IEC 62150-6 Ed1—Basic Test and Measurements, Universal Test Cards for Test and Measurement of Micro Board Photonic Devices) will be published in 2022. This standard defines the generic electro-optic mezzanine board for the test and measurement of micro-optical and microphotonic devices, including a wide diversity of photonic integrated circuit (PIC) technologies spanning technologies from transceivers, switches, sensors, neuromorphic networks and LiDAR to quantum integrated circuits. In addition to European research and development projects FP7 PhoxTroT [27], H2020 Nephele [28] and H2020 COSMICC [6], the Aurora platform with new variants of these mezzanine boards has formed the basis of the United States of America High Density User Packaging Group (HDPuG) Opto-electronics project, which culminated in a comprehensive round-robin between multinational organizations including Cisco, Corning, Nokia, Seagate, Fujitsu, Huber + Suhner, Fraunhofer IZM, R&M, Samtec and TTM Technologies.

7. Nexus—Fully Converged Rack-Scale Data Center Test Platform

A converged rack-scale platform, "Nexus", was developed based on the FP7 PhoxTroT project to support the different fully functional optically enabled data storage and switch platforms and provide an interconnection matrix, which would allow them to be characterized and validated for different lengths and configurations of optical interconnects. In addition to the functional optically enabled systems, there is provision for the universal optical interconnect platform, "Aurora" described in the previous section.

As part of the Nexus platform, the Aurora platform allows for advanced, low technical maturity passive and active interconnect technologies to be directly validated with the optically enabled optical data center demonstrators LightningValley2, ThunderValley and Pegasus in the same rack (Figure 15).

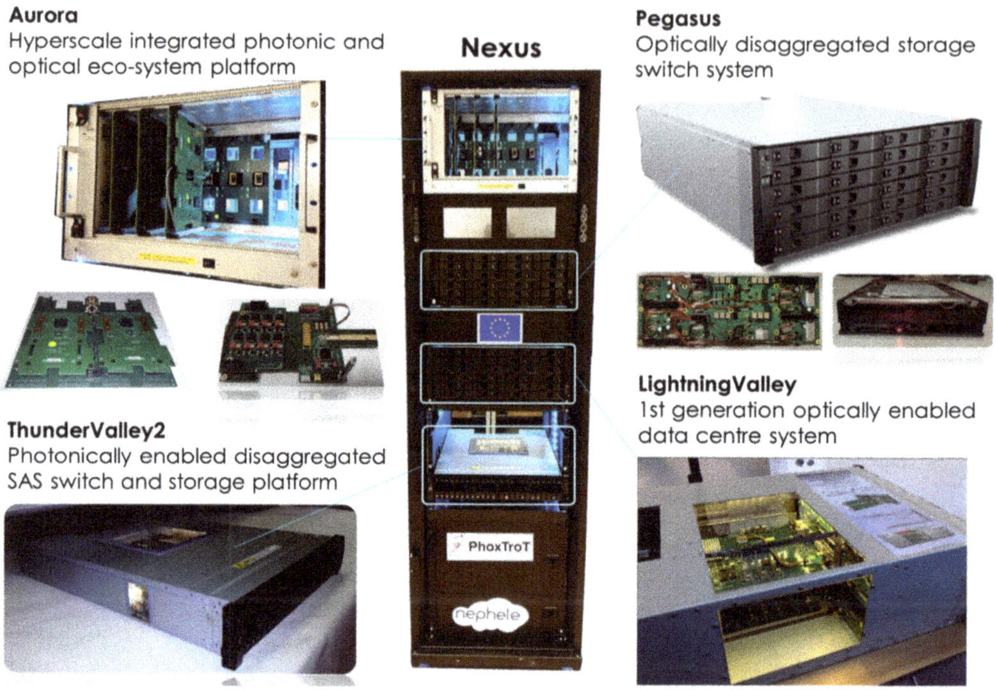

Figure 15. Converged rack-scale data center system [23].

8. Copackaged Optical Assembly—The Ultimate Migration of System-Level Optical Interconnect

Hyperscale data centers are very large data centers typically comprising at least hundreds of thousands of servers and associated storage and networking capacity, and they are run by major internet content providers, such as Google, Microsoft, Amazon and Alibaba, to provide "cloud" services. By 2021 hyperscale data centers have become the dominant form of data center in the world, overtaking private and enterprise data centers, with most organizations outsourcing their data storage requirements to these highly secure facilities with guaranteed quality of service.

Fiber-optic infrastructure is already well established in hyperscale data centers to connect together thousands of racks filled with servers, storage arrays and switch enclosures. External fiber-optic connections are made to optical transceivers, which are plugged onto the "face-plate" of such enclosures, hence the term face-plate pluggable (FPP) modules; however, the exponential increase in data consumption is pushing the bandwidth of data center switch enclosures to the point where there is now not enough space on the face-plate for the transceivers to meet this demand. As shown in Figure 16g, data center switch ASICs of 51 Tbps are expected to be introduced by 2024, and those of 100 Tbps are expected before 2027, which represent substantially higher bandwidths than the bisection bandwidths reported in the demonstration platforms in this paper. There are now massive efforts underway to move optical transceiver cores or "chiplets" into the switch ASIC package itself, in so-called copackaged optical (CPO) assemblies as shown in Figure 16a [29].

As reported in this paper, migration of optical interconnect to the system, board and chip level will be required to substantially mitigate the communications bottlenecks resulting from the increase in both per-lane bandwidth and internal interconnect link lengths. At the board level, this migration will see the development of electro-optical circuit boards comprising detachable fiber shuffles, but in the future, embedded polymer or glass waveguides [30] may play an increasing role if the propagation losses can be reduced to be comparable to those of fiber.

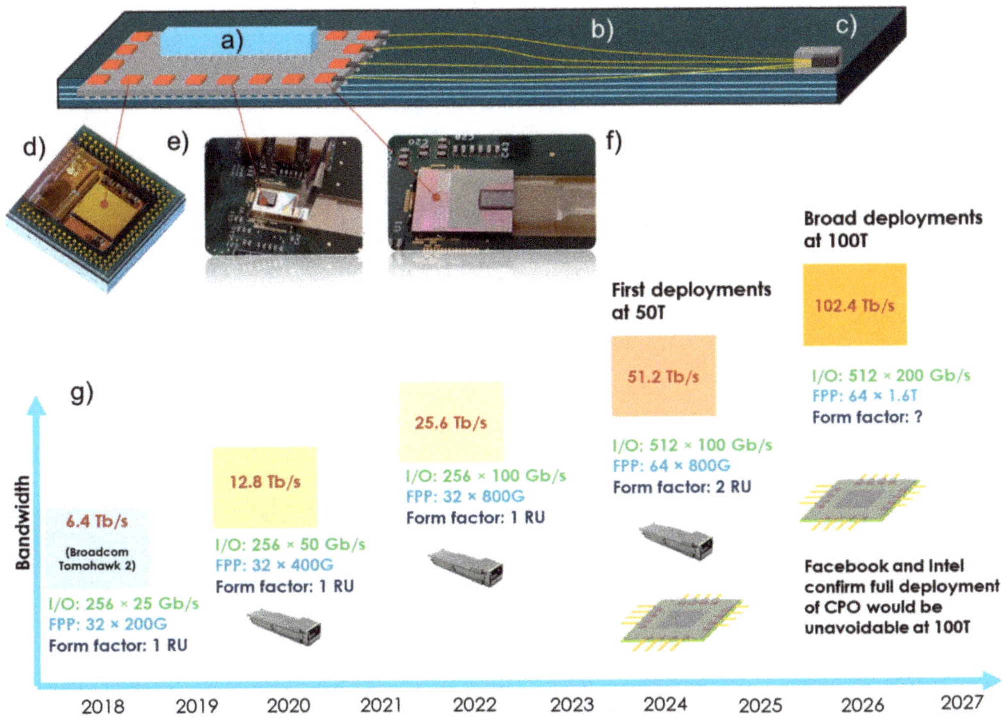

Figure 16. Copackaged optical assembly: (**a**) schematic view of copackaged optical module comprising an ASIC (blue) and transceiver chiplets (red) on common carrier (grey) mounted on PCB (green) of a host card [31], (**b**) optical fiber connections from the copackaged optical module to (**c**) passive optical connector at front face-plate and/or backplane/midplane interface, (**d**) example of transceiver chiplet (courtesy of AIO Core Ltd.) [31,32], (**e**) 1st generation silicon photonics transceiver from H2020 COSMICC, (**f**) 2nd generation of silicon photonics transceiver from H2020 COSMICC project [33], (**g**) projected timeline for data center switch ASICs with increasing aggregate bandwidths.

While front-pluggable and midboard transceivers are evolving, at the chip level, photonic integrated circuit technology (e.g., silicon photonics) has opened the door to the deployment of microtransceiver chiplets (Figure 16d–f) in CPO modules. Front-panel, midplane, backplane and midboard passive optical connectors will be an essential part of this ecosystem connecting system to board to chip.

In this paper we have reported on the first generation of midboard transceivers introduced in 2012; however, by 2020 these fixed modules had become largely obsolete due to lack of field replaceability. The second generation of midboard transceivers specified by the Consortium for On-Board Optics (COBO) are pluggable onto the middle of the board [34]; however, as they are still in separate modules, they would take up a lot of valuable board area. Therefore, it is expected that in such switch enclosures midboard transceivers will be bypassed altogether with a combination of FPPs and CPO modules used to accommodate the huge data center switch bandwidths.

Currently, a 12.8 Tbps switch can be accommodated by 32 400 G face-plate pluggable (FPP) transceiver modules, which fill the face-plate area of a standard 1 RU high enclosure. In order to accommodate a 51.2 Tbps switch with four times the I/O bandwidth of a

12.8 Tbps switch, it is envisaged that sixteen 3.2 Tbps CPO engines [35] will be required with potentially four times the number of fibers at the front face-plate. In addition, most CPO solutions will use an external light source (ELS) to provide the source of continuous-wave light to the modulators in the chiplets, and this ELS module will preferably also be a pluggable module on the face-plate, which will further reduce the available space on the face-plate.

The media by which optical signals are conveyed between the on-board CPO module and the passive optical connectors on the front face-plate will at first be fiber due to its maturity and availability. In applications where the number of fibers becomes prohibitive, sophisticated fiber management solutions will be introduced, such as under-board fiber shuffles. The media on the CPO module itself to provide the possibly hundreds of optical connections between the chiplets and the main board outside the CPO module will also be fiber-based in the short term; however, other media such as polymer waveguides [36] and planar glass waveguides [37,38] have also been investigated.

9. Conclusions and Further Work

We have reported on the evolution of system embedded optical interconnect technologies in optically enabled sub-TOR data center systems. The architectures of three generations of optically enabled data storage, switch and compute platforms have been described; the first two generations demonstrated partially and fully optically enabled data storage and switch capability based on the SAS protocol while the final generation demonstrated a fully optically disaggregated, object-oriented system based on Ethernet and the Seagate OpenStorage protocol. In addition, we have described and comparatively characterized electro-midplanes based on multimode polymer and multimode fiber flexible circuits, described optical interface cards allowing standard edge storage and compute devices to be connected into an optical topology and introduced a proprietary electro-optical connector for object-oriented end nodes.

Looking ahead to the continual evolution of optical interconnects, which are expected to mature as a prerequisite to complete disaggregation of compute, memory and storage, the Aurora and Nexus platforms provide mechanisms by which third-party vendors providing optical hardware can be assessed for performance and compared against competing offerings.

Author Contributions: Conceptualization, R.P., L.O., S.A.S. and A.R.; methodology, R.P. and K.W.; validation, R.P., K.W., A.M. and V.D.; investigation, R.P., K.W., A.M. and V.D.; writing—original draft preparation, R.P. and L.O.; writing—review and editing, A.R., A.J., K.G., S.A.S. and L.O.; visualization, A.R., A.J., K.G. and R.P.; project administration, R.P. All authors have read and agreed to the published version of the manuscript.

Funding: This research was funded by the EU FP7 project "PhoxTrot", for which it has received funding from the European Union Seventh Framework Programme (FP7/2007–2013) under grant agreement No. 318240, the Horizon2020 Nephele project (Grant No. 645212), the Horizon2020 COSMICC project (Grant No. 688516).

Institutional Review Board Statement: Not applicable.

Informed Consent Statement: Not applicable.

Data Availability Statement: Not applicable.

Conflicts of Interest: The authors declare no conflict of interest.

References

1. Mahajan, R.; Li, X.; Fryman, J.; Zhang, Z.; Nekkanty, S.; Tadayon, P.; Jaussi, J.; Shumarayev, S.; Agrawal, A.; Jadhav, S.; et al. Co-Packaged Photonics for High Performance Computing: Status, Challenges and Opportunities. *J. Light. Technol.* **2021**, *40*, e3104725. [CrossRef]
2. Fathololoumi, S.; Nguyen, K.; Mahalingam, H.; Sakib, M.; Li, Z.; Seibert, C.; Montazeri, M.; Chen, J.; Doylend, J.K.; Jayatilleka, H.; et al. 1.6 Tbps Silicon Photonics Integrated Circuit for Co-Packaged Optical-IO Switch Applications. In Proceedings of the Optical Fiber Communication Conference (OFC), San Diego, CA, USA, 13 March 2020; p. T3H.1.
3. Liao, L.; Fathololoumi, S.; Hui, D. High Density Silicon Photonic Integrated Circuits and Photonic Engine for Optical Co-packaged Ethernet Switch. In Proceedings of the 2020 European Conference on Optical Communications (ECOC), Brussels, Belgium, 20–24 September 2020; pp. 1–4. [CrossRef]
4. Schmidtke, K.; Flens, F.; Worrall, A.; Pitwon, R.; Betschon, F.; Lamprecht, T.; Kraehenbuehl, R. 960 Gb/s optical backplane using embedded polymer waveguides and demonstration in a 12G SAS storage array. In Proceedings of the 2013 Optical Interconnects Conference, Santa Fe, NM, USA, 5–8 May 2013; pp. 29–30. [CrossRef]
5. Pitwon, R.; Worrall, A.; Stevens, P.; Miller, A.; Wang, K.; Schmidtke, K. Demonstration of fully enabled data center subsystem with embedded optical interconnect. In Proceedings of the Optical Interconnects Xiv, San Francisco, CA, USA, 8 March 2014; Volume 8991, p. 899110. [CrossRef]
6. Pitwon, R.C.; Immonen, M.; Schroeder, H.; Neitz, M.; Wang, K. Universal test system for system embedded optical interconnect. In Proceedings of the SPIE—The International Society for Optical Engineering, San Francisco, CA, USA, 22 February 2018; Volume 10538, p. 1053804.
7. Pitwon, R.; O'Faolain, L. Modular test system for high-speed silicon photonics transceivers. In Proceedings of the SPIE—The International Society for Optical Engineering, San Francisco, CA, USA, 4 March 2019; Volume 10924, p. 109240I.
8. Pitwon, R.; Wang, K.; Worrall, A. Converged photonic data storage and switch platform for exascale disaggregated data centers. In Proceedings of the SPIE—The International Society for Optical Engineering, San Francisco, CA, USA, 20 February 2017; Volume 10109. [CrossRef]
9. Ghiasi, A.; Nowell, M.; Sommers, S. QSFP-DD/QSFP-DD800/QSFP112 Hardware Specification for QSFP DOUBLE DENSITY 8X AND QSFP 4X PLUGGABLE TRANSCEIVERS. QSFD-DD MSARevision 6.0128 May 2021. Available online: http://www.qsfp-dd.com/wp-content/uploads/2021/05/QSFP-DD-Hardware-Rev6.01.pdf (accessed on 15 October 2021).
10. Schmidtke, K.; Flens, F. Trends and Future Directions in Optical Interconnects for Datacenter and Computer Applications. In *Proceedings of the Optical Fiber Communication Conference Postdeadline Papers*; The Optical Society: San Diego, CA, USA, 2010; p. OTuP4.
11. Schmidtke, K.E.; Flens, F.; Mahgarefteh, D. Taking Optics to the Chip: From Board-mounted Optical Assemblies to Chip-level Optical Interconnects. In Proceedings of the Optical Fiber Communication Conference, San Francisco, CA, USA, 9 March 2014; pp. 1–3.
12. Maniotis, P.; Terzenidis, N.; Siokis, A.; Christodoulopoulos, K.; Varvarigos, E.; Immonen, M.; Yan, H.J.; Zhu, L.X.; Hasharoni, K.; Pitwon, R.; et al. Application-Oriented On-Board Optical Technologies for HPCs. *J. Light. Technol.* **2017**, *35*, 3197–3213. [CrossRef]
13. Schmidtke, K.; Flens, F.; Worrall, A.; Pitwon, R.; Betschon, F.; Lamprecht, T.; Krahenbuhl, R. 960 Gb/s Optical Backplane Ecosystem Using Embedded Polymer Waveguides and Demonstration in a 12G SAS Storage Array. *J. Light. Technol.* **2013**, *31*, 3970–3975. [CrossRef]
14. Flens, F. Fiber Optics Packaging. In *Proceeding of the Optical Fiber Communication Conference, Los Angeles, CA, USA, 22–26 March 2015*; Paper Th1G.4; Optica Publishing Group: Washington, DC, USA, 2015.
15. Telcordia. Generic Reliability Assurance Requirements for Passive Optical Components. 2010. Available online: https://telecom-info.njdepot.ericsson.net/ido/AUX/GR_1221_TOC.i02.pdf (accessed on 15 October 2021).
16. Kruse, K.; Middlebrook, C. Laser-direct writing of single mode and multi-mode polymer step index waveguide structures for optical backplanes and interconnection assemblies. *Photonics Nanostruct.-Fundam. Appl.* **2015**, *13*, 66–73. [CrossRef]
17. Nawata, H.; Oshima, J.; Kashino, T. Organic-inorganic hybrid material SUNCONNECT(R) for photonic integrated circuit. In *Optical Interconnects XVIII*; International Society for Optics and Photonics: Bellingham, WA, USA, 2018; Volume 10538, pp. 70–78. [CrossRef]
18. Pitwon, R.C.A.; Brusberg, L.; Schroder, H.; Whalley, S.; Wang, K.; Miller, A.; Stevens, P.; Worrall, A.; Messina, A.; Cole, A. Pluggable Electro-Optical Circuit Board Interconnect Based on Embedded Graded-Index Planar Glass Waveguides. *J. Light. Technol.* **2014**, *33*, 741–754. [CrossRef]
19. Tokas, K.; Spatharakis, C.; Patronas, I.; Bakopoulos, P.; Landi, G.; Christodoulopoulos, K.; Capitani, M.; Kyriakos, A.; Aziz, M.; Pitwon, R.; et al. Real Time Demonstration of an End-to-End Optical Datacenter Network with Dynamic Bandwidth Allocation. In Proceedings of the 44th European Conference on Optical Communication-ECOC, Rome, Italy, 23–27 September 2018; pp. 1–3.
20. Pitwon, R.; Miller, A. Optical Connectors. U.S. Patent 10,732,360, 4 August 2020.
21. Kurata, K.; Pitwon, R. Short reach, low cost silicon photonic micro-transceivers for embedded and co-packaged system integration. In Proceedings of the Optical Interconnects XX, San Francisco, CA, USA, 1–6 February 2020.
22. Pitwon, R.; O'Faolain, L.; Kurata, K.; Lee, B.; Ninomyia, T. Hyperscale Integrated Optical and Photonic Interconnect Platform. In Proceedings of the 2020 IEEE Photonics Conference (IPC), Vancouver, BC, Canada, 28 September–1 October 2020. [CrossRef]

23. Pitwon, R. System embedded photonic interconnect for mega-data centre environments. In Proceedings of the 2016 IEEE CPMT Symposium Japan (ICSJ), Kyoto, Japan, 7–9 November 2016; pp. 87–90. [CrossRef]
24. Fraunhofer, I.Z.M. Press Release PhoxTroT: Optical Interconnect Technologies Revolutionized Data Centers and HPC Systems. Available online: http://www.phoxtrot.eu/press-release/ (accessed on 23 December 2017).
25. National Technical University of Athens. Nephele Project. Available online: http://www.nepheleproject.eu (accessed on 8 December 2015).
26. CEA-LETI. Horizon 2020 COSMICC Project. Available online: http://www.h2020-cosmicc.com (accessed on 15 October 2021).
27. Hakansson, A.; Tekin, T.; Brusberg, L.; Pleros, N.; Vyrsokinos, C.; Apostolopoulos, D.; Pitwon, R.; Miller, A.; Wang, K.; Tulli, D.; et al. PhoxTroT—A European initiative toward low cost and low power photonic interconnects for data centres. In Proceedings of the 2015 17th International Conference on Transparent Optical Networks, Budapest, Hungary, 5–9 July 2015; pp. 1–5.
28. Bakopoulos, P.; Christodoulopoulos, K.; Landi, G.; Aziz, M.; Zahavi, E.; Gallico, D.; Pitwon, R.; Tokas, K.; Patronas, I.; Capitani, M.; et al. NEPHELE: An End-to-End Scalable and Dynamically Reconfigurable Optical Architecture for Application-Aware SDN Cloud Data Centers. *IEEE Commun. Mag.* **2018**, *56*, 178–188. [CrossRef]
29. Stone, R.; Chen, R.; Rahn, J.; Venkataraman, S.; Wang, X.; Schmidtke, K.; Stewart, J. Co-packaged Optics for Data Center Switching. In Proceedings of the 2020 European Conference on Optical Communications (ECOC), Brussels, Belgium, 6–10 December 2020; pp. 1–3. [CrossRef]
30. Pitwon, R.; Yamauchi, A.; Brusberg, L.; Wang, K.; Ishigure, T.; Neitz, M.; Worrall, A. Planar polymer and glass graded index waveguides for data center applications. *Opt. Interconnects XVI* **2016**, *9753*, 97530. [CrossRef]
31. Kurata, K.; Hagihara, Y.; Kurihara, M.; Yashiki, K.; Kinoshita, K.; Shiba, K.; Kuwata, M.; Muto, T.; Kobayashi, S.; Baba, N.; et al. Short reach, high temperature operation and high reliability silicon photonic micro-transceivers for embedded and co-packaged system integration. In *Optical Interconnects XXI*; SPIE: Bellingham, WA, USA, 2021; Volume 11692, p. 1169204.
32. Kurata, K.; Giorgi, L.; Cavaliere, F.; O'Faolain, L.; Schulz, S.A.; Nishiyama, K.; Hagihara, Y.; Yashiki, K.; Muto, T.; Kobayashi, S.; et al. Silicon Photonic Micro-Transceivers for Beyond 5G Environments. *Appl. Sci.* **2021**, *22*, 10955. [CrossRef]
33. Olivier, S.; Sciancalepore, C.; Hassan, K.; Fowler, D.; Ben Bakir, B.; Ferroti, T.; Duprez, H.; Durel, J.; Abraham, A.; Plantier, S.; et al. Silicon photonic transceivers for beyond 1-Tb/s datacom applications (Conference Presentation). In *Optical Interconnects XVII*; SPIE: Bellingham, WA, USA, 2017; Volume 10109, p. 101090. [CrossRef]
34. Consortium For On-Board Optics. Trends in Optical Networking Communications. 2020. Available online: https://0175c8a8-ec1a-451d-86d7-8291b141a6cb.filesusr.com/ugd/7bc1dc_66d86c38c1d54d04992bec153d6359a0.pdf (accessed on 15 October 2021).
35. 2 Tb/s Copackaged Optics Optical Module Product Requirements Document. 2021. Available online: http://www.copackagedoptics.com/wp-content/uploads/2021/02/JDF-3.2-Tb_s-Copackaged-Optics-Module-PRD-1.0.pdf (accessed on 15 October 2021).
36. Amano, T.; Noriki, A.; Tamai, I.; Ibusuki, Y.; Ukita, A.; Suda, S.; Kurosu, T.; Takemura, K.; Aoki, T.; Shimura, D.; et al. Polymer Waveguide-coupled Co-packaged Silicon Photonics-die Embedded Package Substrate. In Proceedings of the Optical Fiber Communication Conference (OFC) 2021, Washington, DC, USA, 7 June 2021; p. Th4A.1.
37. Brusberg, L.; Grenier, J.R.; Matthies, J.; Miller, A.M.; Terwilliger, C.C.; Clark, J.S.; Zeng, B.; Beneke, P. Passive Aligned Glass Waveguide Connector for Co-Packaged Optics. In Proceedings of the 2021 European Conference on Optical Communication (ECOC), Bordeaux, France, 23–25 September 2021; pp. 1–4.
38. Brusberg, L.; Zakharian, A.R.; Kocabas, S.E.; Yeary, L.W.; Grenier, J.R.; Terwilliger, C.C.; Bellman, R.A. Glass Substrate with Integrated Waveguides for Surface Mount Photonic Packaging. *J. Light. Technol.* **2021**, *39*, 912–919. [CrossRef]

applied sciences

MDPI

Article

Multiple-Valued Logic Modelling for Agents Controlled via Optical Networks

Alexey Yu. Bykovsky

P.N. Lebedev Physical Institute RAS, Leninsky pr. 53, 119991 Moscow, Russia; bykovskiyay@lebedev.ru

Abstract: The methods of data verification are discussed, which are intended for the distant control of autonomous mobile robotic agents via networks, combining optical data links. The problem of trust servers is considered for position verification and position-based cryptography tasks. In order to obtain flexible quantum and classical verification procedures, one should use the collective interaction of agents and network nodes, including some elements of the blockchain. Multiple-valued logic functions defined within discrete k-valued Allen–Givone algebra are proposed for the logically linked list of entries and the distributed ledger, which can be used for distant data verification and breakdown restoration in mobile agents with the help of partner network nodes. A distributed ledger scheme involves the assigning by distant partners of random hash values, which further can be used as keys for access to a set of distributed data storages, containing verification and restoration data. Multiple-valued logic procedures are simple and clear enough for high-dimensional logic modelling and for the design of combined quantum and classical protocols.

Keywords: multiple-valued logic; data verification; quantum key; linked list; ledger

Citation: Bykovsky, A.Y. Multiple-Valued Logic Modelling for Agents Controlled via Optical Networks. *Appl. Sci.* **2022**, *12*, 1263. https://doi.org/10.3390/app12031263

Academic Editors: Fabio Cavaliere and Luca Potì

Received: 30 November 2021
Accepted: 17 January 2022
Published: 25 January 2022

1. Introduction

1.1. Data Protection Problem of Network Robotics

Modern fiber optics networks have become the base for high throughput communication networks combining 5/6G Internet [1,2], big data services [3], quantum key distribution (QKD) lines [4,5], and futuristic quantum computing systems [6]. However, now they are also a good candidate for the massive distant control of autonomous robotic devices, which involves the Internet of unmanned transport vehicles (IoV) [7], industrial systems [8], medicine nursing robots [9], the Internet of Things (IoT) [10], and smart cities [11], supported by mechatronics drives reproducing human movements [12] and computer vision systems [13,14]. Such autonomous robots are planned to be used as separate devices [15] in a collective of collaborating agents commonly called the multiagent system (MAS) [16] or in a global network-centric system [17]. All such designs are based on methods of artificial intelligence (AI), which is the method of imitating the individual and collective work of people [15]. As mobile robotic agents need wireless data links, communication networks should combine fiber optics lines and wired Ethernet pairs with wireless Wi-Fi, Bluetooth, and atmospheric satellite or terrestrial laser lines [4,5,18] (see Figure 1). Modern smartphones [19] supporting AI algorithms can also be used for the distant control of robots.

Thus, modern networks add many new participants of network data exchange and enlarge the number of possible hardware and software platforms for the interaction of robotic agents with distant users. The modelling of such systems becomes very complicated as, besides network communications, it should describe human decision making and collective work procedures [15]. The limiting factor for network robotics is the data protection problem, as modern networks and especially their wireless components cannot provide the necessary level of protection from attacks and data leakages [20,21]. Quantum key distribution systems [4,5,18,22,23] partially improve the situation, but even they do not exclude specific quantum attacks and malicious impacts on supporting classical components. Unfortunately, data leakages due to disloyal personnel have become more of an actual problem than computer viruses and attacks aimed at stationary network nodes [23]. Furthermore, distant mobile robotic agents have many additional vulnerabilities, as current computer vision and decision-making systems [14,15] cannot replace human monitoring. At the same time, mass and unexpensive network robotics need to avoid trusted network nodes, exploiting the special design of servers and intense human control [24,25] that are mainly appropriate for large corporative systems.

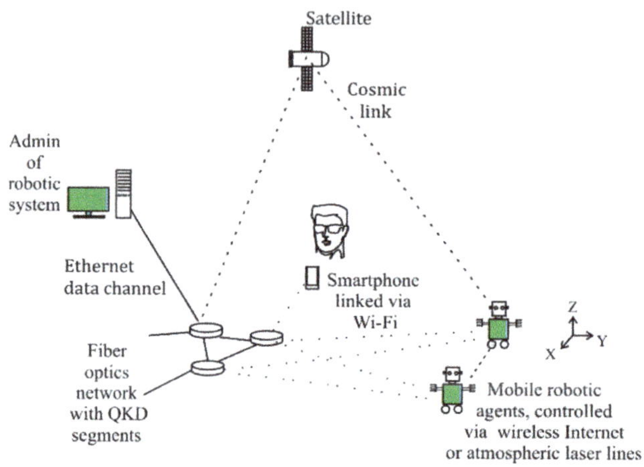

Figure 1. Distant control task of robotic agents via the global communication network includes fiber optics lines, wired Ethernet, and different wireless data links involving cosmic lines and Wi-Fi.

In order to enhance the security level for the distant control of autonomous agents and MASs, it is necessary to combine quantum and traditional data protection methods with new AI algorithms, which now are grounded in concepts of agent and MASs [7–11,15,26–28]. The presented paper is devoted to the original properties of multiple-valued logic (MVL) [29], which provides the possibility to design new schemes of agents based on multiparametrical logic functions and easily provides the modelling of various procedures in a high-dimensional space of parameters [30]. MVL functions can also realize schemes of secret data coding for one-time cipher pads and random keys [31], to accumulate random data into the scheme of random oracle [32], to model the theoretically proven and most protected position-based cryptography protocol by D. Unruh [33], and to realize data classification procedures [30].

That is why the actual task of secured distant work of robotic agents and MASs [34] is the potential field of applications for MVL functions [29,30], which now needs the integration of disparate algorithms and schemes into the whole data-protection agent, combining quantum and classical data coding methods. The present paper discusses the design of "indestructible" and easy-for-verification MVL data structures, which adapt some blockchain schemes [35–39] for collective data verification by means of collective interaction of the agent with loyal network nodes without trusted servers.

1.2. Basic Properties of Agents

The concept of agents and MASs is one of the basic models in modern AI [15,22] and, in general, supports a large set of properties that hardware and software systems should possess for the efficient imitation of human abilities. Purely software agents are now called bots and are being successfully used in Internet data searches and creative activity [26,27], stock trading [28], and cryptocurrencies mining [35,36]. However, the realization of hardware robotic agents is a much more difficult task, as it is necessary to simultaneously control many subsystems [34,40,41]. In general, the model of the universal agent should possess dozens of properties [15,34,36,41–43], which should include at least:

- autonomous activity and self-sustained decision making;
- reactivity to external media and adaptive correction of behavior;
- active impact on the external media, motivated by understanding of the scene;
- interaction with other objects based on goals, desires, collaboration, and trust;
- communicative abilities, including secured data exchange and storage;
- flexible planning of activity and self-analysis of goals, tasks, and resources;
- responsibility to the team, following to general aims and collaboration rules.

Modern MASs can deal with stationary and mobile agents, where stationary ones are mainly represented by software bots for information services [26–28,36] and hardware metering devices for power grid control [37–39]. Initially, the term "mobile agent" was used for software designs capable of expanding their activity from one platform to another [44], but now it is often used for hardware robots for transport, industrial systems, medical surgery, and nursing [7–9]. However, all of them have unresolved issues with data protection.

1.3. Data Protection of Stationary Agents in Classical and Quantum Network Nodes

As a whole, data protection methods for agents and MASs are based on the same principles as any other network nodes, but the unsatisfactory level of modern communication networks [20,21] is reinforced by the complexity of robotic agents. The concept of a trusted network node (or server) and trust modelling [24–26] is actually for both schemes with QKD lines and for traditional data links. The basic idea of a trusted server is that all its subsystems are well-secured ones, including data storages, data channels, devices for the detection of data leakages, and systems for the monitoring of administrative processes. In any situation, data coming from a trusted node are considered as absolutely correct and reliable for every user, and typically trusted nodes are associated with the expensive networks of government organizations and big corporations, which maintain large security staff with adequate administrative procedures and human monitoring. However, even cloud services and big data centers [45,46] can be successfully attacked.

The modern interpretation of trust problems for agents dates back at least to the document published by the National Institute of Standards and Technologies of the USA (NIST) [44]. Most typical problems of illegal activity included unauthorized access and data disclosure or corruption. Main countermeasures [22,24,25,37–40,42,43] included such methods as the verification of path histories, proof carrying code, partial result encapsulation, mutual itinerary recording with optional replication or voting, execution tracing, environmental key generation, and computing with encrypted functions. Trusted modelling in MASs considers concepts of trust and reputation [47], as well as argumentation [48].

As the intrinsic idea of autonomous agents is their working without permanent human monitoring [15], then some verification procedures should be shifted onto other agents and partner network nodes. Illegal activity can then be revealed by other agents by means of the estimation of an inadequate agent's behavior, contradicting the rules used by the majority of participants [37–39]. AI algorithms imitating the work of human experts for the coordination of node activity were designed for stock trading bots [28], systems of e-commerce, and e-business [35,36]. Efforts to raise the trust level were based on trust estimates carried out by neighboring agents [49,50].

The practical design of collectively secured agents was represented by the smart infrastructure of an energy power grid [37–39], using smart metering and control of phasor for devices suspected in illegal activity. Such systems have combined the collective interaction of metering devices and gateways, which have involved the blockchain (BC), data integrity control, and insider attack mitigation methods.

Respectively, the BC method [37–39,51,52] and cryptocurrencies mining [53,54] demonstrate quite useful schemes of collective network data protection. Ledger (or registry) schemes in the BC [37–39,53,54] are based on the principle of linked data lists and have become the popular trend in modern computer technologies. The main idea here is that the distributed ledger of linked entries [53,54] is supported by the decentralized consortium of users, whose integral voting guarantees the correctness of data. However, attempts to apply trust estimation to mobile robotic systems have led to the design of complicated deep-learning schemes based on pre-trained learning with supervisor and acceptance observation history [55].

As modern fiber optics lines have obtained impressive throughput parameters of 115 Tb/s for single-mode fiber and 10 Pb/s for multi-core fiber [56–59], they seem to be excellent platforms for the secured distant control of robotic agents. Although modern QKD systems did not obtain enough parameters for mass stream secret coding by the most secured one-time cipher pad (OTP) coding method and have some quantum and classical vulnerabilities of their optical and computer components, they are hoped to provide large additional technical, competence, and cost barriers for eavesdroppers [22,23,60]. The drawbacks of modern fiber optics QKD lines refer to the real limited length of quantum key transfer (less than 80–100 km), low fan-out coefficients, the possibility of physical damage of a QKD line, and the threat of being suppressed by intense optical noise. Current designs of quantum memory [61] are too far from practical needs, and modern QKD schemes cannot provide reliable data storage and exclude illegal modifications. Moreover, quantum networks together with ordinary ones can suffer from data leakages carried out by disloyal personnel [62]. Nevertheless, the research of QKD schemes continues, and in order to raise the security level of energy power systems [63], the trusted server with "star" architecture was combined with the large number of QKD lines. Other attempts [64,65] also aimed at the realization of trusted nodes by various quantum schemes. The possibility of using atmospheric quantum laser lines for agents was principally demonstrated by key distribution between ground stations and satellites [66,67], as well as by experiments with an airborne platform [68]. However, satellite QKD links can only be a partial solution for mobile agents, as data exchange sessions are scheduled by timetables of orbital flights, but their positioning can be somewhat unpredictable.

Thus, the schemes and methods discussed above do not provide ready solutions for data protection in mass hardware robotic agents and easy ways to avoid expensive trusted servers. That is why one of the opportunities for agents and MASs is to use the principles of the linked list and the distributed ledger taken from BC methods [37–39]. Cryptocurrencies mining and stock trading seem not to be actual tasks for robotic agents, but new verification schemes are necessary for the autonomous confirmation of their rights and powers, licenses, as well as for the verification of technology data, and breakdown restoration. In contrast to cryptocurrencies mining schemes [53,54], the verification of data in the collective robotic protocol can be based on the simple majority voting of participants, which are to be involved based on mutual interests or service payments, but not on mining. Respectively, the term "entry" but not "transaction" is used further.

1.4. Mobile Robotic Agents and Position-Based Cryptography Schemes

Hardware mobile robotic agents are now represented by autonomous vehicles [7,69], collectives of agents [70], and industrial [8] and healthcare systems [9]. The review of applications for wireless sensor networks formed by mobile agents was presented in [71] and included such tasks as required data collection from the field of view by drones, traffic control, biomedical health and environmental monitoring, virtual reality, and industrial systems control, as well as target tracking, urban control, monitoring of events, and intrusion detection.

Another trend here is the research of schemes for position control and verification of mobile agents [72], where quantum optics is involved as the tool to raise the security level. A position-based cryptography (PBC) task [22] is the confidential transmission of messages from one space location to another one, excluding the decoding of this message by any receiver having any other space coordinates. This task dates back to an ongoing problem of position verification of subscribers in mobile telephony [73], which has caused attempts to use QKD and to obtain unconditional security to not depend on the computing resources of the eavesdropper.

The most secured quantum PBC protocol was proven theoretically by D. Unruh [33] and is further tagged as PU. In fact, this protocol tried to prevent an actual "man-in-the-middle" attack [74] for the quantum optics scheme, but it has obtained only a partial result and did not provide unconditional security for the PBC scheme, as it can be deceived by a large enough number of fake qubits randomly emitted by illegal laser sources located between the eavesdropper and verifiers. This protocol [33] involves quite a realistic laser atmospheric QKD line with an Einstein–Podolsky–Rosen (EPR) source of entangled photon pairs [75]. However, in the review [22] protocol, PU was estimated as an interesting but purely theoretical scheme, as it involves an ideal random oracle (RO), which principally cannot be obtained in practice. RO here is the ideal black box, which for an arbitrary input signal generates random outputs with a fixed number of bits [76]. The obligatory conditions for RO are that it should reproduce the output for any repeated input, and there should be no ways to predict its output.

In order to use protocol PU in practice, the scheme for an RO device was proposed in the paper [77] and included a memory device learned by a quantum random number generator [78] with the help of a special algorithm. Such a scheme of an RO device was intended for the verification of a planned route for passing by a mobile autonomous vehicle [79]. This verification scheme is being discussed further, together with the verification scheme from [77], so that both of them are schematically shown in Figure 2.

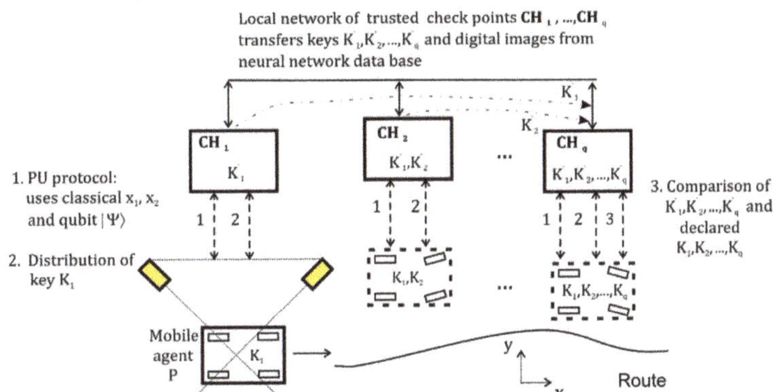

Figure 2. The principal scheme of quantum route verification of a mobile agent M, uniting methods proposed in [32,79].

The task of a verification procedure is to check the prehistory of visits of a mobile agent to checkpoints. According to terms used in [33], the mobile agent P within the verification scheme is regarded as a prover requesting some service and initially approving its space position for this. The idea of protocol PU [33] was discussed in [32,77] for a 2D case. At every checkpoint, agent P passes through the special quantum optics scheme, shown in Figure 2, by "crossed" pairs of verifier modules. Every such "cross" scheme of a verifier is supposed to emit and detect qubits (i.e., quantum bits or single photons characterized by some quantum states $|\Psi>$) and is also supposed to exchange classical laser signals with the prover. The protocol is activated when the prover P sends the initial request signal. The verifier replies by subsequent transfer of random classical data x_1, x_2 and qubit $|\Psi>$. (As there is yet no appropriate quantum memory for qubit storage, it is necessary to transfer it with some time delay after classical ones.) The prover P measures x_1, x_2, calculates the XOR operation $x_1 \oplus x_2$ and further calculates the random hash function $H(x_1 \oplus x_2)$, using the RO module [22,77]. After this, the prover P has the correct basis for the measurement of the classical value $\widetilde{q}$ of the qubit $|\Psi>$. Further, P simultaneously sends measured $\widetilde{q}$ by its EPR source to a pair of verifiers located along one axis (the details of the selection of necessary bits were not set by the protocol). If both photons were sent simultaneously by one entangled photon pair, their classical values obtained by the verifier modules will be equal (for equal distances), accurate to the sign. Then, position verification is successful, and further steps can be performed.

In fact, the choice of EPR scheme by D. Unruh [33] was determined by the necessity to simultaneously emit entangled photons along space axes, as the minimal time interval between the detection of two these photons guarantees the advantage for the protocol PU. That is why the protocol PU is limited by light polarization scheme. However, one should note that although the protocol PU principally used a wireless atmospheric data line, the EPR scheme [75] can also be used in the fiber optics quantum line.

After successful position verification, the second step of the procedure is to generate a random key, secretly distributed by the EPR source between the agent P and the visited checkpoint. Thus, during the passing of the route, agent P will accumulate random keys $K_1, \ldots, K_q$, and each visited checkpoint will have only one of keys $K_1, \ldots, K_q$. In order to approve the route passing to the final checkpoint, the agent P declares its set of keys, as the checkpoint chooses some of keys for verification and requests their copies via the trusted network. Then, the final checkpoint compares the declared keys with their network's replicas. For access to the requested service, agent P should also additionally pass another check step, based on the comparison of the digital image obtained during the route passing and declared by P, with its replica transferred from a distant neural network database [32]. The main result of the paper [32] was that both the quantum protocol PU and template digital image taken from a neural network were represented by multiple-valued logic expressions attributed to time scale. Such a model has considered the joint work of the prover P with the verifier as the interaction of two agents, described by separate AGA functions.

However, the drawback of verification schemes [32,79] was that all involved checkpoints were considered as trusted nodes, because:

- classical data in protocol PU are transferred without errors, although the acceptable quantum channel error in this protocol was estimated as 3.5%;
- checkpoints are supposed to exchange accumulated keys and neural network data without errors and limitations.

Such a scheme, including stationary trusted servers with QKD links, can potentially be reproduced in global IoV and city traffic control systems [7], but the problem is that stationary trusted nodes are not applicable for irregular routes of a small-scale MAS. Thus, the question arises, if one can design a less expensive route verification scheme without trusted checkpoints and intense human control.

Another problem here is that the refusal from trusted nodes and human control in fact means that the level of data storage protection in checkpoints will be reduced substantially, and it is necessary to compensate it somehow. As QKD networks are much more expensive systems than classic ones, they may be equipped with more reliable data storage devices. However, as current designs of quantum memory [61] are far from practical, QKD links can only partially protect traditional memory from illegal modifications carried out directly via data channels.

1.5. Advantages of Data Protection Schemes Based on Multiple-Valued Logic

Discrete k-valued logic calculus of Allen–Givone algebra (AGA) [29] and data coding schemes on their base [23,30–32] have several features that are beneficial for the implementation of data verification schemes for network MASs. First of all, AGA makes the design of multiparametrical logic models easy, containing arbitrary associations of digital and symbolic parameters and aggregated by the appropriate choice of input variables and variants of functions. MVL modelling is also attractive for the integration of quantum schemes, which can be, e.g., easily described by AGA functions [32] for the quantum PU protocol [33]. The principal advantage of discrete k-valued logic is the high data capacity of AGA models, which can describe much more states in compact logic expressions [30] than Boolean logic.

1.6. The Goal of the Presented Paper

The motivation to design an MVL analog of the BC's linked list and distributed ledger [37–39] is determined by the necessity to envisage distributed schemes for autonomous critical data verification, backup, and restoration in a distant MAS. Necessary verification procedures should be grounded in sources of reliable data, distributed in MASs and loyal network nodes. Then, data verification algorithms for MASs will be able to monitor the proper functioning of an agent, to detect attacks, and to restore the work state of an agent.

The aim of the presented paper is to propose an MVL analog of the BC's linked list for a distributed ledger scheme of data protection and for verification of network agents. The task is also to demonstrate the method for combining classical and quantum verification procedures in the logic ledger model.

2. Methods

Although AGA [29] is a very specific calculus [30] with comparatively limited set of intrinsically given tools, multiparametrical logic expressions written for a large number of input variables give the possibility to form a set of logic functions, reproducing the idea of the linked list in BC, and are quite compatible with the methods described in [37–39].

2.1. The Structure of MVL Function

Traditional design of models for robotic agents implies the formation of mathematical expressions intrinsically based on Boolean logic operations *AND*, *OR*, *NOT*, but MVL logic calculus of Allen–Givone algebra (AGA) is based on more general operators *Maximum*, *Minimum*, and *Literal* [29]. The comparison of MVL operators with Boolean ones and other possible logic systems was carried out in [23,30,31]. AGA [29] is one of several known versions of MVL models that operates with discrete MVL functions $y = f(x_1, \ldots, x_n)$, given for n input variables $x_1, \ldots, x_n$ and one output variable y. Specifics of AGA are that instead of two truth levels {0,1} in Boolean logic, all its input variables $x_1, x_2, \ldots, x_n$ and the output variable y can have k discrete truth levels: $x_1, x_2, \ldots, x_n, y \in L = \{0, 1, \ldots, k-1\}$, i.e., they are natural numbers. In earlier designed MVL coding schemes [23,30,31] for 8-bit controllers, k was supposed to be 256, which was enough even for the description of secret coding procedures and communication messages. However, the further proposed MVL model of distributed ledger is to be based on definitions and algorithms of AGA [29] defined for functions with the larger number of truth levels k.

The complete set of non-Boolean operators [29]

$$< 0,\ 1, \ldots,\ k-1,\ X(a,b),\ *\ ,+ > \tag{1}$$

ensures the possibility to represent arbitrary function $y = f(x_1, \ldots, x_n)$ as some combination of logic operators taken from the list given further:

- $0,\ 1, \ldots,\ k-1$ are constants;
- operator *Minimum* (x_i, x_j) or *MIN* is marked by $(*)$ and means the choice of the minimal one in the pair x_i, x_j;
- operator *Maximum* (x_i, x_j) or *MAX* is marked by (+), it means the choice of the maximal value in the pair x_i, x_j;

- operator $X(a,b)$ is called *Literal* and is given by Equation (5):

$$X(a,b) = \begin{cases} 0, & if\, b < x < a \\ k-1, & if\, a \le x \le b, \end{cases} \quad (2)$$

where for any $X(a,b)$, always $b \ge a$, and $a,b \in L = \{0,1,\ldots,k-1\}$.

Equation (2) demonstrates that operator Literal, in fact, resembles the discrete band filter with variable lower and upper limits. This expression is either equal to the maximal truth level $k-1$ or 0, which simplifies the analysis of data.

It was commented in [30] that the general structure of MIN and MAX operators dates back to basic definitions of the set "lattice" and its operators "Infimum" and "Supremum", which namely for the "lattice" were proved to be MAX and MIN. In fact, these binary operators $MAX(x_1,x_2)$ and $MIN(x_1,x_2)$ are the comparison of two logic values x_1, x_2, which can reproduce Boolean operators *AND, OR* if written as appropriate rows into the MVL truth table. However, MVL operators do not possess some specific theorems of Boolean logic, providing its more universal properties. Thus, AGA is a more general but very poor set of tools for data procession, which mainly provides calculation of arbitrarily given MVL functions and their minimization (i.e., simplification), described in detail in [30]. For example, traditional arithmetic operations are not defined in AGA, and they can be reproduced in AGA only by the specially given MVL functions with appropriate truth tables. At the same time, AGA calculus [29] provides primitive logic calculations, whose result is predictable only for a specific numerical data set and does not give appropriate tools for design of trapdoors and schemes to break cryptographic one-way functions and hash ones.

Equivalent representations of MVL functions include truth table, logic expression, and matrix form [29]. Any MVL function $y = F(x_1,\ldots,x_n)$ can always be given by the truth table shown in Table 1, which has the overall number of rows equal to k^n, as in Boolean logic it has only 2^n ones. Thus, for large values of k (say, more than 256) and n (several dozens), the number of rows can be extremely large and, in reality, one can use only some part of the truth table. That is why MVL function is much more attractive [31,32,77] for modelling complicated systems, cryptography schemes, and random hash functions than the Boolean one.

Table 1. Truth table of a MVL function with equivalent product terms for rows with nonzero output value.

N_{row}	Input Variables					Output		
	x_1	x_2	...	x_{n-1}	x_n	$F(x_1,\ldots,x_n)$		
0	0	0	...	0	0	$F(0,0,\ldots,0)$		$F(0,0,\ldots,0) * X_1(0,0) * X_2(0,0) * \ldots * X_n(0,0)$
1	1	0	...	0	0	$F(1,0,\ldots,0)$	$\rightarrow$	$F(1,0,\ldots,0) * X_1(1,1) * X_2(0,0) * \ldots * X_n(0,0)$
2	2	0	...	0	0	$F(2,0,\ldots,0)$		$F(2,0,\ldots,0) * X_1(2,2) * X_2(0,0) * \ldots * X_n(0,0)$
...	...	...	...	...	...	...		...
k^n-1	$k-1$	$k-1$	...	$k-1$	$k-1$	$F(k-1,\ldots,k-1)$		$F(k-1,k-1,\ldots,k-1) * X_1(k-1,k-1) * X_2(k-1,k-1) * \ldots * X_n(k-1,k-1)$

The column for output variable y should be filled in by some set of logic constants $C = \{0, 1,\ldots, k-1\}$. According to [29], any input vector $x_1,\ldots,x_n$, i.e., every row of the truth table with nonzero y, has equivalent product term shown to the right of Table 1, and they are written via logic constants, *Literals*, and operators MIN (marked $*$) and MAX (marked +). In this product term, the value of $F(x_1,\ldots,x_n)$ is taken from the column "Output" in Table 1, and *Literals* are to be filled in by equal parameters, e.g., $X(1,1)$.

Unfortunately, the truth table is a convenient tool only for a small number of input vectors, so that real AGA models should use either formal expression (3) written via AGA operators as a set of product terms, or equivalent matrixes [77]. The function given by the truth table (see Table 1) responds to the set of product terms

$$y = f(0,0,\ldots,0) * X_1(0,0) * X_2(0,0) * \ldots * X_n(0,0) + +f(0,0,\ldots,1) * X_1(1,1) * X_2(0,0) * \ldots * X_n(0,0) + f(k-1,k-1,\ldots,k-1) * X_1(k-1,k-1) * X_2(k-1,k-1) * \ldots * X_n(k-1,k-1). \quad (3)$$

One should especially note that, according to definitions given in [29], Literals written for separate rows of the truth table and shown to the right of Table 1 have equal values $a = b \in L = \{0,1,\ldots,k-1\}$ for product terms. So, for the arbitrary truth table, one can always write formal Equation (3), which has equal lower and higher parameters *a,b* in Literals the structure $f(0,\ldots,1) * X_1(a_1,a_1) * X_2(a_2,a_2) * \ldots * X_n(a_n,a_n)$. This notation means that Literal is nonzero only for one natural number $n = a = b$. However, further simplifications and transformations based on subsuming of product terms can shorten the computing time. Such modification of logic expressions can be held by means of minimization procedure by consensus method [29]. Then, the set of product terms can be shortened, but some parameters *a,b* of its Literals will be modified. In this case, notation *X(a,b)* with different parameters $a \neq b$ means that Literal is nonzero for the discrete set of values a, $a + 1$, $a + 2$, ... $b - 1$, b. It is essential and advantageous for MVL data protection methods that minimization procedure [29,30] strongly depends on the specific set of used data and is "unpredictable" beforehand, which prevents the design of hidden trapdoors. Besides this, any such legal or illegal procedure will change the set of Literals, i.e., expressions for Literal will receive different lower and higher indexed parameters *a,b*. In addition, the advantage of procedures proposed in the present paper is that they do not initially need minimization procedures. Within the framework of the present paper, logic minimization will finally be necessary for the large number of rows in the truth table or for large number of entries in the MVL model. That substantially simplifies processing and verification of MVL models.

Another possible case of transformation of Literals parameters without minimization refers to Section 3.3, where in Equation (9) some fragment of the truth table is subdivided into loyal and prohibited subsets of rows, which is based directly on total estimates of unused (or prohibited) rows for the known numbers of all possible rows and already used ones.

In turn, its consensus minimization procedure [29] is based on extension of initially given function and adding of so-called "don't care states" with constants $k - 1$ substituted into the function. The choice of such states for extended function leads to multi-criteria optimization task that additionally complicates very long minimization procedure. Respectively, if namely MVL function is used as a learned hash or a ledger function, its illegal modification is a very costly task.

In the present paper, for clarity, mainly truth tables are used, but for other modelling stages, one should use more compact equivalent representation of MVL function, written as the set of matrixes in Equation (4), which reproduce parameters a, b and logic constants used in Equation (3):

$$A_u = \begin{pmatrix} a_{11} & \ldots & a_{1n} \\ \ldots & \ldots & \ldots \\ a_{k-1,1} & \ldots & a_{k-1,n} \end{pmatrix}, \quad B_u = \begin{pmatrix} b_{11} & \ldots & b_{1n} \\ \ldots & \ldots & \ldots \\ b_{k-1,1} & \ldots & b_{k-1,n} \end{pmatrix}, \quad C = \begin{pmatrix} c_{11} & \ldots & c_{1v} \\ \ldots & \ldots & \ldots \\ c_{k-1,1} & \ldots & c_{k-1,v} \end{pmatrix} \quad (4)$$

where: $b_{ij} \geq a_{ij}$, n—is the number of input variables, k—is the number of truth levels, and $u \in \{1,2,\ldots,v\}$. Matrixes A_u and B_u in Equation (4) specify indexed sets of parameters a and b in Literals $X_j(a,b)$. Matrix C in (4) shows the set of constants for all product terms, united by operator $Max(+)$, and v is the number of columns of product terms. Initially, due to Table 1, matrix C will have only one column, but further minimization of MVL function [29] can modify it.

2.2. Transformation of Dimension of AGA Function

According to basic definitions [29], the number of input variables for AGA function is not limited, but it has one output variable (see Figure 3a). However, verification procedures need to output many signals for one input vector. In order to avoid modification of AGA calculus, one can simply add one or several input variables in order to obtain the allocated counter cycle for outputs. In Figure 3a, the basic AGA function $y = F(x_1, x_2, \ldots, x_{N_0})$ has the number of input variables equal N_0, but the only output variable can describe only $k - 1$ output values. However, if one will define AGA function $F(x_1, x_2, \ldots, x_{N_0+1})$, shown in Figure 3b, it can produce the extended number of output signals in the external cycle, obtained by the sequential substitution of values for x_{N_0+1}. Such a procedure is reasonable to carry out after the calculation of all *Literals* in product terms with $x_1, x_2, \ldots, x_{N_0}$. Note that the value $x_{N_0+1} = 1$ but not 0 should be used for the first output signal to provide nonzero product term for it. This small limitation seems not to be substantial.

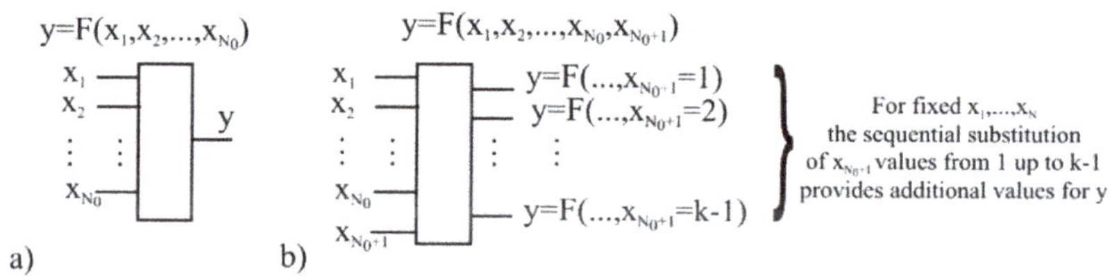

Figure 3. (**a**) The basic scheme to calculate function in AGA. (**b**) The simple scheme to use additional counter x_{N_0+1} in order to receive $k - 1$ output values for the same set of input vectors given for MVL function.

3. Results

The MVL version of collective ledger and the linked list are proposed to be based on the procedure of the collective generation of values of a random hash function by Q participants of the network pool of users or nodes, motivated in mutual service and shown in Figure 4 as $V_1, V_2, \ldots, V_Q$. These nodes are called verifiers, as they verify entries (or messages) coming from P and other participants of the pool. It is supposed for simplicity that only one mobile agent P can represent the "interests" of some MASs and is responsible for transfer of entries between this MAS and network nodes. This agent P is further called the prover, as in [33], because its data should be approved, corrected, or restored with the help of external network nodes $V_1, V_2, \ldots, V_Q$. Verification, authentication, and identification data coming from P to the external nodes are described by entries $\boldsymbol{e} = \{e_1, \ldots, e_n\}$. The entry $\boldsymbol{e} = \{e_1, \ldots, e_n\}$ received at time t has n fields represented as natural numbers and can be interpreted, e.g., as a message such as "(1) vehicle N"—(2) "is certificated for"—"(3) ecology class Euro . . . "— . . . —(n) "date". For simplicity, we do not consider any language structures here.

3.1. Collective Generation of Data for MVL Storage and Ledger Functions

In Figure 4, the simple procedure is proposed to accumulate data collectively, which is necessary for the generation of an MVL logic-linked list and ledger functions. The prover P is considered as the participant of the pool of loyal nodes. The prover agent wants to add some entry $\boldsymbol{e} = \{e_1, \ldots, e_n\}$ into the distributed backup storage, which is being sustained collectively by the pool of network nodes. We suppose that this entry is not confidential and does not need preliminary secret coding, but even this procedure will not disturb the procedures disclosed further. The prover P has the list $N_N = \{N_1, \ldots, N_Q\}$ of identifiers (or simply numbers $1, \ldots, Q$) corresponding to verifiers $V_1, V_2, \ldots, V_Q$, who are agreed to provide the service. Principally, some of the participants can be switched off or busy, so that full list can be unused, but this does not change the scheme, and for simplicity we suppose that all registered participants are taking part in the formation of the entry. The prover P tags the entry $\boldsymbol{e} = \{e_1, \ldots, e_n\}$ as a new one and sends it via the network to one of the participants, which is taken to be V_1. Verifier V_1 assigns to this entry the time stamp t and some randomly chosen natural number h_1. After this, V_1 writes the set of parameters $\{e,\ t,\ h_1\}$ in its memory and sends the copy back to the prover. Further, prover P writes the obtained parameters in memory and sequentially sends the pair $\{e,\ t\}$ to all other nodes from the list. As a result, it accumulates random hash function values $h_1, \ldots, h_Q$ assigned by verifiers $V_1, V_2, \ldots, V_Q$ to the pair $\{e,\ t\}$. After this, P sends the finally formed set $\{e_1, \ldots, e_n,\ t, h_1, \ldots, h_Q\}$ to all involved participants according to the earlier used list. Further, this set of data is to be represented as a separate logic product term in the collective ledger function.

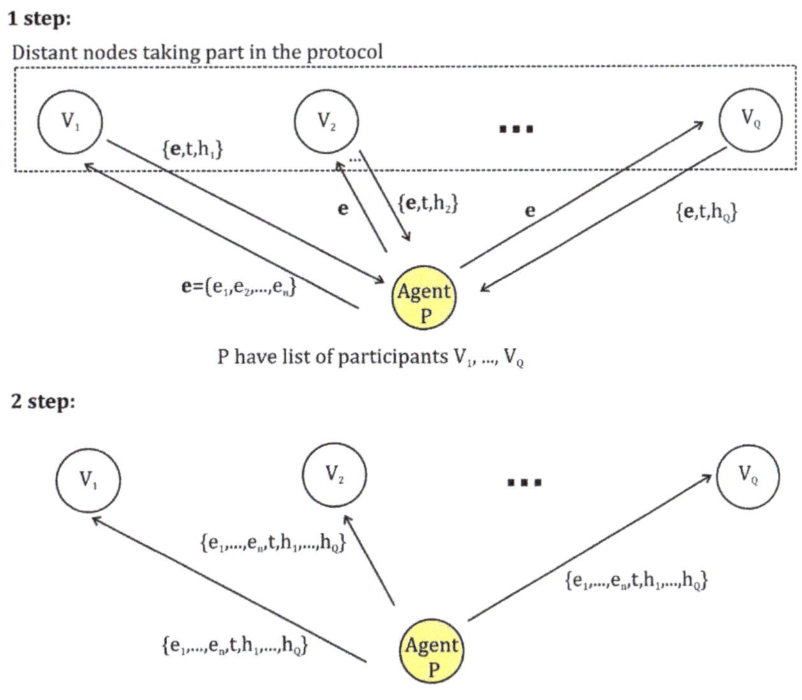

Figure 4. The scheme of assigning random hash values h by verifiers to the entry received from the mobile agent P.

The procedure given above is shown further as Algorithm 1. Its necessary preliminary conditions for entry formation include:

- complete number of involved loyal verifiers Q should respond to $Q \leq k - 1$;
- waiting time t_{wait} is limited for the reply from the verifier V_{j+1}. If it is exceeded, then the zero value should be assigned to the corresponding h_{j+1};
- every node taking part in the protocol should possess either quantum random number generator or RO module with the scheme proposed in [77], which should generate the set of randomly given hash values h_j to make the formation of entries unpredictable and corresponding product terms;
- all the assigned parameters are natural numbers.

Algorithm 1: Collective accumulation of hash values for entry

Input		
	$k \leftarrow$	Number of truth levels
	$Q \leftarrow$	List of adresses for Q participants of protocol
	$t_{wait} \leftarrow$	Waiting time for reply entry $\boldsymbol{e} = \{e_1, \ldots, e_n\}$
	$\boldsymbol{e} = \{e_1, \ldots, e_n\} \leftarrow$	Preliminary prepared entry
1.	Prover P	assigns counter $j = 1$ sends $\boldsymbol{e} = \{e_1, \ldots, e_n\}$ to the verifier V_1 according to the list of addresses
2.	Verifier V_1	assigns time stamp t to $\boldsymbol{e}$ assigns randomly chosen h_j writes the set $\{\boldsymbol{e}, t, h_j\}$ in its memory sends the copy of $\{\boldsymbol{e}, t, h_j\}$ to prover P
3.	Prover P	waits for reply during t_{wait}, if reply $\{\boldsymbol{e}, t, h_j\}$ has come writes it in its memory otherwise writes it with $h_j = 0$ assigns $j = j + 1$ sends pair $\{\boldsymbol{e}, t\}$ to the next verifier V_{j+1}
4.	Verifier V_{j+1}	assigns randomly chosen h_{j+1} writes the set $\{\boldsymbol{e}, t, h_{j+1}\}$ in its memory
5.	Verifier V_{j+1}	sends the copy of $\{\boldsymbol{e}, t, h_{j+1}\}$ to the prover P
6.	Prover P	checks if $j = Q$, if yes goes to step 7, otherwise repeats steps 3–6
7.	Prover P	sends $\{e_1, \ldots, e_n, t, h_1, \ldots, h_Q\}$ to all verifiers $V_1, V_2, \ldots, V_Q$
Output		Ledger entry $\{e_1, \ldots, e_n, t, h_1, \ldots, h_Q\}$ for each of verifiers $V_1, V_2, \ldots, V_Q$

As the prover sends its entry $\boldsymbol{e}$ according to the preliminary formed list of participants, the number of replies does not exceed Q. As the limiting time parameter t_{wait} is assigned in the reply only by the first verifier, its possible further illegal modification in one of verifiers does not influence the other's replies and will contradict the data of the simple majority of verifiers and will be excluded at the stage of decision making. The absence of the verifier's reply is marked by a zero hash, which is fixed in all other procedures and can be further used for analysis of attacks. The waiting time t_{wait} period for reply can be limited by the average and variance values estimated for all involved verifiers.

The accumulated set of user data and random hash values above can be directly used for the formation of the simple version of the logic ledger function, but a more pragmatic variant is to firstly generate the separate backup storage function F_{st}, accessible for distant reconstruction if the prover has declared a correct access key. The simple variant here responds to the reconstruction of entry $\boldsymbol{e_m}$ when the prover declares (i.e., substitutes as input variables) the collectively generated key given by the hash values $h_1, \ldots, h_Q$. The request is sent according to the known list of participants and is considered correct if it is reproduced by the majority of participants. Another simple variant responds to the approval of certificate hash values $h_1, \ldots, h_q$ extracted from the storage for the declared entry $\boldsymbol{e_m}$. Here, the reliability of data also can be estimated according to the majority of replies. More complicated schemes such as the "proof-of-work" one in the BC protocol [37–39] are principally possible, but they seem to be not actual here.

If all participants of the collective protocol possess identical data sets $\{e_1, \ldots, e_n, t, h_1, \ldots, h_q\}$ written for every entry in the storage, then identical storage function F_{st} can be generated in all the involved verifiers using the known scheme. Two simple variants of storage function mentioned above are shown further as the function of entries F_{ste} and the function of hash values F_{sth}. As the key to extracting data from backup storage should be realized by hash values $h_1, \ldots, h_Q$ assigned by Q network verifiers, then namely these hash parameters are to be used as input variables in F_{ste}. In addition, the set of entries $\boldsymbol{e_m} = \{e_{1,m}, \ldots, e_{n,m}\}$ should be considered as the values of output of the variable y. Such a procedure should use the general algorithm given in [29] and Table 1.

The formation of the MVL storage function F_{sth} is shown in Table 2. This truth table follows the general format, so that besides parameters $h_1, \ldots, h_Q$ it contains an additional counter of verifier q. For brevity, Table 2 demonstrates the product term only for final entry $\boldsymbol{e_m}$, and all previously added product terms are not shown. One should especially note that, by default or according to the agreement of participants, the first row in the truth table can be filled in by arbitrarily taken random numbers. If some verifier from the list was switched off and did not take part in the formation of F_{entr}, it is necessary to assign 0 to its hash value in the row. Here, hash values $h_1, \ldots, h_Q$ will be the key to the verification procedure, when some of participants will want to restore the entry with critical data.

Table 2. The supplementation of the newcomer entry $e_m = (e_{1,m}, \ldots, e_{n,m})$ to the truth table of the storage function F_{sth} and the formation of corresponding product terms (shown to the right). Previous entry is $e_{m-1} = (e_{1,m-1}, \ldots, e_{n,m-1})$, the overall number of verifiers is Q. Service variables t and q respond to the time stamp and the counter of verifiers.

Input Variables						Output	
h_1	h_2	...	h_Q	t	q	e_j	
...	...	...	...	...	...	...	
$h_{1,m-1}$	$h_{2,m-1}$	...	$h_{Q,m-1}$	t_{m-1}	Q	$e_{Q,m-1}$	Product terms added for the new entry:
$h_{1,m}$	$h_{2,m}$	...	$h_{Q,m}$	t_m	1	$e_{1,m}$	$e_{1,m} * X_1(h_{1,m}, h_{1,m}) * \ldots * X_n(h_{Q,m}, h_{Q,m}) * X_t(t_m, t_m) * X_q(1,1)$
$h_{1,m}$	$h_{2,m}$	...	$h_{Q,m}$	t_m	2	$e_{2,m}$	→ $e_{2,m} * X_1(h_{1,m}, h_{1,m}) * \ldots * X_n(h_{Q,m}, h_{Q,m}) * X_t(t_m, t_m) * X_q(2,2)$
...	...	...	...	...	...	...	...
$h_{1,m}$	$h_{2,m}$	...	$h_{Q,m}$	t_m	Q	$e_{Q,m}$	$e_{Q,m} * X_1(h_{1,m}, h_{1,m}) * \ldots * X_n(h_{Q,m}, h_{Q,m}) * X_t(t_m, t_m) * X_q(Q,Q)$

The second version of storage function F_{ste} can be based on the same data set $\{e_1, \ldots, e_n, t, h_1, \ldots, h_q\}$ and is shown in Table 3. It is to approve certificate hash values $h_1, \ldots, h_q$ for the declared entry e_m, i.e., it can output numbers of licenses or permission if entries contain data referring agent and MASs. Actually, this function should reproduce collectively assigned random hash values.

A special comment should be given to the assigning of random values to the hash function. Their use is determined by the desire to provide an unpredictability of parameters for the MVL function, the storage function, and the ledger function. As most secured data protection schemes [80] are grounded in the OTP method and random data, AGA verification should also follow this trend, excluding preliminary forecasting of backup functions. The method to assign random hash values in the algorithm described above can use the RO scheme proposed earlier in [77] for the protocol PU. The idea of this scheme supposes the quantum random number generator (QRNG) [78] and learning procedure, forming the MVL function according to the algorithm proposed in [77]. An interesting aspect here is that, instead of a hardware QRNG, it is principally possible to use software for one of the known and tested types of cryptography hash functions such as, e.g., SHA256 [81], which guarantees a high quality of hash values set. The principal advantage of the MVL function here is that it can be taught to reproduce any reliable and tested source of random bit sequences with a fixed length, including both hardware QRNG and software SHA256 data.

Table 3. Creation of new product term (shown to the right) in the storage function F_{ste} for the newcomer entry $e_m = (e_{1,m}, \ldots, e_{n,m})$. Previous entry is $e_{m-1} = (e_{1,m-1}, \ldots, e_{n,m-1})$ and the number of used verifiers is Q. Service variables q and t respond to the time stamp and the counter of verifiers.

Input Variables						Output	
e_1	e_2	...	e_n	t	q	h	
...	...	...	...	...	...	...	
$e_{1,m-1}$	$e_{2,m-1}$	...	$e_{n,m-1}$	t_{m-1}	Q	$h_{Q,m-1}$	Product terms added for the new entry:
$e_{1,m}$	$e_{2,m}$	...	$e_{n,m}$	t_m	1	$h_{1,m}$	$h_{1,m} * X_1(e_{1,m}, e_{1,m}) * \ldots * X_n(e_{Q,m}, e_{Q,m}) * X_t(t_m, t_m) * X_q(1,1)$
$e_{1,m}$	$e_{2,m}$	...	$e_{n,m}$	t_m	2	$h_{2,m}$	→ $h_{2,m} * X_1(e_{1,m}, e_{1,m}) * \ldots * X_n(e_{Q,m}, e_{Q,m}) * X_t(t_m, t_m) * X_q(2,2)$
...	...	...	...	...	...	...	...
$e_{1,m}$	$e_{2,m}$	...	$e_{n,m}$	t_m	Q	$h_{Q,m}$	$h_{Q,m} * X_1(e_{1,m}, e_{1,m}) * \ldots * X_n(e_{n,m}, e_{n,m}) * X_t(t_m, t_m) * X_q(Q,Q)$

In Tables 2 and 3, schemes demonstrate two complementary procedures to represent data as AGA functions and to extract them from a reserve backup storage of entries written by logic product terms. However, these two complementary functions discussed above have no internal tools yet to prevent the modification of data, but the calculus of AGA also provides a logic analog of a linked list used in BC schemes [37–39].

3.2. MVL Function of Two Entries as a Base Model for the Logic-Linked List

In order to prevent illegal modifications of entries written in some format unified by participants, it is proposed to form the logic ledger function F_{ldg} by product terms, containing pairs of entries obtained at different time moments t_m and t_{m-s} (see Figure 5). In such a ledger scheme, the function formation responds to the generation of the AGA truth table, where fields of entries may have repeated values if they respond to the same agents, tasks, documents, permissions, plans, and used resources. In order to split such data into separate rows of the MVL ledger function, every entry receives time stamp parameter t and the number of entry m in the ledger. External network nodes are to assign random hash values h in order to exclude the prediction of the ledger function and to simplify its verification, as the uniform distribution of random numbers is appropriate for revealing possible modifications.

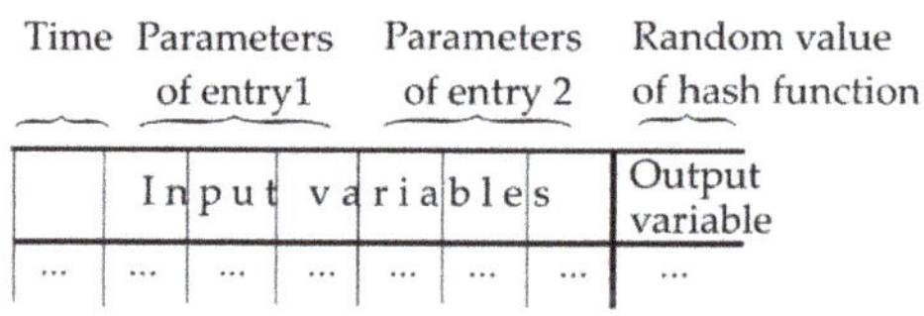

Figure 5. Principle of formation of the ledger function with two linked entries 1 and 2 obtained at different time moments.

The last entry obtained at time moment t_m is written as $\boldsymbol{e}_m = \{e_{1,m}, \ldots, e_{p,m}\}$, where m is the number of the entry in the ledger and p is the chosen number of fields in the entry. An earlier obtained entry $\boldsymbol{e}_{m-s} = \{e_{1,m-s}, \ldots, e_{p,m-s}\}$ corresponds to the time moment t_{m-s}. In order to confirm authenticity for two entries, the ledger should contain the data of both of them and sets of random keys unique for each of them. In Figure 5 this responds to the shift to right of the position of entry $\boldsymbol{e}$ when a new entry arrives.

Definition 1. *Logic ledger function or a linked list of logic entries is given within AGA as a hash function*

$$h^{(m,s)} = F_{ldg}(m, s, t, e_{1,m}, \ldots, e_{p,m}, e_{1,m-s}, \ldots, e_{p,m-s}, h_{1,m}, \ldots, h_{Q,m}, h_{1,m-s}, \ldots, h_{Q,m-s}) \quad (5)$$

In Equation (5), m is the number of the last entry in the ledger, s is the shift of the number for the previous entry, t denotes the assigned time stamp for the last entry, $e_{1,m}, \ldots, e_{p,m}$ are the parameters of the last entry $\boldsymbol{e}_m$; $e_{1,m-s}, \ldots, e_{p,m-s}$ correspond to parameters of the earlier received entry $\boldsymbol{e}_{m-s}$, $h_{1,m}, \ldots, h_{Q,m}$ are hash values assigned by network verifiers to the last entry $\boldsymbol{e}_m$, $h_{1,m-s}, \ldots, h_{Q,m-s}$ are hash values of the earlier received entry $\boldsymbol{e}_{m-s}$, and q is the number of a verifier, $q = 1,.., Q$. Non–indexed parameters m, s, q and indexed parameters t, e, h are natural numbers. For brevity, Equation (5) can be rewritten briefly as:

$$h^{(m,s)} = F_{ldg}(m, t, \boldsymbol{e}_m, \boldsymbol{e}_{m-s}, \boldsymbol{h}_m, \boldsymbol{h}_{m-s}) \quad (6)$$

The specifics of data backup for autonomous robotic agents are that the verification of data should include the possibility of its use both by a distant human expert and by a software bot. That is why a logic-linked list should not be accented only on the procedure of rigid binding of a block of data to the pointer of the next block, as in BC schemes [37–39], which can be realized by the special field-to-code data pointer just in the entry. Certainly, the assigning of data pointers should be adapted to a specific software platform.

Thus, using the same set of basic rules disclosed by Tables 1, 3 and 4, Equation (5) for the logic ledger can be written according to its truth table. As it was mentioned in the previous section, by default storage functions F_{ste} and F_{sth} should have the first row of the ledger filled in by a randomly chosen set of data. The output variable column should contain random hash values for all rows in order to exclude any preliminary prediction of the ledger function. In Table 4, the new entry is tagged by the number m and the time stamp t. For a simple version of the ledger parameter, s can be taken as equal to 1 and Equation (5) for $h^{(m,1)}$ will simply connect the last and the previously received entries such as with blocks of data in BC schemes [37–39,53,54].

Table 4. Truth table of the MVL ledger function $h^{(m,1)} = F_{ldg}(m, t, e_m, e_{m-1}, h_m, h_{m-1})$, whose input variables are the newcomer entry $\boldsymbol{e}_m = (e_{1,m}, \ldots, e_{n,m})$, previous entry is $\boldsymbol{e}_{m-1} = (e_{1,m-1}, \ldots, e_{n,m-1})$, and their corresponding sets of hash values $\boldsymbol{h}_m = (h_{1,m}, \ldots, h_{Q,m})$ and values $\boldsymbol{h}_{m-1} = (h_{1,m-1}, \ldots, h_{Q,m-1})$ at time t_m and t_{m-1}.

Input Variables														Output
m	t	$e_{1,t}$	…	$e_{p,t}$	$h_{1,t}$	…	$h_{q,t}$	$e_{1,t-1}$	…	$e_{p,t-1}$	$h_{1,t-1}$	…	$h_{q,t-1}$	$h^{(\mathbf{m},1)}$
1	t_1	$e_{1,1}$	…	$e_{p,1}$	$h_{1,1}$	…	$h_{Q,1}$	$e_{1,0}$		$e_{p,0}$	$h_{1,0}$	…	$h_{Q,0}$	$h_1^{(1,1)}$
…	…	…	…	…	…	…	…	…	…	…	…	…	…	…
$m-1$	t_{m-1}	$e_{1,m-1}$	…	$e_{p,m-1}$	$h_{1,m-1}$	…	$h_{Q,m-1}$	$e_{1,m-2}$	…	$e_{p,m-2}$	$h_{1,m-2}$	…	$h_{Q,m-2}$	$h_{m-1}^{(m-1,1)}$
m	t_m	$e_{1,m}$	…	$e_{p,m}$	$h_{1,m}$	…	$h_{Q,m}$	$e_{1,m-1}$	…	$e_{p,m-1}$	$h_{1,m-1}$	…	$h_{Q,m-1}$	$h_m^{(m,1)}$

According to definitions given in [29] and commented in Tables 1–3, the last row of the truth table shown in Table 4 for the ledger responds to the product term Equation (7) tagged with the number m and time t_m:

$$\begin{aligned} product\ term_{\ m} = {} & h_m^{(m,1)} \star X_m(m,m) \star X_t(t_m,t_m) \star X_{e,1,m}(e_{1,m},e_{1,m}) \star \ldots \star \\ & X_{e,1,p}(e_{p,m},e_{p,m}) \star \ldots \star X_{h,1,m}(h_{1,m},h_{1,m}) \star \ldots \star X_{h,Q,m}(h_{Q,m},h_{Q,m}) \star \\ & X_{e,1,m-1}(e_{1,m-1},e_{1,m-1}) \star \ldots \star X_{e,p,m-1}(e_{p,m-1},e_{p,m-1}) \star \\ & X_{h,1,m-1}(e_{1,m-1},e_{1,m-1}) \star \ldots \star X_{h,Q,m-1}(e_{Q,m-1},e_{Q,m-1}) \end{aligned} \tag{7}$$

3.3. The Scheme to Exclude Illegal Modification of Non-Specified States of AGA Function

The task to protect the ledger from illegal data modification includes the problem of minimizing the frequency of needed full checks of the overall list of entries and simplifying verification procedures. However, the possible consequences of various illegal modifications in the logic ledger can differ. The illegal deleting of rows or adding of fake rows will distort data written after specific m and t values in the truth table, and it can be revealed by direct access to that ledger. Nevertheless, a more dangerous attack can be carried out if the logic minimization of AGA expressions [29,30] is periodically held in order to shorten the ledger and its computing time. Although this minimization is not an obligatory procedure, participants of the protocol can be interested in it to lower expenses. Namely, this procedure can hide the results of illegal replacement of the true logic constant in a product term by a fake value, as it is disclosed by Example 1.

Example 1. *Let us consider a ledger with k = 32,656 and input variables x_1, x_2, x_3, x_4, where the product term* $567 * X_1(24,24) * X_2(1275,1275) * X_3(317, 331) * X_4(587,1144)$ *is one of the ledger's entries. A possible attack of an eavesdropper to compromise the ledger is to add a fake entry, where the parameters of Literals are not modified but the randomly chosen hash value, assigned as logic constant C = 567, is replaced, e.g., as 830. Further minimization here can delete clear signs of attack and may complicate its verification, and in order to avoid this one should use the basic definition of subsuming of product terms, which is the critical component of simplification procedures* *for AGA functions [29]. Further, the definition used is cited from [29] and was commented on in detail in [30].*

Definition 2. *Product term* $r_1 * X_1(a_1, b_1) * \ldots * X_n(a_n, b_n)$ ***subsumes*** *another product term* $r_2 * X_1(c_1, d_1) * \ldots * X_n(c_n, d_n)$, ***if and only if*** *conditions (1) and (2) are true:*

1. $r_1 \leq r_2$;
2. $c_i \leq a_i \leq b_i \leq d_i$ for all X_i, $i = 1, \ldots, n$.

Using this definition for Example 1, one can see that $567 \leq 830$ is true and $c_i = a_i \leq b_i = d_i$ for all X_i, $i = 1, 2, 3$, respectively. Minimization after illegal replacement of data will provide further that the first product term will subsume the second one and consequently will be deleted:

$$\begin{aligned}
&\cancel{567 * X_1(24{,}24) * X_2(1275{,}1275) * X_3(317, 331) * X_4(587{,}1144)} + \\
&+ 830 * X_1(24{,}24) * X_2(1275{,}1275) * X_3(317, 331) * X_4(587{,}1144) = \\
&= 830 * X_1(2{,}24) * X_2(1275{,}1290) * X_3(317, 331) * X_4(587{,}1144).
\end{aligned}$$

Thus, the ledger function can be compromised by illegal changes, although further comparison of the true number of product terms and values of m and t will disclose the mismatch, but it will be difficult to determine the step when it is carried out.

If participants of the protocol want to use minimization procedures to shorten the ledger and at the same time prefer to exclude possible illegal modifications, then an AGA model of a ledger provides the original method to prevent modifications hidden by minimization. The advantage of this scheme is that it is grounded only in basic definitions [29] and does not need any additional theorems or constraints. In order to apply this method, the participants of the protocol should agree to use constant $k - 1$ only as a "technical" parameter for the minimization procedure [29], which will help to uniformly describe all parts prohibited for modification of the truth table, as illustrated in Figure 6. To obtain a formal logic model for the prohibited part of the truth table, one should define complementary product terms, assigning maximal logic constant $k - 1$ to all unused rows located "between" both of true entries in the multiparametrical space given by the truth table. Such a prohibited product term will be the subsumed term for any fake product terms having less logic constants and various sets of input variables.

The procedure is proposed further to define a complementary set of prohibited product terms for the truth table, which is general enough and can be used not only for a logic ledger function with two entries but also for any storage AGA function. That is why, for brevity, let us consider a logic function containing only three input variables x_1, x_2, x_3, where the last two entries (or rows) were added into the truth table at different time moments t_m and t_{m-1}, such as in Figure 6. Such a segment of the function includes only true entries, it is tagged by "+" (i.e., a "good" segment), and it can be written as Equation (8):

$$\begin{aligned}
\textit{Function segment}\,{}^{+}_{m-1,\ m} = \; & h_{m-1} \star X_1(\overline{x}_1, \overline{x}_1) \star X_2(\overline{x}_2, \overline{x}_2) \star X_3(\overline{x}_3, \overline{x}_3) + \\
& h_m \star X_1\left(\overline{\overline{x}}_1, \overline{\overline{x}}_1\right) \star X_2\left(\overline{\overline{x}}_2, \overline{\overline{x}}_2\right) \star X_3\left(\overline{\overline{x}}_3, \overline{\overline{x}}_3\right).
\end{aligned} \tag{8}$$

Double index $m - 1$, m in the left part of Equation (8) indicates the numbers of entries to be processed.

If the segment of the logic function with true entries is given by Equation (8), then the corresponding segment with prohibited entries also can be designed, as one can describe the total border between two classes (or function segments) given in the multiparametrical space given in AGA [30]. The principle of formation of prohibited product term $pt^{-}_{m-1,m}$, describing all possible potentially fake entries, is very simple: it one should include all possible neighboring entries (or rows in the truth table), i.e., whose values of variables differ by one or more. If only one of the variables provides unused entries, other variables may have values coinciding with true entries.

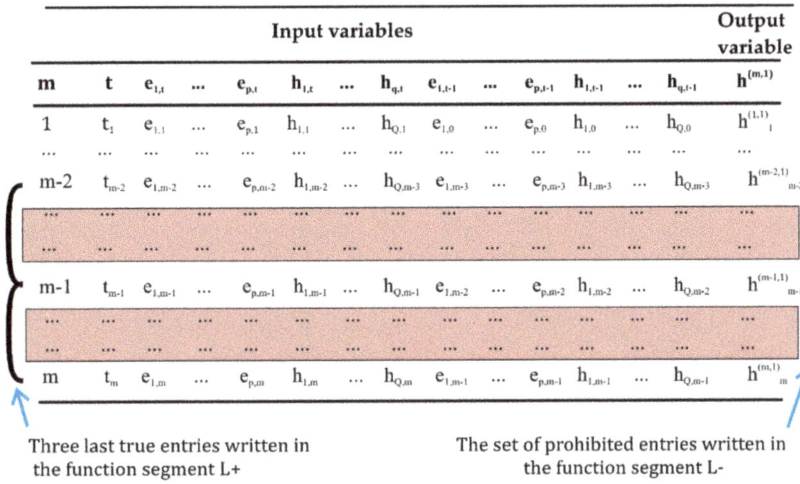

Figure 6. Formation of the part prohibited for modifications of the truth table, corresponding to the segment L-, which is complementary to the true segment L+.

Prohibited product term $pt^{-}_{m-1,m}$ should contain logic constant $k-1$, which will provide subsuming of any illegally added product terms with smaller constants written into the prohibited part of the truth table. Such a property of AGA is being actively used in its minimization method based on "don't care states" [29]. According to it, the idea of minimization in AGA is grounded in the extension of the initially given AGA function. One should initially add unused sets of input variables into the truth table and assign logic constants $k-1$ to them, thus defining by hand the "don't care states". These new product terms will be the subsumed ones for all added fake product terms with smaller constants and equal sets of variables [29], so that one can also exploit this experience of a minimization procedure and intentionally assign constant $k-1$ to all unused product terms, thus excluding their use by eavesdroppers.

In order to obtain formal expression for prohibited product term $pt^{-}_{m-1,m}$, one can firstly guess the appropriate logic expression, propose the algorithm for its realization, and further prove it.

In the further proposed Algorithm 2, a necessary preliminary condition is that one should arrange entries in the truth table according to the sequential enlargement of time variable and entry numbers m. Note that further x_i corresponds to time moment t_{m-1}, and $\overline{\overline{x}}_i$ responds to t_m.

Algorithm 2: Formation of the prohibited segment of an AGA function for the last entry $h_m \star X_1\left(\bar{\bar{x}}_1, \bar{\bar{x}}_1\right) \star X_2\left(\bar{\bar{x}}_2, \bar{\bar{x}}_2\right) \star \ldots \star X_p\left(\bar{\bar{x}}_p, \bar{\bar{x}}_p\right)$ and the previous one $h_{m-1} \star X_1(\bar{x}_1, \bar{x}_1) \star X_2(\bar{x}_2, \bar{x}_2) \star \ldots \star X_p(\bar{x}_p, \bar{x}_p)$.

Input		
	$k \leftarrow$	Number of truth levels
	$Q \leftarrow$	List of addresses for Q participants of protocol
	$t_{wait} \leftarrow$	Waiting time for reply
	$e_m \leftarrow$	Last entry $h_m \star X_1\left(\bar{\bar{x}}_1, \bar{\bar{x}}_1\right) \star X_2\left(\bar{\bar{x}}_2, \bar{\bar{x}}_2\right) \star \ldots \star X_p\left(\bar{\bar{x}}_p, \bar{\bar{x}}_p\right)$
	$e_{m-1} \leftarrow$	Previously received entry $h_{m-1} \star X_1(x_1, x_1) \star X_2(x_2, x_2) \star \ldots \star X_p(x_p, x_p)$
	1.	Set $i = 1$ for the counter of input variables, $i = 1, \ldots, p$
	2.	Calculate $b = Max\left(\bar{x}_i, \bar{\bar{x}}_i\right)$, $a = MIN\left(\bar{x}_i, \bar{\bar{x}}_i\right)$
	3.	If $b_i - a_i \geq 2$ write Literal as $X_i(a+1, b-1)$, otherwise go to step 6
	4.	Set z = 1; where z is the flag of prohibited rows
	5.	Go to step 8.
	6.	If $b_i - a_i < 2$ write Literal as $X_i(a, b)$
	7.	Set flag z = 0
	8.	Set counter $i = i + 1$
	9.	If $i \leq p$, go to step 2, otherwise go to step 10.
	10.	If z = 1, then write prohibited segment as $Function\ segment^{-}_{m-1,\ m} = (k-1) * X_1(\ldots,\ldots) * \ldots * X_p(\ldots,\ldots)$, if z = 0, then write prohibited segment as $Function\ segment^{-}_{m-1,\ m} = 0$.
Output		The set of parameters of the product term, defining the prohibited part between two given entries

Proof of correctness: The prohibited product term according to its idea cannot coincide with the true one, thus its Literals parameters are natural numbers and in the formal expression should differ at least for one. Due to the definition of operator Literal [29], always for any input variable $b_i \geq a_i$. As condition $b_i - a_i < 2$ is equivavent to $b_i < 2 + a_i$, but always $b_i \geq a_i$, then according to Figure 7, possible values of b_i are either a_i or $a_i + 1$. Non-coinciding with true ones, the prohibited segment cannot then be inserted. As condition $b_i - a_i \geq 2$ responds to condition $b_i \geq a_i + 2$, a non-coinciding prohibited product term can include at least one possible value, and briefly this can be written as $X_i(a+1, b-1)$.

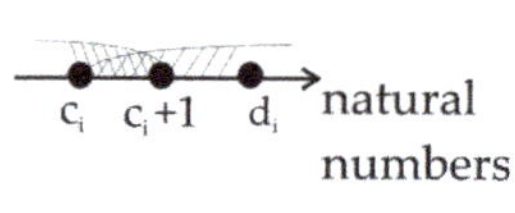

Figure 7. Possible values of Literal parameters.

Based on Algorithm 2, it is more appropriate to design the joint segment of the logic function by combining both true entries and the prohibited segment "between" them. Such a joint expression of Equation (9) tagged by "±" should contain three product terms: the true product term for previous entry pt^{+}_{m-1}, tagged by number of $m-1$, the true product term for the last entry pt^{+}_{m}, tagged by number m, and the prohibited product term $pt^{-}_{m-1,m}$ uniting all unused and potentially dangerous rows located between the last and the previous entries:

$$
\begin{aligned}
Function\ segment^{\pm}_{m-1,\ m} = \; & h_{m-1} \star X_1(\bar{x}_1, \bar{x}_1) \star X_2(\bar{x}_2, \bar{x}_2) \star X_3(\bar{x}_3, \bar{x}_3) + \\
& (k-1) \star X_1\left(Min\left(\bar{x}_1, \bar{\bar{x}}_1\right) + 1, Max\left(\bar{x}_1, \bar{\bar{x}}_1\right) - 1\right) \star \\
& X_2\left(Min\left(\bar{x}_1, \bar{\bar{x}}_1\right), Max\left(\bar{x}_1, \bar{\bar{x}}_1\right)\right) \star X_3\left(Min\left(\bar{x}_3, \bar{\bar{x}}_3\right), Min\left(\bar{x}_3, \bar{\bar{x}}_3\right)\right) + h_m \star X_1\left(\bar{\bar{x}}_1, \bar{\bar{x}}_1\right) \star \\
& X_2\left(\bar{\bar{x}}_2, \bar{\bar{x}}_2\right) \star X_3\left(\bar{\bar{x}}_3, \bar{\bar{x}}_3\right).
\end{aligned} \tag{9}
$$

Thus, according to Algorithm 2, Equation (9) can describe sets of prohibited rows even without directly composing the truth table for all possible states. The prohibited part, if necessary, can also be written as a separate AGA function.

3.4. Representation of Quantum Protocol by AGA Function

The model of the logic ledger $h^{(m,s)} = F_{ldg}(m, t, e_m, e_{m-s}, \boldsymbol{h}_m, \boldsymbol{h}_{m-s})$ proposed above describes a system whose states are linked by a logic expression for entries taken for different time moments. This model principally can be used for description of an arbitrary procedure, including various classical and quantum protocols, as the truth table and its equivalent product terms are simple and visual tools for preparation and debugging of a logic function. This is illustrated further for the case of the route verification scheme cited in Section 1.2.

Verification procedures discussed in [32] and in Section 1.2 suppose that the prover and the verifier in each of the checkpoints interact as two different agents and execute the quantum protocol PU [33], which includes the transfer and measurements of classical and quantum signals.

Given above in the Figure 3, the actions of verifiers in every checkpoint were represented by the logic function $F_{TN}(P, t, V_1, V_2, S_1, S_2, S_3, S_4, J)$ [32], describing the work of crossed measurement modules. Input variable P here describes the initiating signal, t is time, and V_1 and V_2 depict the transfer of two equal values q by entangled photons emitted by the EPR module. Indexed parameters S_1, S_2, S_3, S_4 describe the transfer of classical test signals coming from the RO module via four crossed modules of the verifier, and J is the counter of the involved devices and procedures.

In its turn, actions of the prover P were represented in [32] by logic function F_P $(t, \Psi, S_{p1}, S_{p2}, S_{p3}, S_{p4}, J)$, where t is time, Ψ is the measurement of the photon polarization state $|\Psi>$ in the basis $B = H(x_1 \oplus x_2 \oplus x_3 \oplus x_4)$ responding to the 2D scheme of verification, $S_{p1}, S_{p2}, S_{p3}, S_{p4}$ refer to measurements of classical signals sent to four verifier modules in the crossed scheme, and J is the counter of the devices and procedures. H is the random hash function of the RO [32,77] used by all verifiers and provers.

In order to model the interaction of the two agents commented above in the joint logic model, one should define the united AGA function F_{VP} given by Equation (10) and disclosed in Table 5:

$$y = F_{VP}(t, P, V_1, V_2, S_1, S_2, S_3, S_4, \Psi, S_{p1}, S_{p2}, S_{p3}, S_{p4}, J_v, J_p) \tag{10}$$

Table 5. Truth table of the F_{VP} function, modelling interaction of the prover and the verifier agents in the route verification scheme [79]. Procedure describes execution of protocol PU and involves RO. Logic function can be written as a set of product terms and used as entry in the logic ledger.

Input Variables															Output	Activated Procedure
t	P	V_1	V_2	S_1	S_2	S_3	S_4	J_V	Ψ	S_{p1}	S_{p2}	S_{p3}	S_{p4}	J_P	F_{VP}	
t_0	p	-	-	-	-	-	-	-	-	-	-	-	-	-	C_1	Prover initiates protocol
t_1	-	-	-	s_1	-	-	-	j_2	-	-	-	-	-	-	C_2	Step 1: load s_1 from RO
t_1	-	-	-	-	s_2	-	-	j_3	-	-	-	-	-	-	C_3	Load s_2 from RO
t_1	-	-	-	-	-	s_3	-	j_4	-	-	-	-	-	-	C_4	Load s_3 from RO
t_1	-	-	-	-	-	-	s_4	j_5	-	-	-	-	-	-	C_5	Load s_4 from RO

Table 5. *Cont.*

Input Variables															Output	Activated Procedure
t	P	V_1	V_2	S_1	S_2	S_3	S_4	J_V	Ψ	S_{p1}	S_{p2}	S_{p3}	S_{p4}	J_P	F_{VP}	
t_2	-	-	-	-	-	-	-	-	-	-	-	-	-	j_{p1}	C_{14}	Step 2: transfer initiating signal p
t_2	-	-	-	-	-	-	-	-	ψ	-	-	-	-	j_{p2}	C_{15}	Load $\mid\psi>$ into quantum memory
t_2	-	-	-	-	-	-	-	-	-	s_{p1}	-	-	-	j_{p3}	C_{16}	Measure s_1
t_2	-	-	-	-	-	-	-	-	-	-	s_{p2}	-	-	j_{p4}	C_{17}	Measure s_2
t_2	-	-	-	-	-	-	-	-	-	-	-	s_{p3}	-	j_{p5}	C_{18}	Measure s_3
t_2	-	-	-	-	-	-	-	-	-	-	-	-	s_{p4}	j_{p6}	C_{19}	Measure s_4
t_2	-	-	-	-	-	-	-	j_1	-	-	-	-	-	-	C_6	Re-switching of memory units
t_2	-	-	-	-	-	-	-	j_2	-	-	-	-	-	-	C_7	Transfer of $\mid\psi>$
t_2	-	-	-	-	-	-	-	j_3	-	-	-	-	-	-	C_8	Transfer s_1
t_2	-	-	-	-	-	-	-	j_4	-	-	-	-	-	-	C_9	Transfer s_2
t_2	-	-	-	-	-	-	-	j_5	-	-	-	-	-	-	C_{10}	Transfer s_3
t_2	-	-	-	-	-	-	-	j_6	-	-	-	-	-	-	C_{11}	Transfer s_4
t_3	-	-	-	-	-	-	-	-	-	-	-	-	-	j_{p1}	C_{20}	Step 3: calculate basis $B = H(s_1 \oplus s_2 \oplus s_3 \oplus s_4)$ and measure ψ in basis B
t_4	-	-	-	-	-	-	-	-	-	-	-	-	-	j_{p2}	C_{21}	Step 4: transfer of $\mid\psi>$ by EPR pairs
t_5	-	q_1	q_2	-	-	-	-	j_1	-	-	-	-	-	-	C_{12}	Step 5: measurements and check if $q_1 = q_2$
t_6	-	-	-	-	-	-	-	j_1	-	-	-	-	-	-	C_{13}	Step 6: repeat steps for C1-C12 and estimate if error exceeds 3.5%

This function is defined for the set of all 14 input variables used earlier in F_P and F_{TN}, and the output variable is given by the set of constants $C_1, \ldots, C_{21}$, for simplicity tagged as they are in paper [32]. According to [32,33,79], the used version of route verification protocol PU [33] was slightly adapted for the 2D case, and one checkpoint was considered as one verifier agent, which interacts with the prover agent. Their interaction corresponds to the protocol PU [33] and is subdivided into six steps tagged by time countdowns $t_0, \ldots, t_6$. Here, the value p of variable P obtained at t_0 refers to the initial signal coming from the prover and initiating the verification protocol, clocking scheme, memory device, random oracle, and microcontrollers or FPGAs. As time variable t is the same for both agents in [32,79], it does not need to be repeated twice in the united Equation (10). Two counters tagged J_V and J_p principally can be replaced by one, but for simplicity they respond to the variables used earlier.

A special comment should be given to the other steps necessary for theoretically adapting the proven PU protocol [33], for practice. The choice of EPR scheme is critical here, as namely the simultaneous emission of two entangled photons provides maximal precision to the measurements of distances between the prover and the pair of verifiers (in the crossed detection scheme) in Figure 2. Respectively, the EPR scheme in the PU protocol determines the use of the polarization coding of qubits. However, the EPR scheme can have three variants, concerning the polarization of the pump photon [82,83]. The signal and idler photons can have the same polarization with each other and can coincide with the destroyed pump photon, but their equal polarization can also be orthogonal to the pump photon polarization. One more opportunity is that the signal and idler photons can have perpendicular polarizations. Thus, the truth table in Table 5 can potentially describe more details for these variants.

The choice of input variables leaves some alternatives for the procedure of input of the classical testing signals s_1, s_2, s_3, s_4 generated by the RO module and controlled by variables S_1, S_2, S_3, S_4. These classical signals should be sent to the prover by the verifier, and one can use a separate controller for them and thus replace four signals by one. The same alternative refers to signals $s_{p1}, s_{2p}, s_{p3}, s_{p4}$ emitted by the prover and described by variables $S_{p1}, S_{p2}, S_{p3}, S_{p4}$.

It is necessary to emphasize that mobile agents are mainly associated with the atmospheric line for entangled data exchange during route passing by agents, but principally modifications of the protocol PU can be adapted for a fiber optics scheme with an EPR source of qubits [75].

Special commentary should also be given to the "-" symbols used in the rows of the truth table of the function F_{VP} for the shortened description of equivalent product terms. For the ideal physical set-up, one may fill in the truth table by the set of variables written in the table and replace all "-" with zeros, as there should be no signals for these variables. These zeros in the truth table correspond to Literals $X_i(0,0)$ written in the product terms, as it was carried out for Tables 1–3. However, real signals may have nonzero levels due to transient processes and noise. That is why, for the real system, an ideal logic model may not work correctly, and noise deviations should be taken into account. This fact leads to the necessity to add into the truth table all possible rows containing the expected spectrum of all possible noise signals, which is a very wasteful method, but such information may be not accessible at the initial stage of design. At the stage of composing the truth table, one can suppose that input variables tagged by the symbol "-" are not observed and may have arbitrary transient values. This may be written by Literal $(0, k-1) \equiv k-1$, which can always be correctly "installed" into any shortened product term for omitted variables. This can be disclosed by Equation (11), where all variables tagged by "-" in Table 5 are omitted:

$$
\begin{aligned}
F_{VP} = \;& C_1 * X_t(t_1,t_1) * X_P(p,p) * X_{jp}(j_{p1},j_{p1}) + \\
& C_2 * X_t(t_1,t_1) * X_{s1}(s_1,s_1) * X_j(j_2,j_2) + \\
& C_3 * X_t(t_1,t_1) * X_{s2}(s_2,s_2) * X_j(j_3,j_3) + \\
& C_4 * X_t(t_1,t_1) * X_{s3}(s_3,s_3) * X_j(j_4,j_4) + \\
& C_5 * X_t(t_1,t_1) * X_{s4}(s_4,s_4) * X_j(j_5,j_5) + \\
& C_6 * X_t(t_2,t_2) * X_j(j_1,j_1) + \\
& C_7 * X_t(t_2,t_2) * X_j(j_2,j_2) + \\
& C_8 * X_t(t_2,t_2) * X_j(j_3,j_3) + \\
& C_9 * X_t(t_2,t_2) * X_j(j_4,j_4) + \\
& C_{10} * X_t(t_2,t_2) * X_j(j_5,j_5) + \\
& C_{11} * X_t(t_2,t_2) * X_j(j_6,j_6) + \\
& C_{12} * X_t(t_5,t_5) * X_{v1}(q_1,q_1) * X_{v2}(q_2,q_2) * X_j(j_1,j_1) + \\
& C_{13} * X_t(t_6,t_6) * X_j(j_1,j_1) + \\
& C_{14} * X_{tp}(t_0,t_0) * X_{jp}(j_{p1},j_{p1}) + \\
& C_{15} * X_{tp}(t_2,t_2) * X_\psi(\psi,\psi) * X_{jp}(j_{p2},j_{p2}) + \\
& C_{16} * X_{tp}(t_2,t_2) * X_{sp1}(s_{p1},s_{p1}) * X_{jp}(j_{p3},j_{p3}) + \\
& C_{17} * X_{tp}(t_2,t_2) * X_{sp2}(s_{p2},s_{p2}) * X_{jp}(j_{p4},j_{p4}) + \\
& C_{18} * X_{tp}(t_2,t_2) * X_{sp3}(s_{p3},s_{p3}) * X_{jp}(j_{p5},j_{p5}) + \\
& C_{19} * X_{tp}(t_2,t_2) * X_{sp4}(s_{p4},s_{p4}) * X_{jp}(j_{p6},j_{p6}) + \\
& C_{20} * X_{tp}(t_3,t_3) * X_{jp}(j_{p1},j_{p1}) + \\
& C_{21} * X_{tp}(t_4,t_4) * X_{jp}(j_{p1},j_{p1}).
\end{aligned}
\tag{11}
$$

However, the first product term $C_1 * X_P(p,p) * X_t(t_1,t_1) * X_j(j_1,j_1)$ in Equation (11) should be rewritten equivalently as:

$$
\begin{aligned}
& C_1 * X_P(p,p) * X_t(t_1,t_1) * X_j(j_1,j_1) = \\
& \quad C_1 \; *X_t(t_1,t_1) * X_P(p,p) * X_{V1}(0,k-1) * X_{V2}(0,k-1) * X_{S1}(0,k-1) \\
& \quad *X_{S1}(0,k-1)*X_{S2}(0,k-1)*X_{S3}(0,k-1)*X_{S4}(0,k-1) \\
& \quad *X_\psi(\psi,\psi)*X_{Sp1}(0,k-1)*X_{Sp2}(0,k-1)*X_{Sp3}(0,k-1) \\
& \quad *X_{Sp4}(0,k-1)*X_{IV}(j_1,j_2)*X_{IP}(j_1,j_2)
\end{aligned}
\tag{12}
$$

The other 20 product terms also should be appropriately rewritten.

After the design and clarification of the involved hardware components, one should specify the appropriate noise parameters and reduce the band of indexed parameter b in Literals with $(0,\ k-1)$ from $k-1$ to some determined by noize value $\bar{\bar{b}}$. In other words, function $y = F_{VP}$ given by Equation (12) should be redefined after clarification of its noise parameters. Namely, this procedure will be correct, as there are no formal reasons to mix 0 and $k-1$ constants in one logic expression, and this should be carried out by hand in the final truth table for the new function $y = F_{VP}$.

Using the procedures discussed above, one can write the function, describing the interaction of prover and verifier agents, into the ledger function as a reference procedure for tests or restoration. Position-based verification with protocol PU [33] here is not the obstacle, and both quantum and classical procedures can be united in the AGA logic model.

One more substantial note should be given, referring the further necessity to consider the model of trust for the interaction between the prover and the verifier. The joint logic function $y = F_{VP}$ discussed above has included the prover and the verifier as two different agents, according to their signals used in the verification protocol. However, their real interaction will also depend on the honest and dishonest behavior of the prover. In order to obtain a more detailed and realistic description of the verification procedure, it is further necessary to add a model of trust for their interaction.

Another interesting aspect of a fiber optics network segment is that advances in this field create the possibility to complement the PBC task and its more secured but specific PU by a less protected but more massive QKD protocol for the delivery of confidential digital data, such as photos, from the neural network database. At least, the recent research of the continuous-variable QKD scheme [84] can potentially be used for the design of k-valued truth models of AGA. Moreover, this paper discusses the trusted component of the phase noise model, so that the agent model for such hardware schemes can directly include several different noise models into the control scheme of the agent. That means there is a necessity to design more complicated models of trust, directly including physical parameters into agent models. It seems that multiparametrical models of AGA can be quite appropriate for the integration of such physical processes. The frequency polarization multiplexing system used in [85] additionally enlarges the interest in the continuous-variable QKD scheme. Besides this, the equalization enhanced phase noise [86] is attributed to wavelength division multiplexing schemes of fiber optics, which seems to be the most reliable scheme for the realization of multi-level logic. Moreover, the phase-coded noise compensation tools in QKD schemes [86] are attributed to dispersion compensating fiber, chirped Bragg gratings, dispersion-shifted fiber, and reduction of the chromatic dispersion effect by optimal initial pulse duration. All these schemes can potentially be considered as components for AI control agents, but the agent's architecture and MVL model can differ substantially for them.

3.5. Computing Aspects of AGA Functions

As the ledger verification scheme for robotic agents is finally designated for data protection protocols and RO design, possible schemes of MVL procession should be obligatorily adapted to low-level programming and distributed schemes of microcontrollers and memory chips located inside a hardware agent. The two-controller testing circuit board was designed earlier for a fuzzy logic modelling device, which is shown schematically in the Figure 8, where input bus commutators used ADC in agents, but here it was used for the run of MVL function calculations. This external link controller, chips at the output bus, and some of clocking signals are not discussed here and are not shown, for brevity. Such a laboratory circuit board has a lower response in comparison with modern ARDUINO controllers [87], which are good for conjugation with modern computer ports and for quick access to embedded memory with limited capacity, but they are not appropriate for experiments with brunched multi-port schemes with active interaction of several controllers and high-capacity external memory chips, which is actually for the design of hardware agents.

The 8-bit circuit board that was used included 1 MB SRAM, 512 KB ROM, and two 24 MHz ATMEL microcontrollers, possessing four input/output banks. The microcontroller MC1 was intended for data acquisition from external and network devices, and the second one, MC2, calculated the MVL ledger test function. Trigger registers Rg1, 2, and 3 are necessary here for the clocking of external memory chips. Addressing of the necessary SRAM cell is performed via bank (i.e., port) two of MC2 by means of trigger registers and inversed pins CE OE, WE. The data from SRAM and ROM are inputted via bank P0 of MC2.

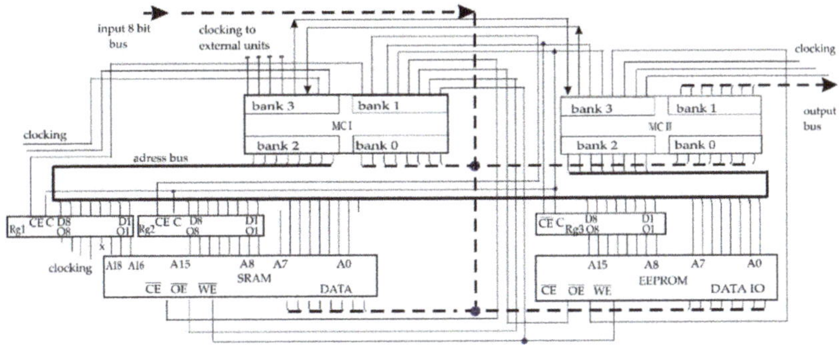

Figure 8. Scheme of the testing circuit board for AGA microassembler software (shortened).

The scheme shown in Figure 8 with the microassembler program for MC2 calculates the fit of the arbitrary external entry to the MVL ledger model with assigned quasi random hash value $h^{(m,1)}$. For clarity, the hash value $h^{(m,1)}$ assigned to the last ledger entry is written in ROM, as in fact the RO device is to be the combination of the memory and random number generator, reproducing random hash for any repeated input signal. Respectively, it can be emulated by outputting value $h^{(m,1)}$ taken from the set $h^{(1,1)}, \ldots, h^{(m-1,1)}, h^{(m,1)}, \ldots h^{(k-1,1)}$ written in cells with addresses from, for example, #1 up to #255.

The test task included the processing of eight Literals shown in Table 2. In order to shorten the task, only one external hash value approves the entry, and every entry e consists of two components, e_{11} and e_{12}. According to Table 4, the MVL template ledger function has emulated the numerical test data coming from the external agent and written them in fixed memory cells, as shown in Table 6.

Table 6. Structure of template and input test data to be written in SRAM for entry verification.

Template	m	t	e11	e12	h1	e21	e22	h2
Addresses	#1,#2	#3,#4	#11,#12	#13,#14	#15,#16	#17,#18	#19,#20	#21,#22
Test data	mt	tt	et11	et12	ht1	et21	et22	ht2
Addresses	#101	#103	#111	#113	#115	#117	#119	#121

The value of the calculated product term is to be written in SRAM cell #150.

The results of running of the program for microcontroller MC2 were also emulated by the simplest microassembler simulator AVSIM. Given a further short fragment of software demonstrates the calculation of only one Literal in a scheme with external memory addressing, in order to show the role of the agent's architecture. Note that this program processes a general case of Literals with different low and higher indexed parameters of a, b in separate cells.

Fragment of the program for calculation of Literal X(a,b) in MC2.

```
ORG 0H
AJMP START
ORG 30H
...
LITERALS: MOV R6,#1; begin cycle for calculation of 8 Literals
SJMP PT
NEXTLIT: INC R6
PT: CLR C;
... addressing cell #1 in SRAM ...
MA: MOV A,P0; read a for variable m
MOV R5,A; assign a to register R5
... addressing cell #2 in SRAM ...
MB: MOV A,P0; read b for variable m
MOV R4,A; assign b to register R4
... addressing cell #101 in SRAM ...
MT: MOV A, P0; read mt
MOV R3,A
COMPMA: SUBB A, R5; compare mt with a for Xm
JNC LITMB; go to LITMB if bit C = 0 and mt is greater than a for Xm
AJMP PT_0; mt does not fit to Literal and the whole product term is equal to zero
COMPMB: CLR C
MOV A,R4; assign b to accumulator A
SUBB A, R3; compare mt with b for Xm
JNC CHECKLIT; if bit C = 0 i.e., mt less than b for Xm, then process next Literal
AJMP PT0; mt does not fit to Literal and the whole product term is zero
CHECKLIT: CJNE R6,#8,PT;
PT0: ... SRAM cell #150 addressing ...
MOV P0,#0; write 0 for product term into SRAM
...
END
```

A further example is shown for the shortest designed versions of MIN, MAX operators possible, only without external memory addressing.

Fragments of the program for calculationof MAX(x,y) or MIN(x,y) in MC2

	MINIMUM (#N1,#N2): choose smaller of two values	MAXIMUM (#N1,#N2): choose greater of two values
INPUT #N1 ← Natural number #N2 ← Natural number		

	MOV A,#N1
	MOV R7,#N2
	CLR C
	SUBB A,R7
	JNC MIN_N2
	MOV R0,#N1
	SJMP NEXT
	MIN_N2:MOV R0,#N2
	NEXT
		MOV A,#N1
		MOV R7,#N2
		CLR C
		SUBB A,R7
		JNC MAX_N1
		MOV R0,#N2
		SJMP NEXT
		MAX_N1:MOV R0,#N1
		NEXT:

OUTPUT:	MIN value in register R0	OUTPUT: MAX value in register R0

Results of microassembler modeling.

- Microassembler 8-bit modeling of MVL operators and functions does not detect any additional problems in comparison with traditional microcontroller applications such as step motors or control or sensors, but the choice of agent architecture influences greatly on the time response and clocking. It should be chosen before the full-scale modeling of agent interaction;
- For test circuit board shown above, the calculation of basic operators for AGA function is slow enough as, e.g., $k = 256$ truth levels и $n = 30$ variables the calculation will need $\approx 210,000$ work cycles, which responds to 0.1 s at 24 MHz, as minimally possible parameters are $t_{\text{MIN-MAX}} \approx 4$ µs and $t_{\text{Literal}} \approx 9$ µs. However, addressing of external memory devices can additionally enlarge the number of cycles, at least for ≈20–25%. That complicates the design of the universal PC's program for interaction with the hardware agent;
- FPGA, at for example 500 MHz, are more appropriate for practical work with AGA, as due to [88] its typical work cycle responds to four clocking cycles and calculations of the AGA function will take

$$t = N_{truth\ levels} \cdot 4T_{work\ cycle} \cdot N_{operators\ MIN\&MAX} \tag{13}$$

  and can be lowered to the ~ms scale appropriate for real technical tasks. However, parallel schemes of AGA calculations are not investigated for such platforms.
- In order to solve the problem of non-compatibility of instructions for computer and microcontroller platforms using the AGA logic ledger, a modern RISC-V platform [89] with open code seems to be the possible solution, as it proposes universal short instructions for devices with different throughput.

For further conjugation of microassembler and PC procedures in MASs, one can use as the base the program for the computing of the AGA function given in C language, which was published as a supplement to the recent paper [30]. As for C++, since the standard ANSI C [90] does not define library functions corresponding to operators $MAX(x,y)$ and $MIN(x,y)$, then for the prototype stdlib.h traditional cycles or macro, *max*( ) and *min*( ) can be used for calculations of these operators. The program's main { . . . } part should then use instructions such as:

printf "max/min of x, y is %d\n", max/min (x, y));
return 0;.

For such a design, the final architecture of the hardware agent should be chosen, as the data exchange in the distributed scheme directly emulating procedures for the proposed logic ledger will need high-capacity external memory chips, substantially modifying time response estimations.

3.6. Discussion of Prospectives for Further Research of AGA Functions

Modern fiber optics networks provide a possibility to transfer large volumes of digital information to distant nodes, and QKD schemes enhance the protection level for data transfer channels, which creates new possibilities for massive distant control of robotic MASs. However, there are currently no ready and well-formed concepts to provide reliable autonomous monitoring and self-restoration of a MAS in case of attacks and breakdowns. That is why there are several actual tasks to continue the MVL designs presented above.

The logically linked list for a distributed ledger proposed above can be further used as a simple scheme for backup storage, aimed at autonomous verification and restoration of valuable and critical data for distant agents. AGA models are attractive here due to their ability to form correct multiparametrical logic expressions for arbitrary sets of parameters. They can combine parameters of classical and quantum lines for collective verification using quantum keys, RO, and "bit commitment schemes".

For future robotic system tasks, data concerning rent, repairs, and licenses for a small-scale MAS seems to be not confidential, but technical schemes of objects, prehistory of the agent's work represented by terrestrial maps, biometrics keys, verification digital schemes, and images need additional secret coding. The simplest way here is to use the encapsulation of an additional protocol into a standard one, and the AGA analog of OTP secret coding scheme [31] can potentially be adapted for the large number of truth levels in a ledger.

The proposed schemes of a logic-linked list of entries and distributed logic ledger allow the formal logic scheme to design a vaster model of trust for the interaction of agents with other ones and network nodes. The time-attributed model of linking entries, necessary for the quantum protocol PU and other classical verification schemes, principally can be extended from space coordinates to other variables. A multi-stage procedure resembling a Merkle tree in the BC [37–39] can be used here based on the same random hash functions as for RO-based schemes [22]. Estimations of trust parameters can be gathered in a separate AGA function, which can be used the same as the MVL classification scheme proposed earlier [30] for decision making.

As noise models for fiber optics systems [84] can now involve independent trust estimates, more detailed schemes for the integration of such models into robotic agents are interesting for more efficient control of fiber optics networks. Verification schemes with enhanced trust models can also describe pairs of bot–hardware agents, where the software bot is used as a "deputy admin" at a distant MAS.

The task to protect memory storage in unmanned network robots from illegal modifications may be considered further as the creation of self-sustained robotic "collaborative media" for the partner interaction of inexpensive small-scale robotic systems, providing individual and collective verification, approval, and restoration of data. Unfortunately, a trusted server is too expensive of a way for mass MASs. Then, a set of small distant backups with branched addressing of entries, muddy structure, and additional secret coding can be the alternative, thus extending the trend for collective interaction of agents, demonstrated by IoV, IoT, BC, and smart grids.

Experiments for MVL models are reasonably based on the unified structure of short instructions at RISC-V platforms [89], which are interesting for the design of RO versions based on known versions of the SHA256 cryptography hashing function and QRNG.

Conjugation with computer vision systems and neural networks for autonomous agents are considered as the obligatory step for the realization of unmanned robots without trusted server control. That is why the route verification scheme [32,79] is only a preliminary attempt to work out schemes for the integration of quantum verification schemes with digital images.

4. Conclusions

Modern fiber optics can become the base for the distant control of unmanned robotic agents and MASs. However, mass and inexpensive unmanned mobile hardware agents can scarcely be realized as trusted nodes; moreover, they should interact with network nodes that have different levels of trust. As a result, traditional memory data storage becomes the vulnerable component of a mass autonomous robot, and the situation cannot be radically changed for the modern level of quantum memory technologies. That is why, for the reliable work of autonomous mass agents, they should more closely interact with partner network nodes and agents, which can be used for mutual monitoring and as distributed backup storages of critical data necessary in case of breakdowns and attacks. The trend is also to involve different estimates of trust and various verification procedures. Thus, the problem of illegal modifications of data in mass unmanned robots makes the actual design of complicated verification procedures based on AI and multiagent models difficult.

In the present paper, the use of some of the methods designed earlier for the blockchain and distributed ledgers and to applying logic analogs of the linked list and a distributed ledger scheme are proposed. As a robotic version of a ledger is aimed at technical tasks, the approval of data can be realized as simple majority voting instead of expensive proof-of-work schemes from the blockchain.

In order to integrate classical and quantum verification protocols with digital images and other possible data structures, the AGA logic model of a linked list is proposed for agents, autonomously interacting with network nodes. This model uses the AGA function for the logic modelling of two entries, attributed to the time parameter and random hash values assigned collectively by partner nodes. Such a scheme simplifies the integration of complicated quantum and classical verification protocols, such as the protocol PU for position-based verification, and makes the earlier proposed RO schemes for hashing more interesting.

Keys for access to backup storages in external nodes are proposed to be sets of random hash values, assigned by network partners involved with the pool as verifiers, or by entries written in the special format of the data. Such functions that include entries and hash values can be directly used for data extraction from distributed storages. Entries in these functions are intended to approve licenses, credentials, and other critical data for the unmanned work of agents.

The method of "prohibited" product terms is proposed in order to prevent illegal modification of the logic ledger by the minimization of logic expressions, which may be applied for the shortening of the ledger's length and time computing.

Initial testing was held for MVL programs necessary for agents and logic ledger design, stimulating the choice of specific architecture for involved agents, which is necessary for full-scale testing of the proposed method.

Funding: This research received no external funding.

Institutional Review Board Statement: Not applicable.

Informed Consent Statement: Not applicable.

Data Availability Statement: Not applicable.

Conflicts of Interest: The author declares no conflict of interest.

References

1. Hui, R. Passive optical components. In *Introduction to Fiber-Optic Communications*, 1st ed.; Academic Press: Cambridge, MA, USA, 2020; pp. 209–297.
2. Chi, N.; Zhou, Y.; Wei, Y.; Hu, F. Visible Light Communication in 6G: Advances, Challenges, and Prospects. *IEEE Veh. Technol. Mag.* **2020**, *15*, 93–102. [CrossRef]
3. Venkatram, K.; Geetha, M.A. Review on Big Data & Analytics—Concepts, Philosophy, Process and Applications. *Cybern. Inf. Technol.* **2017**, *17*, 3–27. [CrossRef]
4. Xu, F.; Ma, X.; Zhang, Q.; Lo, H.-K.; Pan, J.-W. Secure quantum key distribution with realistic devices. *Rev. Mod. Phys.* **2020**, *92*, 025002. [CrossRef]
5. Huang, D.-J.; Zhong, W.-Z.; Zhong, J.; Jiang, D.; Wu, H. Optimization and Implementation of Efficient and Universal Quantum Key Distribution. *J. Electr. Comput. Eng.* **2020**, *2020*, 1–9. [CrossRef]
6. Nimbe, P.; Weyori, B.A.; Adekoya, A.F. Models in quantum computing: A systematic review. *Quantum Inf. Process.* **2021**, *20*, 1–61. [CrossRef]
7. Sharma, S.; Kaushik, B. A survey on internet of vehicles: Applications, security issues & solutions. *Veh. Commun.* **2019**, *20*, 100182. [CrossRef]
8. Yan, Z.; Ouyang, B.; Li, D.; Liu, H.; Wang, Y. Network Intelligence Empowered Industrial Robot Control in the F-RAN Environment. *IEEE Wirel. Commun.* **2020**, *27*, 58–64. [CrossRef]
9. Christoforou, E.G.; Avgousti, S.; Ramdani, N.; Novales, C.; Panayides, A.S. The Upcoming Role for Nursing and Assistive Robotics: Opportunities and Challenges Ahead. *Front. Digit. Health* **2020**, *2*, 39. [CrossRef]
10. Kumar, S.; Tiwari, P.; Zymbler, M. Internet of Things is a revolutionary approach for future technology enhancement: A review. *J. Big Data* **2019**, *6*, 1–21. [CrossRef]
11. Samih, H. Smart cities and internet of things. *J. Inf. Technol. Case Appl. Res.* **2019**, *21*, 3–12. [CrossRef]
12. Masumori, A.; Maruyama, N.; Ikegami, T. Personogenesis Through Imitating Human Behavior in a Humanoid Robot "Alter3". *Front. Robot. AI* **2021**, *7*, 165. [CrossRef] [PubMed]
13. Aydin, M.E.; Fellows, R. Building collaboration in multi-agent systems using reinforcement learning. *Lect. Notes Artif. Intell.* **2018**, *11056*, 201–212. [CrossRef]
14. Jamil, S.; Rahman, M.; Haider, A. Bag of Features (BoF) Based Deep Learning Framework for Bleached Corals Detection. *Big Data Cogn. Comput.* **2021**, *5*, 53. [CrossRef]

15. Russell, S.; Norvig, P. *Artificial Intelligence: A Modern Approach*; Prentice Hall: Englewood Cliffs, NJ, USA, 1995; Chs.2,6,7,11.
16. Seeja, G.; Arockia, S.; Berlin, H. A Survey on Swarm Robotic Modeling, Analysis and Hardware Architecture. *Procedia Comput. Sci.* **2018**, *133*, 478–485.
17. Alberts, D.S.; Garstka, J.; Stein, F.P. *Network Centric Warfare: Developing and Leveraging Information Superiority*, 2nd ed.; CCR Press: Boca Raton, FL, USA, 2017; ISBN 1-57906-019-6.
18. Agrell, E.; Karlsson, M.; Chraplyvy, A.R.; Richardson, D.; Krummrich, P.M.; Winzer, P.; Roberts, K.; Fischer, J.K.; Savory, S.J.; Eggleton, B.J.; et al. Roadmap of optical communications. *J. Opt.* **2016**, *18*, 063002. [CrossRef]
19. Lai, M.; Wang, J.; Song, T.; Qi, Z.; Zhou, W.; Liu, N. VSP: A Virtual Smartphone Platform to Enhance the Capability of Physical Smartphone. In Proceedings of the 2016 IEEE Trustcom/BigDataSE/ISPA, Tianjin, China, 23–26 August 2016; pp. 1434–1441. [CrossRef]
20. Camenisch, J.; Doğan Kesdoğan, D. *Open Problems in Network Security. Revised Selected Papers*; IFIP WG 11.4 International Workshop; iNetSec: Zurich, Switzerland, 2015.
21. Dastres, R.; Soori, M. A Review in Recent Development of Network Threats and Security Measures. *Int. J. Inf. Sci. Comp. Eng.* **2021**.
22. Broadbent, A.; Schaffner, C. Quantum cryptography beyond quantum key distribution. *Des. Codes Cryptogr.* **2015**, *78*, 351–382. [CrossRef]
23. Bykovsky, A.Y.; Kompanets, I.N. Quantum cryptography and combined schemes of quantum cryptography communication networks. *Quantum Electron.* **2018**, *48*, 777–801. [CrossRef]
24. Trusted Node Security. Available online: https://www.ibm.com/docs/en/scfz/5.2.0?topic=security-trusted-node (accessed on 18 November 2021).
25. Brogan, C.; Smith, J. *Trust Agents: Using the Web to Build Influence, Improve Reputation, and Earn Trust, 10th Anniversary Edition*; John Wiley: Hoboken, NJ, USA, 2020.
26. Hwang, A.H.-C.; Won, A.S. IdeaBot: Investigating Social Facilitation in Human-Machine Team Creativity. In Proceedings of the 2021 CHI Conference on Human Factors in Computing Systems, Yokohama, Japan, 8–13 May 2021; pp. 1–16. [CrossRef]
27. Burr, C.; Cristianini, N.; Ladyman, J. An Analysis of the Interaction Between Intelligent Software Agents and Human Users. *Minds Mach.* **2018**, *28*, 735–774. [CrossRef]
28. Mathur, M.; Mhadalekar, S.; Mhatre, S.; Mane, V. Algorithmic Trading Bot. *ITM Web Conf.* **2021**, *40*, 03041. [CrossRef]
29. Allen, C.M.; Givone, D.D. The Allen-Givone Implementation Oriented Algebra. In *Computer Science and Multiple-Valued Logic: Theory and Applications*; Rine, D.C., Ed.; North Holland: Amsterdam, The Netherlands, 1984; pp. 262–283.
30. Bykovsky, A.Y. Heterogeneous network architecture for integration of AI and quantum optics by means of multiple-valued logic. *Quantum Rep.* **2020**, *2*, 126–165. [CrossRef]
31. Antipov, A.L.; Bykovsky, A.Y.; Vasiliev, N.A.; Egorov, A.A. Multiple-valued logic-protected coding for an optical non-quantum communication line. *J. Russ. Laser Res.* **2006**, *27*, 492–505. [CrossRef]
32. Bykovsky, A.Y. Multiple-Valued Logic and Neural Network in the Position-Based Cryptography Scheme. *J. Russ. Laser Res.* **2021**, *42*, 618–630. [CrossRef]
33. Unruh, D. Quantum Position Verification in the Random Oracle Model. In *Advances in Cryptology–CRYPTO 2014*; Garay, J.A., Gennaro, R., Eds.; Lecture Notes in Computer Science; Springer: Berlin/Heidelberg, Germany, 2014; pp. 1–18. [CrossRef]
34. Bräunl, T. *Embedded robotics. Mobile Robot Design and Applications with Embedded Systems*, 2nd ed.; Springer: Berlin/Heidelberg, Germany, 2006.
35. OCDE/CAF/CEPAL. *Perspectivas Económicas de América Latina 2018*; OECD Publishing: Paris, France, 2018.
36. Krishnan, H.; Saketh, S.; Tej, V. Cryptocurrency Mining—Transition to Cloud. *Int. J. Adv. Comput. Sci. Appl.* **2015**, *6*, 115–124. [CrossRef]
37. Singh, J.; Sinha, A.; Goli, P.; Subramanian, V.; Shukla, S.K.; Vyas, O.P. Insider attack mitigation in a smart metering infrastructure using reputation score and blockchain technology. *Int. J. Inf. Secur.* **2021**, 1–20. [CrossRef]
38. Prabadevi, B.; Deepa, N.; Pham, Q.-V.; Nguyen, D.C.; Praveen Kumar Reddy, M.; Thippa Reddy, G.; Pathirana, P.N.; Dobre, O. Toward Blockchain for Edge-of-Things: A New Paradigm, Opportunities, and Future Directions. *IEEE Internet Things Mag.* **2021**, *4*, 102–108. [CrossRef]
39. Zhuang, P.; Zamir, T.; Liang, H. Blockchain for Cybersecurity in Smart Grid: A Comprehensive Survey. *IEEE Trans. Ind. Inform.* **2020**, *17*, 3–19. [CrossRef]
40. Abbas, H.; Saha, I.; Shoukry, Y.; Ehlers, R.; Fainekos, G.; Gupta, R.; Majumdar, R.; Ulus, D. Special Session: Embedded Software for Robotics: Challenges and Future Directions. In Proceedings of the 2018 International Conference on Embedded Software, Turin, Italy, 30 September–5 October 2018; pp. 1–10. [CrossRef]
41. Kamdar, R.; Paliwal, P.; Kumar, Y. A State of Art Review on Various Aspects of Multi-Agent System. *J. Circuits Syst. Comput.* **2018**, *27*, 1830006. [CrossRef]
42. Hedin, Y.; Moradian, E. Security in Multi-Agent Systems. *Procedia Comput. Sci.* **2015**, *60*, 1604–1612. [CrossRef]
43. Rashvand, H.; Salah, K.; Calero, J.; Harn, L. Distributed security for multi-agent systems—Review and applications. *IET Inf. Secur.* **2010**, *4*, 188–201. [CrossRef]
44. Karygiannis, T.; Yansen, W. *Mobile Agent Security*; Technical Report NIST SP 800-19; National Institute of Standards and Technology: Boulder, CO, USA, 1999.

45. Most Infamous Cloud Security Breaches. StorageCraft, Arcserve Company. Available online: https://blog.storagecraft.com/7-infamous-cloud-security-breaches/ (accessed on 18 November 2021).
46. Dontov, D. The Future of Ransomware Attacks on Cloud Services. 23 April 2021. Available online: https://www.networkcomputing.com/cloud-infrastructure/future-ransomware-attacks-cloud-services (accessed on 18 November 2021).
47. Cohen, R.; Schaekermann, M.; Liu, S.; Cormier, M. Trusted AI and the contribution of trust modelling in multiagent systems. In Proceedings of the 18th International Conference on Autonomous Agents and Multiagent Systems (AAMAS 2019), Montreal, QC, Canada, 13–17 May 2019; pp. 1644–1648.
48. Sklar, E.I.; Parsons, S.; Li, Z.; Salvit, J.; Perumal, S.; Wall, H.; Mangels, J. Evaluation of a trust-modulated argumentation-based interactive decision-making tool. *Auton. Agents Multi-Agent Syst.* **2015**, *30*, 136–173. [CrossRef]
49. Rishwaraj, G.; Ponnambalam, S.G.; Kiong, L.C. An efficient trust estimation model for multi-agent systems using temporal difference learning. *Neural Comput. Appl.* **2016**, *28*, 461–474. [CrossRef]
50. Cheng, M.; Yin, C.; Zhang, J.; Nazarian, S.; Deshmukh, J.; Paul Bogdan, G. A General Trust Framework for Multi-Agent Systems. In Proceedings of the 20th International Conference on Autonomous Agents and Multiagent Systems, London, UK, 3–7 May 2021; pp. 332–340.
51. Deepa, N.; Pham, Q.-V.; Nguyen, D.C.; Bhattacharya, S.; Prabadevi, B.; Reddy Gadekallu, T.; Kumar Reddy Maddikunta, P.; Fang, F.; Pathirana, P.N. A Survey on Blockchain for Big Data: Approaches, Opportunities, and Future Directions. *arXiv* **2020**, arXiv:2009.00858v2.
52. Jabbar, A.; Dani, S. Investigating the link between transaction and computational costs in a blockchain environment. *Int. J. Prod. Res.* **2020**, *58*, 3423–3436. [CrossRef]
53. Ankalkoti, P. A Relative Study on Bitcoin Mining. *Imperial J. Interdiscip. Res.* **2017**, *3*, 5. Available online: http://www.onlinejournal.in (accessed on 18 November 2021).
54. Buterin, V. A Next-Generation Smart Contract and Decentralized Application Platform. White Paper 3.37. 2014. Available online: https://ethereum.org/en/whitepaper/ (accessed on 12 November 2020).
55. Balas, C.; Karlsen, R.; Muench, P.; Mikulski, D.; Al-Holou, N. Deep-learning trust estimation in multi-agent systems. *Unmanned Syst. Technol. XIX* **2017**, *10195*, 1019510. [CrossRef]
56. Winzer, P.J.; Neilson, D.T.; Chraplyvy, A.R. Fiber-optic transmission and networking: The previous 20 and the next 20 years [Invited]. *Opt. Express* **2018**, *26*, 24190–24239. [CrossRef]
57. Renaudier, J.; Meseguer, A.C.; Ghazisaeidi, A.; Tran, P.; Muller, R.R.; Brenot, R.; Verdier, A.; Blache, F.; Mekhazni, K.; Duval, B.; et al. First 100-nm Continuous-Band WDM Transmission System with 115Tb/s Transport over 100km Using Novel Ultra-Wideband Semiconductor Optical Amplifiers. In Proceedings of the 2017 European Conference on Optical Communication (ECOC), Gothenburg, Sweden, 17–21 September 2017; pp. 1–3. [CrossRef]
58. Soma, D.; Wakayama, Y.; Beppu, S.; Sumita, S.; Tsuritani, T.; Hayashi, T.; Nagashima, T.; Suzuki, M.; Takahashi, H.; Igarashi, K.; et al. 10.16 Peta-bit/s Dense SDM/WDM transmission over Low-DMD 6-Mode 19-Core Fibre Across C+L Band. In Proceedings of the 2017 European Conference on Optical Communication (ECOC), Gothenburg, Sweden, 17–21 September 2017; pp. 1–3. [CrossRef]
59. Liu, X. Evolution of Fiber-Optic Transmission and Networking toward the 5G Era. *iScience* **2019**, *22*, 489–506. [CrossRef]
60. Jain, N.; Stiller, B.; Khan, I.; Elser, D.; Marquardt, C.; Leuchs, G. Attacks on practical quantum key distribution systems (and how to prevent them). *Contemp. Phys.* **2016**, *57*, 366–387. [CrossRef]
61. Gyongyosi, L.; Imre, S. Optimizing High-Efciency Quantum Memory with Quantum Machine Learning for Near-Term Quantum Devices. *Sci. Rep.* **2020**, *10*, 135. [CrossRef]
62. Wagh, K.S. A Survey: Data Leakage Detection Techniques. *Int. J. Electr. Comput. Eng.* **2018**, *8*, 2247–2253. [CrossRef]
63. Hughes, R.J.; Nordholt, J.E.; McCabe, K.P.; Newell, R.; Peterson, C.G.; Somma, R.D. Network-Centric Quantum Communications with Application to Critical Infrastructure Protection. In Proceedings of the 3rd International Conference Quantum Cryptography, Waterloo, ON, Canada, 1–3 October 2014; LA-UR-13-22718. Available online: http://2013.qcrypt.net/program/#invited (accessed on 18 November 2021).
64. Salvail, L.; Peev, M.; Diamanti, E.; Alléaume, R.; Lütkenhaus, N.; Länger, T. Security of trusted repeater quantum key distribution networks. *J. Comput. Sec.* **2010**, *18*, 61–87. [CrossRef]
65. Evans, P.; Peterson, G.; Morgan, T.; Jones, K.; Morrison, S.; Newell, R.; Peters, N. Demonstration of a Quantum Key Distribution Trusted Node on an Electric Utility Fiber Network. In Proceedings of the 2019 IEEE Photonics Conference, San Antonio, TX, USA, 29 September–3 October 2019; pp. 1–2. [CrossRef]
66. Sidhu, J.S.; Brougham, T.; McArthur, D.; Pousa, R.G.; Oi, D.K. Key generation analysis for satellite quantum key distribution. *Proc. SPIE* **2021**, *11881*, 1188106. [CrossRef]
67. Le, H.T.; Pham, H.T.T.; Le, H.-C.; Dang, N.T. Satellite Quantum Key Distribution for Vehicular Visible Light Communication Networks. In Proceedings of the 2020 IEEE Eighth International Conference on Communications and Electronics (ICCE), Phu Quoc Island, Vietnam, 13–15 January 2021; pp. 45–50. [CrossRef]
68. Pugh, C.; Kaiser, S.; Bourgoin, J.-P.; Jin, J.; Sultana, N.; Agne, S.; Anisimova, E.; Makarov, V.; Choi, E.; Higgins, B.L.; et al. Airborne demonstration of a quantum key distribution receiver payload. *Quantum Sci. Technol.* **2017**, *2*, 024009. [CrossRef]
69. Moshref-Javadi, M.; Winkenbach, M. Applications and Research avenues for drone-based models in logistics: A classification and review. *Expert Syst. Appl.* **2021**, *177*, 114854. [CrossRef]

70. Konolige, K.; Fox, D.; Ortiz, C.; Agno, A.; Eriksen, M.; Limketkail, B.; Ko, J.; Morisett, B.; Schultz, D.; Stewart, B.; et al. Centibots: Very Large Scale Distributed Robotic Team. In *Experimental Robotics IX: The 9th International Symposium on Experimental Robotics;* Marcelo, H., Ang, M.H., Khatib, O., Eds.; Springer: Berlin/Heidelberg, Germany, 2006; pp. 131–140.
71. Derakhshan, F.; Yousefi, S. A review on the applications of multiagent systems in wireless sensor networks. *Int. J. Distrib. Sens. Networks* **2019**, *15*, 1550147719850767. [CrossRef]
72. Wang, X.; Poikonen, S.; Golden, B. The vehicle routing problem with drones: Several worst-case results. *Optim. Lett.* **2016**, *11*, 679–697. [CrossRef]
73. Rao, R.U.; Veeraiah, D.; Mandhala, V.N.; Kim, T.-H. Neighbor Position Verification with Improved Quality of Service in Mobile Ad-hoc Networks. *Int. J. Control. Autom.* **2015**, *8*, 83–92. [CrossRef]
74. Mallik, A.; Ahsan, A.; Shahadat, M.M.Z.; Tsou, J.-C. Man-in-the-middle-attack: Understanding in simple words. *Int. J. Data Netw. Sci.* **2019**, *3*, 77–92. [CrossRef]
75. Karan, S.; Aarav, S.; Bharadhwaj, H.; Taneja, L.; De, A.; Kulkarni, G.; Meher, N.; Jha, A.K. Phase matching in β-barium borate crystals for spontaneous parametric down-conversion. *J. Opt.* **2020**, *22*, 083501. [CrossRef]
76. Boneh, D.; Dagdelen, M.; Fischlin, M.; Lehmann, A.; Schaffner, C.; Zhandry, M. Random oracles in a quantum world; *Advances in Cryptology*. In Proceedings of the 17th International Conference on the Theory and Applied of Cryptology and Information Security, Seoul, Korea, 4–8 December 2011; pp. 41–69. [CrossRef]
77. Bykovsky, A.Y. Multiple-Valued Logic for The Implementation of Random Oracle and Position-Based Cryptography. *J. Russ. Laser Res.* **2019**, *40*, 173–183. [CrossRef]
78. Ma, X.; Yuan, X.; Cao, Z.; Qi, B.; Zhang, Z. Quantum random number generation. *npj Quantum Inf.* **2016**, *2*, 16021. [CrossRef]
79. Bykovsky, A.Y. Position-Based Cryptography with Quantum and Classical Schemes Using Multiple-Valued Logic Computing. *Bull. Russ. Acad. Sci. Phys.* **2020**, *84*, 289–293. [CrossRef]
80. Rabah, K. Implementation of One-Time Pad Cryptography. *Inf. Technol. J.* **2004**, *4*, 87–95. [CrossRef]
81. Beebe, N.H. *The Mathematical-Function Computation Handbook*; Springer: Berlin/Heidelberg, Germany, 2017. [CrossRef]
82. Lerch, S.; Bessire, B.; Bernhard, C.; Feurer, T.; Stefanov, A. Tuning curve of type-0 spontaneous parametric down-conversion. *J. Opt. Soc. Am. B* **2013**, *30*, 953. [CrossRef]
83. Boyd, R. *Nonlinear Optics*, 3rd ed.; Academic Press: New York, NY, USA, 2008; pp. 79–88.
84. Shao, Y.; Wang, H.; Pi, Y.; Huang, W.; Li, Y.; Liu, J.; Yang, J.; Zhang, Y.; Xu, B. Phase noise model for continuous-variable quantum key distribution using a local local oscillator. *Phys. Rev. A* **2021**, *104*, 032608. [CrossRef]
85. Jin, C.; Shevchenko, N.A.; Li, Z.; Popov, S.; Chen, Y.; Xu, T. Nonlinear Coherent Optical Systems in the Presence of Equalization Enhanced Phase Noise. *J. Light. Technol.* **2021**, *39*, 4646–4653. [CrossRef]
86. Kiselev, F.; Samsonov, E.; Goncharov, R.; Chistiakov, V.; Halturinsky, A.; Egorov, V.; Kozubov, A.; Gaidash, A.; Gleim, A. Analysis of the chromatic dispersion effect on subcarrier wave QKD system. *Opt. Express* **2020**, *28*, 28696–28712. [CrossRef]
87. Arduino. Available online: https://www.arduino.cc (accessed on 18 November 2021).
88. Sulaiman, N.; Assi Obaid, Z.; Marhaban, M.H.; Hamidon, M.N. Design and Implementation of FPGA-Based Systems—A Review. *Aust. J. Basic Appl.Sci.* **2009**, *3*, 3575–3596.
89. Lee, J.; Chen, H.; Young, J.; Kim, H. RISC-V FPGA Platform Toward ROS-Based Robotics Application. In Proceedings of the 30th International Conference on Field-Programmable Logic and Applications, Dresden, Germany, 31 August–4 September 2020; p. 370. [CrossRef]
90. ISO/IEC 9899:1999 Programming Languages—C. Available online: https://www.iso.org/standard/29237.html (accessed on 18 November 2021).

applied sciences

MDPI

Article

Planning of Optical Connections in 5G Packet-Optical xHaul Access Network

Mirosław Klinkowski * and Marek Jaworski

National Institute of Telecommunications, Szachowa 1, 04-894 Warsaw, Poland; M.Jaworski@il-pib.pl
* Correspondence: M.Klinkowski@il-pib.pl

Abstract: One of the main challenges in dense 5G radio access networks (RANs) is provisioning of low-cost connectivity between a large number of antennas, located at remote sites, and a central site (hub) in which baseband processing functions are performed. Packet-switched Ethernet and wavelength division multiplexing (WDM) are two principal transport network technologies enabling the reduction of the demand for direct optical fiber connections between the antennas and the hub. Whereas Ethernet allows for statistical multiplexing of multiple xHaul (fronthaul/midhaul/backhaul) flows and their aggregation in a high-capacity transmission link, WDM makes it possible to establish a number of such links (using different wavelengths) in a single optical fiber. Additional savings in the amount of fibers required can be achieved by means of optical add-drop multiplexers (OADMs) that allow for obtaining access to unused wavelengths by intermediate remote nodes, whenever the capacity on the WDM system is not fully utilized by the end remote node. In this work, we focus on the problem of planning optimal fiber connections, including the placement of OADMs for a set of wavelength demands at remote sites, with the aim of minimizing the amount of fibers used in a packet-optical xHaul access network carrying 5G traffic. We consider a passive WDM system in which the maximum transmission distance, estimated using an optical power-budget model, depends on the number of OADMs that are present on the transmission path. To formulate and solve the optimization problem, we make use of integer linear programming (ILP). We apply the ILP model in network analysis. In particular, by means of numerical experiments performed for two different network topologies, we study the impact of traffic load (in terms of the number of requested wavelengths) and optical multiplexer loss on the number of transmission paths that have to be established in the network. Obtained results show that the savings in fiber connections of up to 65% can be achieved in a packet-optical xHaul network if OADMs are used when compared to the scenario without OADMs.

Keywords: 5G networks; centralized radio access network; packet-switched xHaul network; optical network; WDM; network planning; network optimization; ILP modeling

Citation: Klinkowski, M.; Jaworski, M. Planning of Optical Connections in 5G Packet-Optical xHaul Access Network. *Appl. Sci.* **2022**, *12*, 1146. https://doi.org/10.3390/app12031146

Academic Editors: Amalia Miliou and Fabio Cavaliere

Received: 30 November 2021
Accepted: 20 January 2022
Published: 22 January 2022

1. Introduction

Telecommunication networking is undergoing a profound transformation related to the deployment of 5G networks [1]. This transformation is accompanied by evolution of the mobile network towards centralized and virtualized radio access network (C-RAN/vRAN) architectures [2]. Already in centralized 4G/long term evolution (LTE) network implementations, the base station has been disaggregated into a remote radio head (RRH), located close to the antenna at a remote site (cell), and a baseband unit (BBU) placed at a central site (hub). Distributed RRHs and centralized BBUs are connected using the common public radio interface (CPRI) protocol in such networks. In 5G, the radio frequency processing functions performed by a BBU are realized by a distributed unit (DU) and a central unit (CU), whereas RRH is replaced by a radio unit (RU), which performs low-level physical functions [3]. The DUs and CUs may be placed at different network locations in accordance with particular requirements of diverse 5G services, which can be classified as enhanced mobile broadband (eMBB), ultra-reliable and low-latency communications (URLLC), and massive machine-type communications (mMTC). For instance, the whole radio processing stack (i.e., DU/CU) might be placed at a remote site in mMTC applications, whereas the DU processing for eMBB services may be performed at a hub site, as mentioned in Reference [4]. The disaggregation and distributed placement of 5G RAN functions, results in multiple data flows that differ in terms of bandwidth and latency requirements. These flows, related to fronthaul (FH—between RU and DU), midhaul (MH—between DU and CU), and backhaul (BH—between CU and a 5G core network) connections, should be carried using a convergent xHaul transport network, as mentioned in Reference [5]. In particular, the transport of traffic from remote sites and its aggregation into a hub site is realized through an xHaul access network.

5G networks are expected to make use of a much larger amount of installed antennas and access points than previous generations of mobile networks, which is frequently referred to as network densification. To decrease the capacity requirements of transport links in RANs, different techniques for compression of radio data have been proposed [6–8]. Still, the use of conventional transport solutions, based on dedicated point-to-point CPRI links, is neither scalable nor cost-effective in dense 5G access networks as it results in huge demand for high-bandwidth links between the antennas and the hub site. Therefore, to assure a convergent, scalable, and low-cost transport of radio traffic, the adaptation of well-known packet-based Ethernet technology has been proposed for xHaul transport networks in the IEEE standards 802.1CM [9] and 1914.1 [4]. Ethernet enables statistical multiplexing of data flows and, hence, increased utilization of link bandwidth. The encapsulation of 5G radio data into Ethernet frames is achieved by means of the enhanced CPRI (eCPRI) protocol [10]. The use of Ethernet in xHaul allows for support, as well other services not related to 5G, such as legacy 4G, enterprise, and residential services. In particular, the CPRI data related to 4G services is encapsulated and mapped into Ethernet frames using Radio over Ethernet (RoE) protocol specified in the IEEE 1914.3 standard [11]. Eventually, time-sensitive networking (TSN) features specified in Reference [9] enable prioritized, low-latency transmission of Ethernet frames carrying latency-sensitive fronthaul traffic.

The Ethernet switches located at remote sites, where traffic from local antennas/access points/other sources is aggregated, and a hub site, where certain baseband processing functions are performed and traffic is forwarded towards the network core, will be principally connected by means of high capacity and low delay links. Optical fiber technologies are a first choice for deployment of such links as they satisfy both requirements and provide some additional capabilities, such as wavelength division multiplexing (WDM). WDM increases significantly the capacity of an optical fiber thanks to multiplexing of signals transmitted on different wavelengths using either passive or active WDM equipment installed at the link ends [12]. Passive WDM reduces about 4–6 times the cost of active WDM since it does not use signal amplification and dispersion compensation components, in addition to the fact that it utilizes less expensive WDM equipment [13]. Therefore, passive WDM solutions are often preferred in limited-distance applications (up to 20 km), such as 5G access networks.

WDM enables aggregation of traffic from remote sites by means of OADM devices, which combine selected wavelengths at particular intermediate sites into an optical transmission path going from an end remote site through intermediate sites to a hub site, as shown in Figure 1. The application of an OADM at an intermediate site increases utilization of the WDM link, and, at the same time, it releases from the need to establish a dedicated transmission path between the hub and the site. It leads to the reduction of the number of fiber connections and installed WDM equipment, which translates into lower network deployment costs [14–16]. Recently, a commercial solution available under the trade name flexiHaul has been proposed for a 5G packet-optical xHaul access network. This solution consists of a TSN Ethernet switch [17] and a passive WDM system xWave 400G [16], which allows for the aggregation of wavelengths on optical paths by means of OADMs.

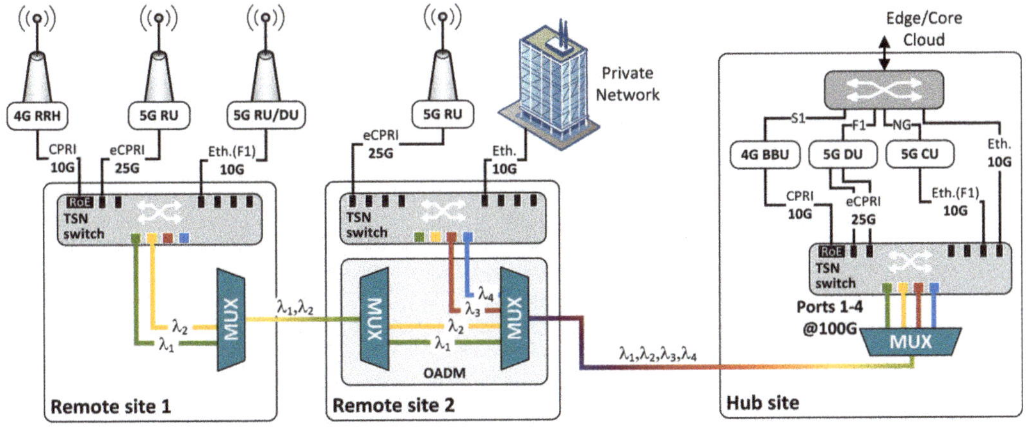

Figure 1. Example of packet-optical xHaul access network carrying 4G, 5G, and private network traffic between a hub and two remote sites (one transmission direction depicted in the optical layer).

In a 5G access transport network based on a transmission system with the above discussed features, a basic network design problem concerns planning of optical connections between remote sites and a hub site. The problem consists of the selection of routes for optical transmission paths over the network, together with the selection of intermediate nodes at which OADMs are located and traffic is aggregated. These decisions are constrained by the optical power budget of the system, which determines the maximum length of optical paths (i.e., transmission reach/distance). Moreover, the use of OADMs introduces losses of optical power, which results in a shortened transmission reach. As a consequence, there is a certain tradeoff that limits the use of many OADMs on longer paths, although such paths might pass through and allow for gathering traffic from a larger number of remote sites. As aggregation of traffic is desirable for decreasing network cost; hence, proper routing and OADM placement decisions are required when planning optical connections in the network. The length of routing paths and resulting propagation delays may be limited additionally due to low-latency requirements of specific 5G services.

In this work, we focus on modeling and optimal solving the discussed connection planning problem in a 5G packet-optical xHaul access network. The main contributions of this work are the following:

1. development and application of a physical-layer transmission model, based on optical power budget calculations and assuming the properties of a passive WDM system, for estimating the reach of transmission paths,
2. formulation of an ILP optimization problem for generation of optimal solutions to the connection planning problem in the 5G packet-optical xHaul access network considered, and
3. assessment of the impact of wavelength aggregation (by means of OADMs) on network performance in different scenarios assuming realistic transmission system parameters.

To the best of our knowledge, the optimization problem addressed and the ILP formulation proposed have not been considered in the literature yet. In addition, we are not aware of a similar work in which a physical-layer transmission model was included to optimization of a 5G packet-optical xHaul access network.

The remainder of this article is organized as follows. In Section 2, we discuss related works. In Section 3, we present main assumptions concerning the transmission model and latency constraints. In Section 4, we formulate the optimization problem and model it as an ILP problem. In Section 5, we report the results of numerical experiments. Finally, in Section 6, we conclude this work.

2. Related Works

Optimization of C-RANs connected using optical fiber networks has been frequently addressed jointly with the problem of placement of BBU/DU/CU processing resources. In Reference [14,15], the authors studied a BBU placement problem in a C-RAN connected using a WDM optical network. The optimization problem was formulated as an ILP problem with the objective to minimize the total network cost represented by either the number of active BBU sites or the number of fibers used to transport the traffic in the network. In the study, the underlying optical network was given, and a generic WDM system was considered, in which transmission distance of optical paths (lightpaths) was not constrained by a physical-layer transmission model. The authors of Reference [18] proposed an ILP formulation for the problem of dimensioning of BBU processing and optical transponder resources in a CPRI-based C-RAN connected using an active WDM optical network equipped with sliceable bandwidth-variable transponders. In that work, dedicated point-to-point connections were assumed, without traffic aggregation, and there were not any transmission reach-related constraints. The problem of designing dense WDM (DWDM) rings in metro and access segments of a survivable 5G transport network was studied in Reference [19]. The authors proposed different schemes for survivable 5G transport and made use of a heuristic approach for planning and dimensioning of ring-based fiber connections. In Reference [20], the BBU location problem with planning of survivable (i.e., primary and backup) lightpath connections in a 5G fronthaul network was addressed. In that work, a generic WDM optical network without any constraints on transmission distance was assumed. The authors of Reference [21] focused on ILP modeling of the problem of DU and CU placement with lightpath provisioning in a ring-based WDM metro/aggregation network. Similar as in other aforementioned works, a generic WDM system was assumed, and transmission reach was not modeled. A literature survey on resource allocation-related problems and solutions in centralized RANs can be found [22].

Different solutions have been considered for passive WDM networks. Coarse wavelength division multiplexing (CWDM) has been widely used in local and metropolitan networks. In CWDM, the channel spacing equals 20 nm, which enables the use of 18 channels in the wavelength range from 1271 nm to 1611 nm. Due to the wide range of dispersion, CWDM has limited use in systems with an extreme bit rate of 100 Gb/s per wavelength. Zero dispersion of the G.652D fiber [23] lies in the range of 1312 $\pm$ 12 nm. For this reason, in the IEEE 802.3cu standard [24], the wavelength range is limited to 4 channels: 1271, 1291, 1311, and 1331 nm. A two times denser spacing, i.e., 10 nm, was proposed in the MWDM system [13]—12 channels were located in the range from 1267.5 to 1374.5 nm. In addition, 12-channel LAN-WDM system (LWDM) with an 800 GHz inter-channel spacing is used, which, in the O transmission window, corresponds to a spacing of about 4.5 nm [13] and the center of channels passband equals, respectively, 1269.23, 1273.55, 1277.89, 1282.26, 1286.66, 1291.10, 1295.56, 1300.05, 1304.58, 1309.14, 1313.73, and 1318.35 nm. Recently, an 8-channel system with a 400 GHz channel spacing, i.e., about 2.25 nm in the O transmission window (called nWDM—narrow WDM) has been proposed [16], in which the center of channels passband equals, respectively, 1295.56, 1297.80, 1300.05, 1302.31, 1304.58, 1306.85, 1309.14, and 1311.43 nm. This system, under the trade name of flexiHaul xWave 400G, offers a 20 km transmission distance and a 17 dB power budget. Four-level amplitude modulation (PAM-4) is used, which enables two bits of information to be encoded in one code symbol. Symbols are transferred at 53.125 Gbaud. Forward error correction RS (544,514) is applied to allow error-free transmission for the input bit error rate less than 2.4×10^{-4} [25]. In the mentioned WDM systems, multiplexers (MUX), demultiplexers (DMUX), and add/drop modules (OADM), utilizing optical thin-film filters (TFF), are used to extract a channel of a specific wavelength. They have better transmission properties than Bragg grids (FBG) and array waveguide gratings (AWG), especially with inter-channel spacing greater than 200 GHz [26].

3. Main Assumptions

In this section, we present the details of the optical transmission system and transmission-reach model considered in this study. Moreover, we discuss the assumptions that we take concerning the latencies of transported radio data and their impact on the transmission path length.

3.1. Transmission Model

In this article, we assume the properties of WDM transmission system xWave 400G [16] applied in the xHaul 5G optical-packet access network. As discussed in Section 2, the system is characterized by a narrow inter-channel spacing. The benefit of a small inter-channel spacing is the reduction of signal distortion caused by dispersion, while the drawback is the need to use lasers with low chirp, temperature drift in the transmitter, and greater nonlinear distortions. In this case, the limitation of the transmission distance is caused by the power budget. Therefore, we present a detailed model for estimating this budget in the following section.

Table 1 shows the system parameters required to calculate the transmission distance L expressed by Equation (1), determined on the basis of the power and attenuation budget. The number of OADMs introduced into the optical path is denoted by N.

$$L = \min\left(\frac{P_{\mathrm{TXoma}} - P_{\mathrm{RXoma}} - 4NA_{\mathrm{C}} - 2NA_{\mathrm{MUX}} - A_{\mathrm{MM}}}{\alpha}, L_{\max}\right) \text{ [km]}. \quad (1)$$

Table 1. Transmission system parameters.

Fiber loss coefficient (α) [dB/km]	0.5
Connector loss (A_{C}) [dB]	0.25
Maximum number of carried wavelengths (W)	4
MUX4 loss (A_{MUX}) [dB]	$1.8 \times \{1.0, 0.9, 0.8\}$
Maintenance margin (A_{MM}) [dB]	1.0
Minimal transmitter outer optical modulation amplitude (P_{TXoma}) [dBm]	3.0
Receiver sensitivity (P_{RXoma}) [dBm]	−14.0
Maximum transmission distance ($L_{\max}$) [km]	20.0

Minimal transmitter outer optical modulation amplitude (P_{TXoma}) is defined as the minimal difference between two outer (maximal and minimal) optical power levels in PAM-4 signal at transmitter output. Receiver sensitivity (P_{RXoma}) is defined as the difference between two outer (maximal and minimal) optical power levels in PAM-4 signal at receiver input sufficient to obtain bit error rate less than 2.4×10^{-4} before FEC correction. Power budget is the difference between P_{TXoma} and P_{RXoma} (see Reference [25], Figure 2-1).

The G.652D fiber loss coefficient value of 0.40 dB/km given in Reference [23] is increased here by 0.10 dB/km taking into account additional splices losses and environmental margin. To calculate the influence of connectors on attenuation budget, we exploit a statistical approach due to a large number of connectors utilized ($4N$), and, as a consequence, mean value represents connector loss. The maximum transmission distance ($L_{\max}$) of the flexiHaul xWave 400G system is limited to 20 km [16].

MUX and DEMUX modules always operate in pairs, so, if they have the same order of filters, the total attenuation we estimated is at approximately 2/3 of combined maximum MUX/DEMUX attenuations (e.g., 3.6 dB instead of 5.4 dB) because the channel added in the MUX farthest from the output is the most attenuated, while, in the DMUX, it is dropped first and the least attenuated. This applies to both the MUX/DEMUX pair which is part of OADM (see Figure 1) and to the first/last MUX/DEMUX pair in each optical path. Attenuation of cut-through OADM optical path is increased by patchcord attenuation equal to $2 \times A_{\mathrm{C}}$.

Figure 2 shows the system configuration with indicated parameters necessary to estimate the transmission distance from Equation (1). A minimum of one wavelength is dedicated to a single remote site, so the maximum remote sites number equals W. In the information available to us about the parameters of the flexiHaul xWave 400G system, there is no exact data for the attenuation of MUX and OADM modules. Therefore, the budget calculations are based on the parameters of similar, off-the-shelf components, made in the same TFF technology, available from various manufacturers [27,28]; hence, we use several variants for values of A_{MUX} adopted in the analysis.

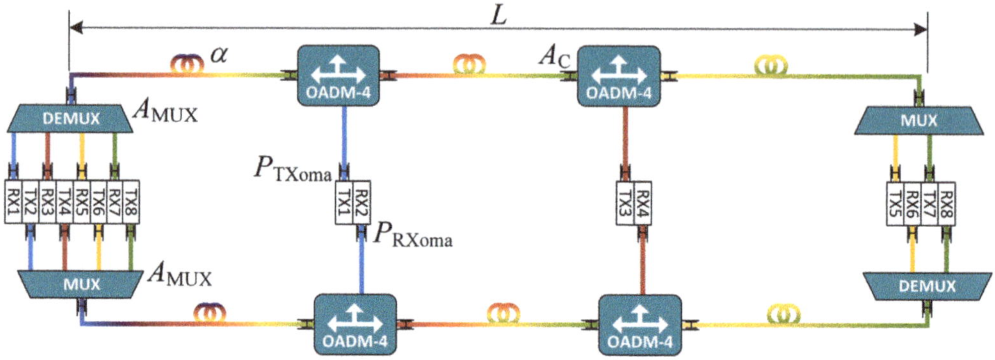

Figure 2. System configuration with denoted parameters necessary to estimate the transmission distance L.

Eventually, in Table 2, we show the transmission distance (L) values obtained using Equation (1) and system parameters presented in Table 1, for different values of MUX loss and the number of OADMs present on the optical transmission path. The obtained values of L are used in the optimization model presented in the next section to determine the maximum number of OADMs that can be installed on a transmission path of a given length.

Table 2. Transmission distance L as a function of MUX loss and number of installed OADMs.

MUX Loss	Number of OADMs (N)			
	0	1	2	3
1.80 dB	20.00 km	13.60 km	4.40 km	–
1.62 dB	20.00 km	15.04 km	6.56 km	–
1.44 dB	20.00 km	16.48 km	8.72 km	0.96 km

3.2. Latency Constraints

The technical specification of the Ethernet switch utilized in the flexiHaul system [17] mentions the implementation of the TSN mechanisms defined by standards IEEE 802.1CM and IEEE 802.1Qbu. We assume that these TSN mechanisms are applied and they assure the prioritized transmission of Ethernet frames carrying latency-sensitive data, in particular, related to 5G fronthaul traffic. Still, under heavy fronthaul traffic loads, the latencies of higher priority frames may be increased due to the need for their buffering since they compete each other for transmission resources at the switch output link. To account for this effect, we consider that the length of transmission paths between remote sites and the hub is constrained additionally by the latency requirements of latency-sensitive data flows.

Namely, in the analysis performed in this work, we consider that the one-way latency budget of latency-sensitive flows is equal to 100 μs (according to Reference [29]). Within this budget, we assume that 50 μs is reserved, among others, for the store-and-forward operations in the switch (a few μs, as mentioned in Annex B in Reference [9]), the transmission time of frames (not exceeding a few μs in a 100 Gbps optical link [30]), and queuing latency (up to some tens of μs), whereas the remaining 50 μs is left for signal propagation in the optical fiber. Note that, in the transmission system studied in this work, the buffering of frames in the xHaul network occurs once for each transmission direction, namely during aggregation of radio data flows into wavelength connections at the remote TSN switch and at the switch located in the hub, respectively, for uplink and downlink direction. Assuming a periodic traffic model [31], in which the radio data frames are transmitted periodically and within a certain transmission window of the size of up to some tens of µs, the queuing latency does not exceed the transmission window duration in the worst case, in which all HP frames arrive at the same moment, and the link capacity is fully utilized.

The above assumed 50 µs propagation delay limit corresponds to the maximum optical path length equal to 10 km. The numerical results presented in Section 5 are obtained considering this limit.

4. Optimization Problem

The packet-optical xHaul access network planning problem addressed in this paper consists of finding a set of transmission paths between a given set of remote sites and a hub site. A transmission path is realized either as a direct fiber connection between the end sites or a fiber connection going through OADM devices installed at some remote sites (as shown in Figures 1 and 2). Each transmission path carries a number of wavelengths. An OADM allows for joining into an existing transmission path the wavelengths originated at the remote node at which this device is installed (see λ_3 and λ_4 in remote site 2 in Figure 1). The connections are constrained by: (a) the allowable number of OADMs traversed by a transmission path, which is limited by the optical power budget of the path (as discussed in Section 3.1), and (b) the maximum number of carried wavelengths that must not exceed the capacity of given WDM system. The use of OADMs allows for decreasing the demand for fiber connections in the network. In this work, we aim at optimizing the configurations of transmission paths, namely their routes and placement of OADMs, to minimize the number of fiber connections required in the network.

We begin by introducing the notions and notation used in problem modeling. Afterwards, we formulate the connection planning problem as an ILP optimization problem.

4.1. Notation

The xHaul transport network is modeled by a graph $\mathcal{G} = (\mathcal{V}, \mathcal{E})$, where $\mathcal{V}$ denotes the set of network nodes, representing remote and hub sites, and $\mathcal{E}$ is the set of links representing fiber connectivity between the nodes. We assume that the capacity of each link in terms of available fibers is not a bottleneck and is sufficient to carry the traffic between remote and hub nodes. Let W denote the capacity of the WDM system, i.e., the maximum number of wavelengths carried in a fiber.

Let $\mathcal{D}$ be the set of demands. Each demand $d \in \mathcal{D}$ represents a number of wavelength connections to be established in the xHaul network between a remote node and a hub node. We assume that each remote node generates a demand. Let $W(d)$ be the number of requested wavelengths of demand d.

Let $\mathcal{P}(d)$ denote the set of candidate transmission paths of demand $d \in \mathcal{D}$. Transmission path $p \in \mathcal{P}(d)$ is a fiber route through the network between the remote node corresponding to demand d and a hub node; in particular, p is a subset of network links ($p \subseteq \mathcal{E}$). Let $D(p)$ denote the physical length of transmission path p. We assume that path p may either: (a) realize a direct fiber connection between the end nodes of a demand or (b) be a fiber connection traversing OADM devices installed at some intermediate remote nodes. Let $N(p)$ denote the number of OADMs allowable on path p. $N(p)$ depends on path length $D(p)$ and is determined by the transmission distance values presented in Table 2, namely $N(p) = \max\{N : D(p) \le L(N)\}$, where $L(N)$ is the value of L for given N in Table 2. Let $\mathcal{Q}(d, \bar{d})$ be the set of candidate transmission paths of demand $\bar{d}$ (i.e., $\mathcal{Q}(d, \bar{d}) \subseteq \mathcal{P}(\bar{d})$) which may go through an OADM installed at the remote node of demand d. The notation is summarized in Table 3.

Table 3. Notation.

Sets and Parameters	
$\mathcal{V}$	set of remote and hub nodes
$\mathcal{E}$	set of links connecting the nodes
$\mathcal{D}$	set of demands
$\mathcal{P}(d)$	set of candidate transmission paths of demand $d \in \mathcal{D}$
$\mathcal{Q}(d,\bar{d})$	set of candidate transmission paths of demand $\bar{d}$ which can serve demand d
$D(p)$	length of transmission path p
$N(p)$	allowable number of OADMs on transmission path p
$W(d)$	number of requested wavelengths of demand d
W	capacity of WDM system (maximum number of wavelengths carried in a fiber)
Variables	
x_{dp}	binary, $x_{dp} = 1$ when path p is used by demand d; $x_{dp} = 0$ otherwise
y_d	binary, $y_d = 1$ when demand d is served by a path belonging to other demand; $y_d = 0$ otherwise
$y_{d\bar{d}}$	binary, $y_{d\bar{d}} = 1$ when demand d is served by a path belonging to demand $\bar{d}$; $y_{d\bar{d}} = 0$ otherwise

4.2. ILP Formulation

To model the optimization problem, we introduce a set of decision variables. Namely, binary variable $x_{dp}, d \in \mathcal{D}, p \in \mathcal{P}(d)$ indicates whether transmission path p is established for demand d. Binary variable $y_d, d \in \mathcal{D}$ indicates whether demand d is served by a transmission path established for some other demand, which is achieved by means of an OADM device introduced into the path. Eventually, binary variable $y_{d\bar{d}}, d \in \mathcal{D}, \bar{d} \in \mathcal{D}$ indicates whether demand d is served by a transmission path established for demand $\bar{d}$. The meaning of variables is also summarized in Table 3.

The ILP formulation of the connection planning problem considered is the following:

$$\text{minimize } z = A \cdot \sum_{d \in \mathcal{D}} x_{dp} + \sum_{d \in \mathcal{D}} \sum_{p \in \mathcal{P}(d)} D(p) \cdot x_{dp}, \tag{2}$$

$$\sum_{p \in \mathcal{P}(d)} x_{dp} + y_d = 1, \qquad \forall d \in \mathcal{D}, \tag{3}$$

$$\sum_{\bar{d} \in \mathcal{D}, \bar{d} \neq d} y_{d\bar{d}} = y_d, \qquad \forall d \in \mathcal{D}, \tag{4}$$

$$\sum_{p \in \mathcal{Q}(d,\bar{d})} x_{\bar{d}p} \geq y_{d\bar{d}}, \qquad \forall d, \bar{d} \in \mathcal{D}, \bar{d} \neq d, \tag{5}$$

$$W(d) \cdot (1 - y_d) + \sum_{\bar{d} \in \mathcal{D}, \bar{d} \neq d} W(\bar{d}) \cdot y_{\bar{d}d} \leq W, \qquad \forall d \in \mathcal{D}, \tag{6}$$

$$\sum_{\bar{d} \in \mathcal{D}, \bar{d} \neq d} y_{\bar{d}d} \leq \sum_{p \in \mathcal{P}(d)} N(p) \cdot x_{dp}, \qquad \forall d \in \mathcal{D}. \tag{7}$$

Optimization objective (2) aims at minimizing the number of transmission paths, which is a primary goal imposed by coefficient A (we consider $A = 1000$), and the overall length of used fibers (a secondary optimization goal). Constraints (3) assure that either a transmission path, selected from the set of candidate paths, is established for a demand ($y_d = 0$) or the demand is served by the transmission path established for some other demand by means of an OADM device ($y_d = 1$). Constraints (4) determine which other demand serves on its transmission path the demand that has not established its own path (i.e., when $y_d = 1$). Constraints (5) assure that, for demands supporting some other demands, their own transmission paths should exist. Constraints (6) assure that the sum of requested wavelengths of a demand for which a transmission path is established (when $y_d = 0$) and the requests wavelengths of other demands served by this transmission path (when $y_{\bar{d}d} = 1$) must not exceed the fiber capacity (expressed by W). Finally, Constraints (7) assure that the number of other demands served by an established transmission path does not exceed the allowable number of OADMs traversed by the path (expressed by $N(p)$).

5. Numerical Results

In this section, we apply the ILP optimization model presented in Section 4 in network analysis. The evaluation is performed in two network topologies of different size: a 17-node city network (WRO17) and a 38-node mesh network (MESH38), shown in Figure 3. Topology WRO17 was developed based on a subset of real antenna locations (marked by triangles in Figure 3) in the center of city Wroclaw in Poland, where remote sites (marked by circles) are placed in proximity of antennas and connected using links driven along streets. The lengths of links in WRO17 reflect real physical lengths of depicted connections. In reference topology MESH38, which was used in C-RAN studies in Reference [20], we consider that link lengths are uniformly distributed between 1 and 3 kilometers. In both topologies, the hub site is denoted by a hexagonal. The routes of candidate transmission paths between remote nodes and the hub site have been generated using a k-shortest path algorithm. The paths of the length exceeding the maximum allowable path length (i.e., 10 km as discussed in Section 3.2) were excluded from the generated sets of candidate paths. In Table 4, we present some link and path-related statistics corresponding to the topologies.

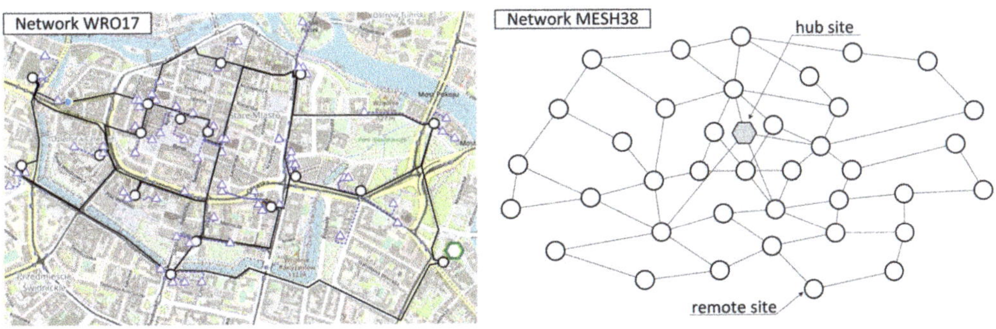

Figure 3. WRO17 and MESH38 network topologies showing fiber connectivity between remote sites (circles) and hub site (hexagon); the triangles in WRO-17 represent real antenna locations.

Table 4. Characteristics of analyzed networks; SP denotes the shortest path.

Network	Remote Sites	Links	Link length [km]		SP length [km]		SP hops	
			Mean	Max	Mean	Max	Mean	Max
WRO17	17	25	0.54	1.55	2.03	3.01	3.7	6
MESH38	38	64	2.12	2.92	4.38	8.42	2.5	4

As mentioned in Section 4, each remote site requests a certain number of wavelengths to be carried towards the hub site. The number of wavelengths requested by a remote site is generated randomly with a uniform distribution between N^{min} and N^{max}. Traffic load, denoted as ρ, is defined as the average number of requested wavelengths per remote site, namely $\rho = (N^{min} + N^{max})/2$. We evaluate different traffic scenarios, where $N^{min} \geq 1$ and $N^{max} \leq 4$. In particular, we have $\rho = 1.0$ for $N^{min} = N^{max} = 1$; $\rho = 1.5$ for $N^{min} = 1, N^{max} = 2$; $\rho = 2.0$ for $N^{min} = 1, N^{max} = 3$, etc. In each traffic scenario, the results are obtained and averaged over 10 randomly generated demand sets.

We assume the transmission model and system parameters discussed in Section 3.1. In particular, we evaluate the impact of MUX loss A_{MUX} on network performance, where $A_{\text{MUX}} \in \{1.8, 1.62, 1.44\}$ dB. As a reference scenario, denoted as no-OADMs, we consider the network in which OADMs are not used and dedicated transmission paths are established between every remote node and the hub. In particular, we have 17 and 38 such transmission paths in WRO17 and MESH38, respectively.

The numerical experiments are performed on a 3.7 GHz 32-core Ryzen Threadripper-class machine with 64 GB RAM. To solve the ILP model, we use CPLEX v.12.9 solver [32]. All the results are optimal and computation times of CPLEX do not exceed 80 s in the most demanding scenario.

In Figure 4, we illustrate optimal transmission paths found in network WRO17 for two selected traffic scenarios with loads $\rho = 1.0$ and $\rho = 2.0$, assuming MUX attenuation $A_{\text{MUX}} = 1.8$ dB and $k = 3$ candidate routing paths (no impact on results were observed for $k > 3$ in WRO17). The paths, as well as the remote nodes making use of the paths, are marked with different colors. For instance, one of the paths in the left-side figure begins in node 8 and goes through intermediate nodes 3 and 0, where some wavelengths are introduced into the path by means of OADMs, and finally terminates in the hub site attached to node 2. We can see that six optical paths are sufficient to carry traffic load $\rho = 1.0$, whereas two more paths (eight paths in total) are required to support scenario $\rho = 2.0$. This difference is a result of a higher number of wavelengths requested at certain sites in the latter scenario which cannot be served due to a limited capacity of the WDM transmission system (4 wavelengths). Note that there are also some differences in the placement of OADMs (i.e., the assignment of remote nodes to the paths) in both scenarios. For instance, node 12 is either an intermediate node of the orange path in scenario $\rho = 1.0$ or an end node of the light blue path in scenario $\rho = 2.0$.

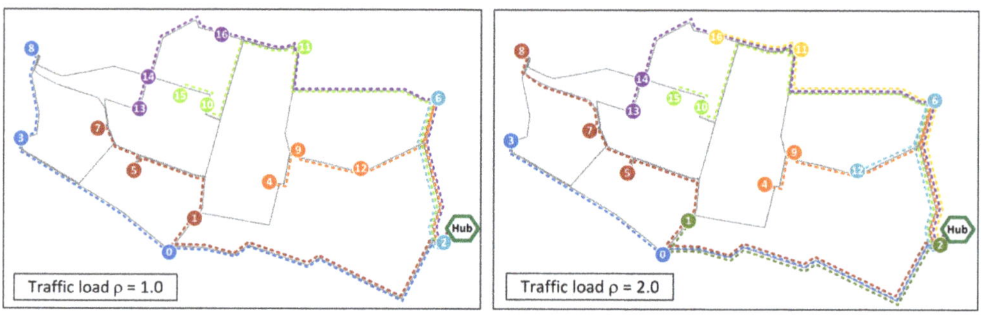

Figure 4. Optimal solutions obtained in WRO17 for traffic load $\rho = 1.0$ (**left**) and $\rho = 2.0$ (**right**).

In Figure 5, we present averaged results of the number of transmission paths (left chart) and overall length of transmission paths (right chart) in a function of traffic load (ρ) in WRO17, assuming MUX loss $A_{\text{MUX}} = 1.8$ dB and $k = 3$ candidate paths. Additionally, in both figures, we show a relative difference (gain) in the obtained results when compared to the noOADM reference scenario.

We can see that 6 transmission paths are sufficient to serve all traffic in the network when each remote node generates a 1-wavelength demand (i.e., $\rho = 1.0$). This is achieved by aggregation of wavelengths (using OADMs) from different remote sites onto WDM optical transmission paths. In this case, the reduction of the number of required transmission paths (i.e., relative gain) versus the scenario without OADMs reaches about 65%. Increasing the traffic load, the number of transmission paths increases (and the relative gain decreases) up to the moment when all remote nodes need dedicated transmission paths. It happens when 3.5 wavelengths, on average, are requested by each remote node ($\rho = 3.5$), and there is no use of OADMs due to saturation of the WDM system, which capacity is 4 wavelengths. The overall length of optical transmission paths (shown in the right chart) follows a similar trend. In particular, the length of dedicated fiber connections required in the noOADM scenario is about 34.5 km, which can be reduced by up to 60% (to about 14 km) in a network with a low load ($\rho = 1.0$). We report that similar results were obtained for other considered values of A_{MUX}.

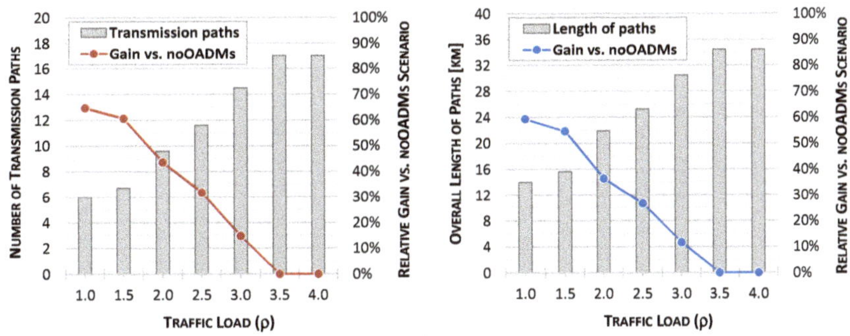

Figure 5. The number of transmission paths (**left**) and overall length of paths (**right**) in a function of traffic load (ρ) in WRO17; relative gain versus the network without OADMs is shown as lines.

In Figure 6, we analyze the impact of the number of candidate routing paths k on obtained results in network MESH38. We focus on the number of optical transmission paths (left chart) in a function of k and traffic load, $\rho \in \{1.0, 1.5, 2.0\}$, for different values of MUX loss (A_{MUX}). Moreover, in the right chart, we show a relative difference in results when compared to a single (shortest) path scenario ($k = 1$). In each scenario, we can see that provisioning of a higher number of candidate routes ($k > 1$) allows for reducing the number of optical transmission paths required in the network when compared to the shortest-path case. The gain in performance increases with k and ranges between 16% and 32%, depending on traffic and MUX scenario, for $k \geq 5$. We can also see that the results stabilize for $k \geq 5$ and that the improvement for $k = 7$ is either none or irrelevant (as for $\rho = 2.0$). Finally, higher differences in obtained results are observed for lower MUX loss values at lower traffic loads. It can be explained by higher transmission distances and numbers of allowable OADMs (as shown in Table 2) in low MUX loss scenarios, which translates into a higher chance to inject single wavelengths into transmission paths at intermediate nodes, especially if several alternative candidate paths are available.

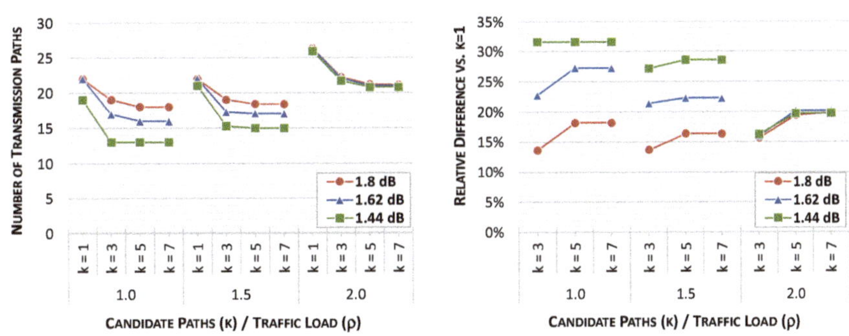

Figure 6. The number of transmission paths (**left**) and relative difference versus a single-path scenario (**right**) in a function of traffic load (ρ) and the number of candidate routing paths (k) for different MUX loss values in MESH38.

Eventually, in Figure 7, we show the results of the number of transmission paths (left chart) and the relative performance gain versus reference scenario noOADM (right chart) in a function of traffic load (ρ) in MESH38 for different MUX loss values (A_{MUX}) and assuming $k = 7$ candidate paths. We can see that, under low traffic loads ($\rho \leq 1.5$), the use of MUXs with lower attenuation results in a lower number of required transmission paths. In this case, higher transmission distances are allowable and more OADMs can be used on the paths (as for $A_{\text{MUX}} = 1.44$ dB), which, under a low demand for wavelengths (at most 2 wavelengths per remote node), allows for higher aggregation of demands, as also discussed in the above remarks concerning Figure 6. At higher traffic loads ($\rho \geq 2.0$), the gains from using low-loss MUXes are either none or negligible in the 4-channel WDM system considered, due to its saturation with the carried wavelengths. The gains from using OADMs are between 52% and 65% for $\rho = 1.0$, depending on the MUX scenario, and are decreasing up to 0%, which is reached at a high load for $\rho = 3.5$.

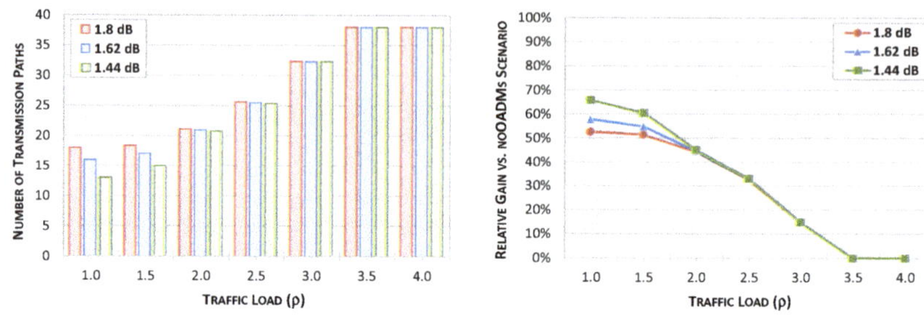

Figure 7. Number of transmission paths (**left**) and relative gain versus the network without OADMs (**right**) in a function of traffic load (ρ) for different MUX loss values in MESH38.

6. Concluding Remarks

We have studied the problem of planning optimal transmission paths, realized using optical fiber connections and optical add-drop multiplexers, in a 5G packet-optical xHaul access network. The planning problem has concerned the selection of routing paths in the network, between a set of remote nodes and a hub site, and placement of OADMs onto the paths for traffic aggregation with the goal to minimize the number of transmission paths in the network. The mentioned optimization problem was formulated as an integer linear programming problem. To estimate maximum transmission distances and determine the maximum number of OADMs allowable on particular paths, we have developed a transmission model based on optical power budget calculation. In network analysis, we have considered the functionality and parameters of a commercial packet-optical xHaul system and optical components available on the market. The evaluation was performed in two different network topologies, including an urban typology developed based on real antenna locations and realistic transmission distances.

We have shown that the application of OADMs allows to decrease the number of required transmission paths significantly when compared to the network in which aggregation of traffic from remote sites by means of OADMs is not realized. The reduction in the number of transmission paths in both evaluated networks can reach up to 65% under a low load, which translates into a lower demand for fiber resources and, consequently, cost savings. Even at higher traffic loads, with an average demand corresponding to 75% of the WDM system capacity, i.e., for 3 wavelengths requested on average per remote site, the savings of about 15% can be achieved. Provisioning of multiple candidate routing paths, instead of using the shortest path only, is beneficial as it allows for a more effective traffic aggregation and, hence, reduction of the number of transmission paths, between 16% and 32%, depending on traffic and MUX loss scenario. Finally, the application of MUXs with lower attenuation can be advantageous for low load scenarios; however, under moderate and higher traffic, they have not offered any performance gains in the analyzed networks.

In future works, we will address packet-optical xHaul network scenarios in which diverse types of OADMs are applied instead of a fixed OADM assumed in this work. In addition, we will extend the optimization models to account for WDM systems in which the power loss levels are not the same for particular wavelengths. Eventually, we plan to study multi-layer xHaul scenarios in which routing decisions concern both the packet and optical layer.

Author Contributions: Conceptualization, M.K.; Funding acquisition, M.K.; Investigation, M.K. and M.J.; Methodology, M.K. and M.J.; Project administration, M.K.; Software, M.K.; Supervision, M.K.; Visualization, M.K.; Writing—original draft, M.K. and M.J.; Writing—review & editing, M.K. and M.J. All authors have read and agreed to the published version of the manuscript.

Funding: This research was funded by National Science Center, Poland under grant number 2018/31/B/ST7/03456.

Conflicts of Interest: The author declares no conflict of interest.

References

1. Agiwal, M.; Roy, A.; Saxena, N. Next Generation 5G Wireless Networks: A Comprehensive Survey. *IEEE Commun. Surv. Tutor.* **2016**, *18*, 1617–1655. [CrossRef]
2. Peng, M.; Sun, Y.; Li, X.; Mao, Z.; Wang, C. Recent Advances in Cloud Radio Access Networks: System Architectures, Key Techniques, and Open Issues. *IEEE Commun. Surv. Tutor.* **2016**, *18*, 2282–2308. [CrossRef]
3. The 3rd Generation Partnership Project (3GPP). Available online: http://www.3gpp.org/ (accessed on 29 November 2021).
4. IEEE. *1914.1-2019—IEEE Standard for Packet-Based Fronthaul Transport Networks*; IEEE: Piscataway, NJ, USA, 2019.
5. Wypiór, D.; Klinkowski, M.; Michalski, I. Open RAN—Radio Access Network Evolution, Benefits and Market Trends. *Appl. Sci.* **2022**, *12*, 408. [CrossRef]
6. Aguerri, I.E.; Zaidi, A. Lossy Compression for Compute-and-Forward in Limited Backhaul Uplink Multicell Processing. *IEEE Trans. Commun.* **2016**, *64*, 5227–5238. [CrossRef]
7. Park, S.H.; Simeone, O.; Sahin, O.; Shamai, S. Robust and Efficient Distributed Compression for Cloud Radio Access Networks. *IEEE Trans. Vehic. Technol.* **2013**, *62*, 692–703. [CrossRef]
8. Aguerri, I.E.; Zaidi, A.; Caire, G.; Shitz, S.S. On the Capacity of Cloud Radio Access Networks with Oblivious Processing. *IEEE IEEE Trans. Inf. Theory* **2019**, *65*, 4575–4596. [CrossRef]
9. IEEE. *802.1CM-2018—IEEE Standard for Local and Metropolitan Area Networks—Time-Sensitive Networking for Fronthaul*; IEEE: Piscataway, NJ, USA, 2018.
10. Common Public Radio Interface: eCPRI V1.2 Interface Specification. 2018. Available online: http://www.cpri.info/downloads/eCPRI_v_1_2_2018_06_25.pdf (accessed on 29 November 2021).
11. IEEE. *1914.3-2018—IEEE Standard for Radio over Ethernet Encapsulations and Mappings*; IEEE: Piscataway, NJ, USA, 2018.
12. Alimi, I.A.; Teixeira, A.; Monteiro, P. Towards an Efficient C-RAN Optical Fronthaul for the Future Networks: A Tutorial on Technologies, Requirements, Challenges, and Solutions. *IEEE Commun. Surv. Tutor.* **2018**, *20*, 708–769. [CrossRef]
13. O-RAN Alliance. WDM-Based Fronthaul Transport Tech. Spec. v1.0. 2021. Available online: https://www.o-ran.org/ (accessed on 29 November 2021).
14. Carapellese, N.; Tornatore, M.; Pattavina, A.; Gosselin, S. BBU Placement over a WDM Aggregation Network Considering OTN and Overlay Fronthaul Transport. In Proceedings of the 2015 European Conference on Optical Communication (ECOC), Valencia, Spain, 27 September–1 October 2015.
15. Musumeci, F.; Bellanzon, C.; Tornatore, N.C.M.; Pattavina, A.; Gosselin, S. Optimal BBU Placement for 5G C-RAN Deployment over WDM Aggregation Networks. *IEEE J. Lightw. Technol.* **2016**, *34*, 1963–1970. [CrossRef]
16. HFR Networks. flexiHaul xWave 400G Data Sheet. Available online: https://hfrnetworks.com/wp-content/uploads/2021/04/HFR-Networks_flexiHaul-xWave-400G_DataSheet_04-13-2021-1.pdf (accessed on 29 November 2021).
17. HFR Networks. flexiHaul Packet M-Series TSN Switch Data Sheet. Available online: https://hfrnetworks.com/wp-content/uploads/2020/10/M6424-Packet-TSN-RoE-Switch-DS_HFR-Networks_102420.pdf (accessed on 29 November 2021).
18. Velasco, L.; Castro, A.; Asensio, A.; Ruiz, M.; Liu, G.; Qin, C.; Proietti, R.; Yoo, S.J.B. Meeting the Requirements to Deploy Cloud RAN Over Optical Networks. *OSA/IEEE J. Opt. Commun. Netw.* **2017**, *9*, B22–B32. [CrossRef]
19. Wong, E.; Grigoreva, E.; Wosinska, L.; Machuca, C.M. Enhancing the Survivability and Power Savings of 5G Transport Networks based on DWDM Rings. *OSA/IEEE J. Opt. Commun. Netw.* **2017**, *9*, D74–D85. [CrossRef]
20. Khorsandi, B.M.; Raffaelli, C. BBU location algorithms for survivable 5G C-RAN over WDM. *Comput. Netw.* **2018**, *144*, 53–63. [CrossRef]
21. Yu, H.; Musumeci, F.; Zhang, J.; Xiao, Y.; Tornatore, M.; Ji, Y. DU/CU Placement for C-RAN over Optical Metro-Aggregation Networks. In Proceedings of the International IFIP Conference on Optical Network Design and Modeling, Athens, Greece, 13–16 May 2019.
22. Ejaz, W.; Sharma, S.K.; Saadat, S.; Naeem, M.; Anpalagan, A.; Chughtai, N.A. A Comprehensive survey on Resource Allocation for CRAN in 5G and Beyond Networks. *J. Net. Comput. Appl.* **2020**, *160*, 102638. [CrossRef]
23. ITU-T. Recommendation G.652: Characteristics of a Single-Mode Optical Fibre and Cable. 2016. Available online: https://www.itu.int/rec/T-REC-G.652/en (accessed on 29 November 2021).
24. IEEE. *Standard 802.3cu-2021, Amendment 11: Physical Layers and Management Parameters for 100 Gb/s and 400 Gb/s Operation over Single-Mode Fiber at 100 Gb/s per Wavelength*; IEEE: Piscataway, NJ, USA, 2021.
25. 100G Lambda MSA. 100G-LR1-20, 100G-ER1-30 and 100G-ER1-40 Tech. Spec. Rev 1.1. Available online: http://100glambda.com/ (accessed on 21 January 2022).

26. Grobe, K. *Handbook of Fiber Optic Data Communication*, 4th ed.; Academic Press: Cambridge, MA, USA, 2013.
27. FS. 4ch Single Fiber 1295.56-1309.14nm LAN-WDM MUX DEMUX Technical Data. Available online: https://www.fs.com/ (accessed on 21 January 2022).
28. EDGE Technologies Ltd.. 5G Fronthaul WDM. Available online: https://edgeoptic.com/product/5g-fronthaul-wdm/ (accessed on 21 January 2022).
29. Common Public Radio Interface: eCPRI V1.2 Requirements for the eCPRI Transport Network. 2018. Available online: http://www.cpri.info/downloads/Requirements_for_the_eCPRI_Transport_Network_V1_2_2018_06_25.pdf (accessed on 29 November 2021).
30. Klinkowski, M. Optimization of Latency-Aware Flow Allocation in NGFI Networks. *Comp. Commun.* **2020**, *161*, 344–359. [CrossRef]
31. Perez, G.O.; Larrabeiti, D.; Hernandez, J.A. 5G New Radio Fronthaul Network Design for eCPRI-IEEE 802.1CM and Extreme Latency Percentiles. *IEEE Access* **2019**, *7*, 82218–82229. [CrossRef]
32. IBM. CPLEX Optimizer. Available online: https://www.ibm.com/analytics/cplex-optimizer (accessed on 29 November 2021).

applied sciences

MDPI

Article

Silicon Photonic Micro-Transceivers for Beyond 5G Environments

Kazuhiko Kurata [1,*], Luca Giorgi [2], Fabio Cavaliere [2], Liam O'Faolain [3,4], Sebastian A. Schulz [3], Kohei Nishiyama [5], Yasuhiko Hagihara [1], Kenichiro Yashiki [1], Takashi Muto [1], Shigeru Kobayashi [1], Makoto Kuwata [1] and Richard Pitwon [1,3,6,*]

1 AIO Core Ltd., 47-12-301, Sekiguchi 1-Chome, Bunkyo-ku, Tokyo 104-0014, Japan; y-hagihara@aiocore.com (Y.H.); k-yashiki@aiocore.com (K.Y.); t-muto@aiocore.com (T.M.); s-kobayashi@aiocore.com (S.K.); m-kuwata@aiocore.com (M.K.)
2 Ericsson Research, Via Moruzzi 1 c/o CNR, Ericsson, 56124 Pisa, Italy; luca.giorgi@ericsson.com (L.G.); fabio.cavaliere@ericsson.com (F.C.)
3 School of Physics and Astronomy, University of St Andrews, North Haugh, St. Andrews KY16 9SS, UK; William.Whelan-Curtin@mtu.ie (L.O.); sas35@st-andrews.ac.uk (S.A.S.)
4 Centre for Advanced Photonics and Process Analysis (CAPPA), Munster Technology University, T12P928 Cork, Ireland
5 I-PEX Inc., Machida ST BLD., 1-33 Morino, Tokyo 104-0014, Japan; nishiyama.kohei@i-pex.com
6 Resolute Photonics, Northover House, 132a Bournemouth Road, Chandlers Ford, Eastleigh SO53 3AL, UK
* Correspondence: k-kurata@aiocore.com (K.K.); rpitwon@resolutephotonics.com (R.P.)

Abstract: Here, we report on the design and performance of a silicon photonic micro-transceiver required to operate in 5G and 6G environments at high ambient temperatures above 105 °C. The four-channel "IOCore" micro-transceiver incorporates a 1310 nm quantum dot laser system and operates at a data rate of 25 Gbps and higher. The 5 × 5 mm micro-transceiver chip benefits from a multimode coupling interface for low-cost assembly and robust connectivity at high temperatures as well as an optical redundancy scheme, which increases reliability by over an order of magnitude.

Keywords: silicon photonics; 5G; 6G; co-packaged optics; data centers; integrated photonics; micro-transceiver; high-performance computer (HPC)

Citation: Kurata, K.; Giorgi, L.; Cavaliere, F.; O'Faolain, L.; Schulz, S.A.; Nishiyama, K.; Hagihara, Y.; Yashiki, K.; Muto, T.; Kobayashi, S.; et al. Silicon Photonic Micro-Transceivers for Beyond 5G Environments. *Appl. Sci.* **2021**, *11*, 10955. https://doi.org/10.3390/app112210955

Academic Editor: Domenico de Ceglia

Received: 16 October 2021
Accepted: 11 November 2021
Published: 19 November 2021

1. Introduction

The conditions required to accommodate the increasing bandwidth densities in 5G+, hyperscale data centers and high-performance computer (HPC) environments are becoming more extreme. Given the requirement that larger numbers of optical transceiver I/O ports are moved closer to the signal source chip in the system (e.g., ASIC or FPGA) in embedded or co-packaged optical assemblies, it follows that transceivers will need to be packed more tightly into very confined spaces. Consequently, transceivers will need to operate at higher ambient temperatures than those to which commercial transceivers have historically been subjected, in many cases exceeding 105 °C as we will outline later on. In order to be deployed in these hyperscale environments, optical transceivers will need to provide higher aggregate bandwidths, a reduced footprint and higher reliability at operating temperatures exceeding 100 °C. The design and performance of a silicon photonic micro-transceiver are presented, which can operate under high ambient temperature regimes exceeding 105 °C. The quad (4 Tx + 4 Rx) silicon photonic micro-transceiver "IOCore" incorporates a 1310 nm Fabry-Perot Quantum Dot laser, the output of which is modulated at data rates of 25, 32 and 50 Gbps per lane. The compact 5 × 5 mm micro-transceiver chip features a multimode optical coupling interface and an optical redundancy circuit. Together, these enable robust, low-cost operation and connectivity at high temperatures and increase reliability by over an order of magnitude compared to non-redundant integrated laser schemes.

The first generation of 100G silicon photonics micro-transceiver comprised four bidirectional channels operating at 25 Gbps [1]. AIO Core collaborated with partners of the European H2020 COSMICC project [2] to demonstrate its first generation of silicon photonics micro-transceivers on mezzanine test cards [3], which have recently been standardized by the IEC [4]. The COSMICC hyperscale optical interconnect ecosystem demonstration platform "Aurora" was exhibited at ECOC 2019 in Dublin in September 2019 (Figure 1).

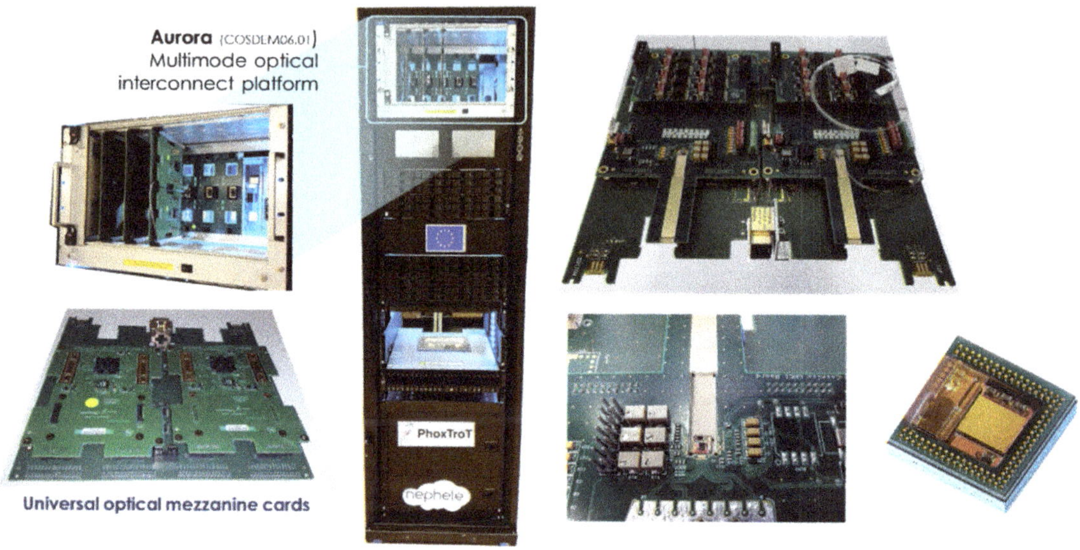

Figure 1. Hyperscale rack-scale demonstrator with H2020 COSMICC "Aurora" platform and test daughtercards with I/O core mezzanine cards connected [5–7].

Silicon photonics-based transceivers were initially developed for data center interconnects (DCIs), where the high requisite volumes of optical transceivers enabled critical investment in new technologies.

Silicon photonics has evolved over the past five years from a high-cost margin telecom technology to a commoditized datacom transceiver technology capable of transmitting bit rates of hundreds of Gbps at transmission costs of an ever decreasing fraction of a US dollar per Gbps and a power consumption of less than 15 picojoules per bit. A relevant example is the specification recently published by the Co-Packaged Optics Collaboration [8] of a 3.2 Tbit/s co-packaged transceiver module [9]. The advent of 5G has given rise to the requirement for high data rates in the fronthaul segment [10], and this trend is expected to increase with the introduction of advanced antenna systems (AAS) [11] and radio transmission in the millimeter wave range [12]. However, the interconnection speed remains lower in radio communications than in DCI: 100 Gbps links are being introduced in the current generation, with the requirement expected to increase to 400 Gbps within the next few years. A possible scenario is depicted in Figure 2.

Two ASICs (depicted in blue in Figure 2)—for example, the ASIC used for digital baseband processing functions—are connected to each of four 400 Gbps optical modules through four 50 Gbps electrical lanes. Each optical module receives inputs from both ASICs in a dual star redundant configuration for protection and computation offload purposes. These four optical modules connect to another four modules across an optical link, the length of which can range from a few centimeters for intra-board optical interconnects to hundreds of meters for split radio equipment. Each of the optical modules on the right-hand side is connected to an ASIC (green), which could, in an advanced antenna system, be a radio frequency processing ASIC connected to an RF antenna element. This architecture can be easily scaled to higher capacities by adding further optical modules or to connect a higher number of antenna elements, thus accommodating various product variants with different bandwidths and antenna array sizes. Radio equipment generally requires a lower aggregate capacity than DCIs [13] but also the capability to scale up capacity and power consumption more effectively. Moreover, even if the transceiver volumes in radio are comparable or even higher than in DCI applications, the target cost is lower. This makes it more difficult for transceiver manufacturers to assume the risk of investing in new technologies while maintaining a sustainable business model. The adoption of standardized solutions and the definition of a limited number of implementation options can definitely mitigate this issue avoiding excessive market fragmentation.

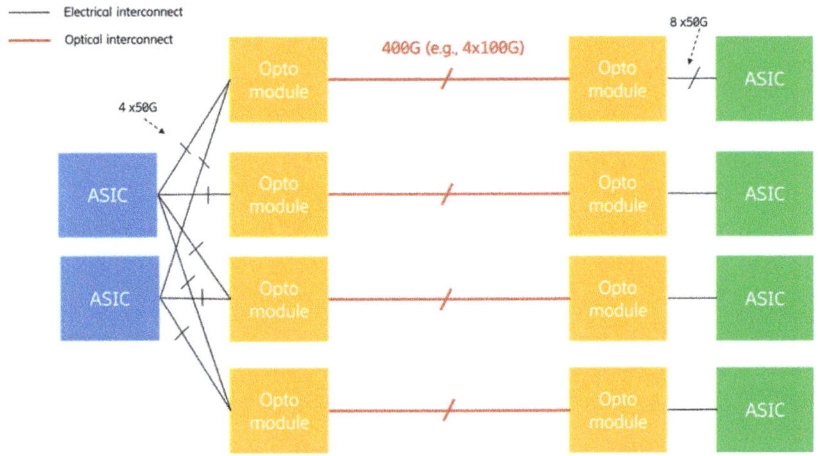

Figure 2. Possible interconnection scenario in radio systems.

Power consumption constraints are also more demanding in radio access networks (RANs) than in DCIs. Data center equipment operates indoors in environments with controlled temperature and humidity. In addition, active cooling systems, e.g., line fans, are commonly implemented, with more advanced solutions such as liquid cooling gaining prominence. Constraints in weight, power supply and installation practices make these solutions impractical or not feasible for radio equipment deployed outdoors or at the top of an antenna pole. Therefore, current RANs require transceivers capable of operating in the so-called industrial temperature (I-Temp) range, from −40 to 85 °C, but the increase in component density in radio boards, due to miniaturization and increase in processing capabilities, is pushing the upper bound of the temperature range above 100 °C. While silicon photonic devices, such as modulators and photodetectors, are tolerant to high temperatures, lasers are very sensitive to them. Transceiver operation beyond 85 °C requires the use of either external laser sources, placed outside the transceiver [14], or quantum dot lasers, which are very promising to withstand high temperatures [15]. Variations in temperature are particularly problematic for wavelength division multiplexing (WDM) systems.

In 5G (and beyond) radio systems operating in the millimeter range, radiating antenna elements and photonic interconnects will be co-integrated in a single hardware unit. This will not only increase the operational temperature but also calls for miniaturized devices—for example, co-packaged transceivers integrated into single multi-chip modules with digital integrated circuits (ICs) and analogue radio frequency (RF) front-ends. This scenario needs more accurate fiber attachment techniques for silicon photonic chips [16] as well as a radical decrease in the power consumption, from the current tens of picojoules per bit to a few picojoules per bit [17]. The IPSR-I integrated photonics roadmap for transceivers [18] makes clear that photonic packaging technology must improve substantially to achieve better energy efficiency. Integrated photonic solutions for radio applications can benefit from the current DCI technology, such as single-mode silicon photonic transceivers; however, operating temperatures, power consumption and cost margins will be much more challenging in 5G+ environments going forward and will require new solutions. In the next section, we provide a detailed technical description of a multimode silicon photonics micro-transceiver designed to operate in high-temperature environments with an integrated laser for lower overall power consumption that can meet the stringent cost targets of future 5G+ environments due to multimode packaging.

2. Silicon Photonics Micro-Transceiver with Multimode Interface

The "IOCore" silicon photonics micro-transceiver chip operates in the 1310-nanometer wavelength regime (O-band) with on-off keying (OOK) modulation data rates per lane of 25, 32 and 50 Gbps. An IOCore chip incorporates four optical transmission and four optical receiver channels under a proprietary multimode optical waveguide or "optical pin" array interface [1] on a 5 × 5 mm footprint (Figure 3).

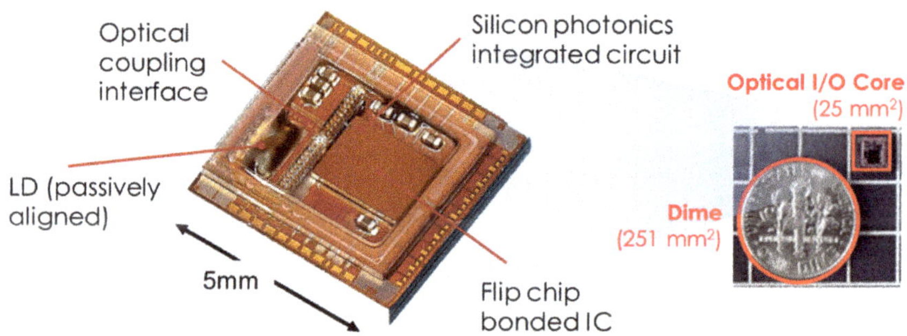

Figure 3. IOCore micro-transceiver footprint of 5 × 5 mm [3].

The IOCore transceiver serves as a transceiver building block that can be assembled into a variety of different standard or bespoke packages to address diverse high-volume, high-temperature and space-restricted applications including co-packaged optics in 5G/6G, hyperscale data centers, HPCs and automotive systems.

We will now introduce the overall structure of the IOCore transceiver, before going on to describe its individual components in more detail.

2.1. Functional Overview of Transceiver Operation

Figure 4 shows the schematic cross-section and optical interfaces of the IOCore micro-transceiver.

Figure 4b shows a top view of the complete IOCore transceiver chip, which is split into a transmitter section on the left-hand side and a receiver section on the right-hand side.

In the transmitter section (Figure 4d), an edge emitting a Fabry-Perot Quantum Dot laser diode is passively aligned within a compliant recess in the silicon substrate such that it couples 1310-nanometer continuous wave (CW) light into four single-mode silicon waveguides in the substrate. Each waveguide conveys the CW light to a Mach–Zehnder interferometer (MZI). High-speed differential electronic signals passed to the chip through electrical signal pads drive the MZI modulator circuit generating the corresponding modulated 1310-nanometer optical signal, which is then coupled out of the chip through a vertical grating coupler.

The vertical grating coupler couples the optical signal out of the silicon photonic substrate at an angle of 82° from the plane of the silicon substrate into a short untapered optical waveguide called an "optical pin". Figure 4a shows a photo of a proprietary array of twelve optical pins on the transmitter section, of which only four are used in this design.

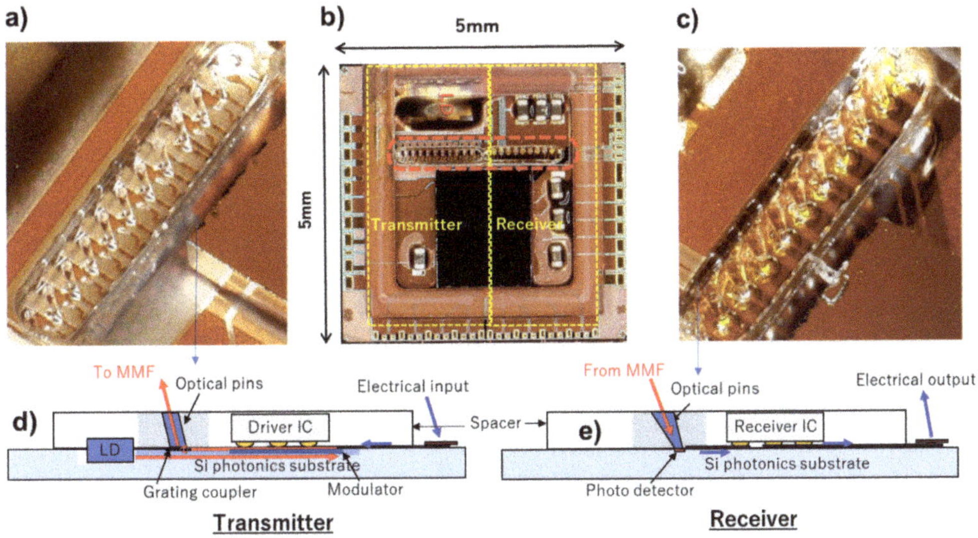

Figure 4. IOCore micro-transceiver: (**a**) photo of proprietary transmitter multimode optical interface [3]; (**b**) top photo of complete IOCore micro-transceiver; (**c**) photo of proprietary receiver multimode optical interface; (**d**) cross-section of transmitter section with untapered waveguides ("optical pins"); (**e**) cross-section of receiver section with tapered waveguides ("optical pins").

In the receiver section (Figure 4e), optical signals are coupled from multimode fibers into optical pins on the same array datum as the transmitter optical pins. The receiver optical pins, however, are tapered to both confine and guide the light directly to integrated silicon germanium photodetectors arranged under the optical coupling interface area. This eliminates the requirement for intermediary fiber-to-waveguide coupling structures, such as lossy polarization splitting grating couplers, and moves the light straight to the photodetector, which is directly connected to transimpedance amplifier (TIA) and limiting amplifier (LA) ICs, thus converting the received optical signals into differential voltages.

The optical pin arrays in the transmitter (Figure 4a) and receiver (Figure 4c) sections are embedded in a glass plate mounted onto the silicon substrate. These form the multimode optical interface, which allows bidirectional communication between a multimode fiber array, such as an MT ferrule, and the micro-transceiver, thus permitting a lower cost and more robust assembly solution with respect to temperature and vibration. This differentiates IOCore from most other commercial silicon photonic transceivers, which have single-mode interfaces.

Simulations by AIO Core predict the insertion loss of the transmit optical pin—i.e., the insertion loss between the vertical grating coupler and the output facet of the optical pin at the point of contact with the MMF—to be 0.54 dB. Furthermore, simulations predict the insertion loss on the receiver optical pin—i.e., the insertion loss between the MMF and the receiving photodetector—to be 0.68 dB.

2.2. Laser Integration, Reliability and Performance

2.2.1. External Light Source vs. Integrated Light Source

The reliability of the laser source is the most important requirement for the high-temperature operation of silicon photonic optical transceivers, yet the integration of reliable light sources is also one of the most challenging aspects of silicon photonic optical transceivers. The three main options for optical sources are (a) an integrated laser; (b) a board-mounted source, but external to the transceiver; and (c) a front-pluggable laser source, as shown in Figure 5. In order to minimize power consumption and cost, the laser should be integrated into or onto the silicon photonic chip as shown in Figure 5a. The disadvantage of this is that the laser is not field-replaceable and therefore must be extremely reliable in the densely populated component environment of the chip. In order to reduce the risk of transceiver failure due to failure of the laser source, many silicon photonics solutions opt for an external light source (ELS) scheme, whereby the CW laser is not integrated but rather located outside the transceiver assembly. This allows the laser to be in a more temperature-controlled environment. Some higher margin solutions may use Peltier coolers to ensure low-risk and stable operation of the laser source, but uncooled on-board (Figure 5b) or front-pluggable (Figure 5c) modules are likely to become more commercially viable, especially if they are field-replaceable.

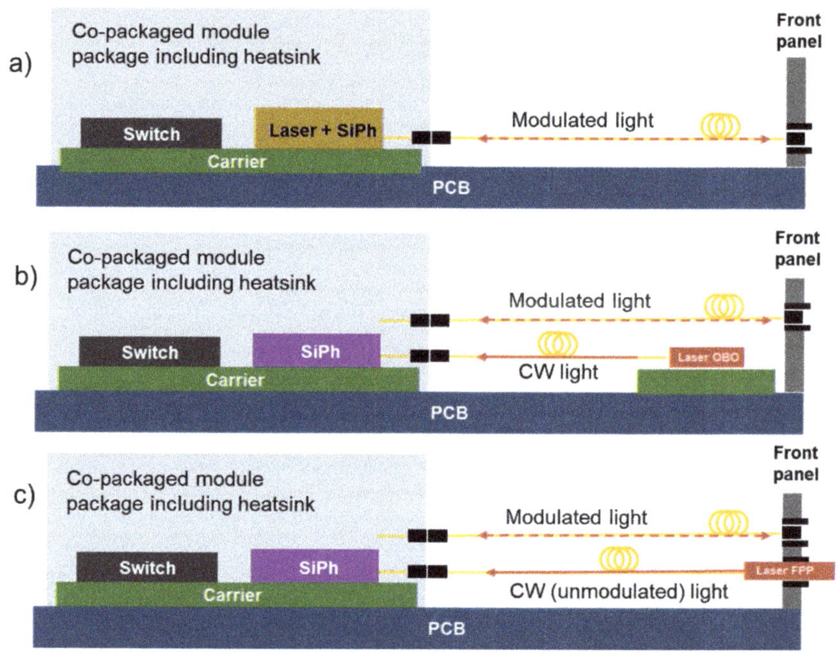

Figure 5. Silicon photonics chiplets in a co-packaged assembly with (**a**) integrated laser source, (**b**) external laser source mounted on the board and (**c**) front-pluggable and completely field-replaceable external light source on faceplate.

Figure 6 compares a multimode silicon photonic micro-transceiver with an integrated laser (Figure 6a) to a single-mode silicon photonic transceiver with an external laser source (Figure 6b). External laser source systems typically require polarization-maintaining fibers (PMFs) to ensure low loss coupling of CW light to the transceiver modulator and will include the additional components of a second package for the laser; therefore, they represent a more expensive solution with higher power consumption.

2.2.2. Quantum Dot Fabry-Perot Laser Diode Performance at High Temperature

In order to accommodate an integrated laser solution, the laser must be exceptionally reliable at higher temperatures. For this purpose, a Quantum Dot Fabry-Perot Laser Diode (QD-LD) with an operational wavelength of 1310 nm was integrated in the IOCore micro-transceiver.

As shown in the L-I curves of Figure 7a, the QD-LD has a characteristically stable optical output even at high temperatures, with little deviation in output power over a wide temperature range. In addition, as shown in the RIN characteristics of Figure 7b, the QD-LD exhibits strong suppression of noise due to back-reflected light; therefore, they can be coupled into a silicon photonic circuit without an isolator, allowing further packaging and cost reduction.

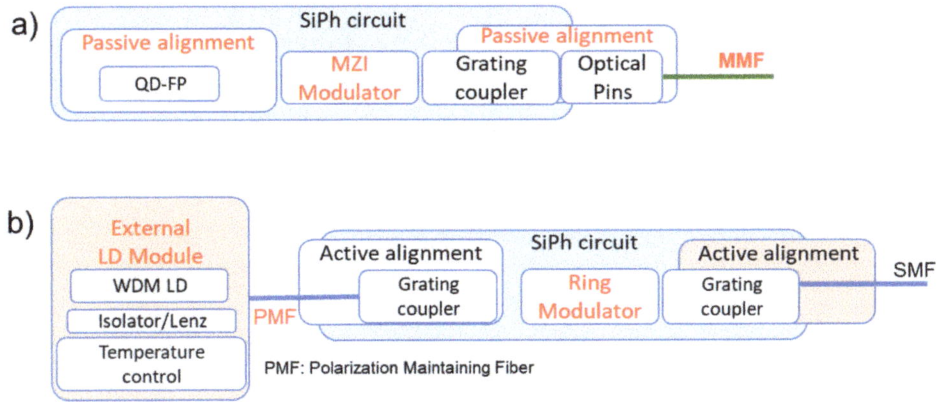

Figure 6. Functional block comparison of different laser integration schemes. (**a**) Integrated laser scheme in single wavelength IOCore micro-transceiver with multimode fiber; (**b**) high-end WDM silicon photonics transceiver with external multiwavelength laser source.

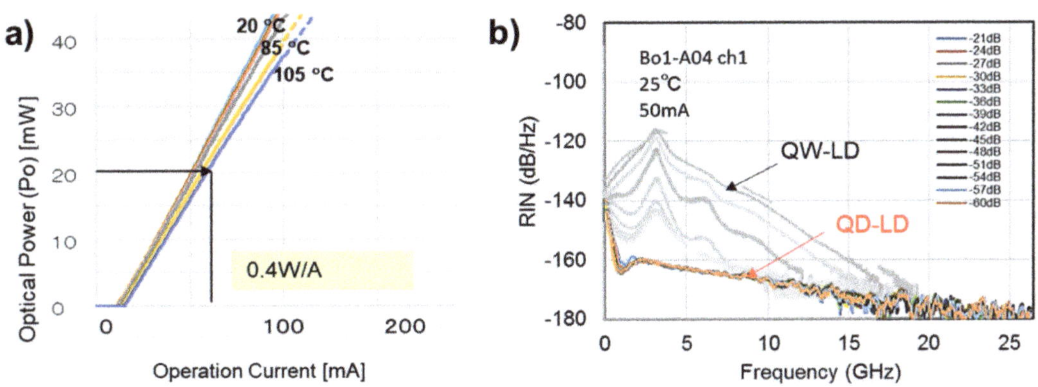

Figure 7. QD-LD performance: (**a**) L-I characteristics of QD-LD at different temperatures; (**b**) RIN for back reflection.

2.2.3. Redundant Laser Scheme for Increased Reliability

The first-generation IOCore micro-transceiver incorporated a single QD-LD with four CW 1310 nm outputs, each of which was modulated onto each transmitter channel directly (Figure 8a). However, this provided the lowest reliability as any single failure in any one of the outputs or the laser module would have resulted in the failure of the overall system.

In order to address this, a redundant optical circuit was introduced in the second-generation micro-transceiver (Figure 8b), wherein the original single four-channel QD-LD array was replaced by two independent two-channel QD-LD arrays. One QD-LD array is the primary device, with both its optical outputs passed to 3 dB power couplers which split the light equally into two branches, resulting in four CW streams, each of which is modulated onto a transmitter channel. The second QD-LD array is held on stand-by, only to be activated upon the failure of the primary QD-LD array. An additional Mach–Zehnder switch was designed between the outputs of both the primary and secondary QD-LD arrays and the transmitter channels. As shown in Figure 8b, in the event of failure of the primary QD-LD array, the secondary QD-LD array is activated and its two outputs are switched to the 3 dB couplers, which in turn split them into four transmitter channels.

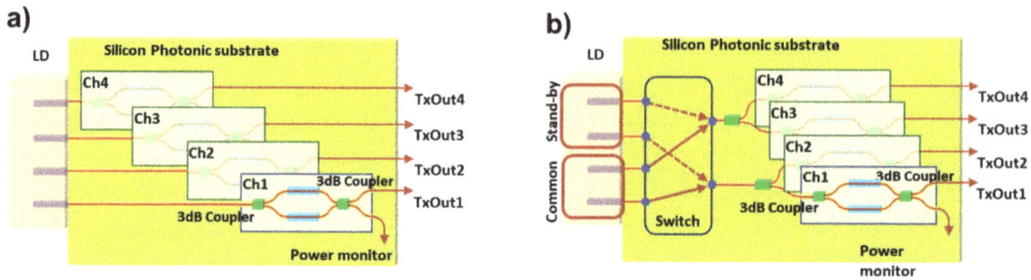

Figure 8. IOCore laser redundancy circuit: (**a**) conventional silicon photonic circuit scheme with single QD-LD array with four optical outputs; (**b**) second-generation IOCore micro-transceiver with two QD-LD arrays, each with two CW optical outputs [3].

Using the parameters obtained from reliability testing of the QD-LD, the time within which the chance of failure was 1% was increased from 10,000 to 100,000 h when moving from the non-redundant laser circuit scheme to the redundant scheme.

The qualification of the redundancy circuit scheme is underway and expected to be deployed in commercial devices by the start of 2022.

2.3. Passive Assembly of QD-LD into Silicon Photonics Chip

As shown in Figure 9a, the QD-LD array (Figure 9b) is passively mounted onto the silicon photonics substrate with silicon pedestals providing vertical alignment, and a fiducial is formed on the substrate to enable visual placement with respect to corresponding alignment marks on the QD-LD chip (Figure 9c) enabling horizontal alignment. Once the QD-LD is aligned, AuSn solder is applied to fix the array in place.

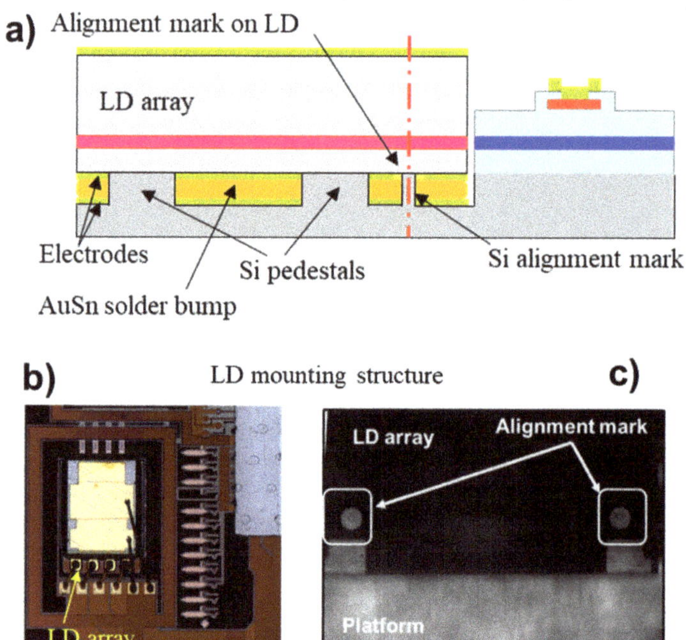

Figure 9. Passive assembly of QD-LD onto silicon photonics substrate. (**a**) Side cross-sectional schematic showing passive assembly features of QD-LD array onto silicon photonics substrate; (**b**) photo of QD-LD array; (**c**) photo of alignment marks used to visually align QD-LD array.

This assembly process can be automated with a pick-and-place machine, allowing high-volume, high-yield production.

Figure 10a shows the statistical distribution of mounting accuracy before and after solder attachment, with 99% (σ = 2.3) of mountings falling within an accuracy of $\pm$3 µm.

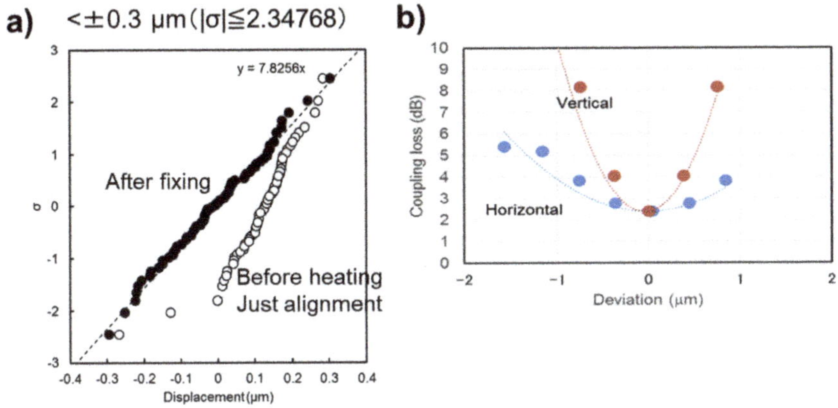

Figure 10. Laser mounting accuracy and coupling loss. (**a**) Distribution of mounting accuracy; (**b**) coupling loss variation with horizontal and vertical misalignment.

The QD-LD aperture is aligned to a tapered spot size converter (SSC) in the silicon photonic substrate, which allows laser light to couple to the silicon waveguides. Figure 10b shows the coupling loss variation with horizontal and vertical misalignment. The minimum coupling loss between QD-LD laser aperture and the silicon waveguide was measured to be 2 dB. The vertical misalignment tolerance is tighter than the horizontal misalignment tolerance, with a vertical misalignment of 0.5 µm resulting in ~4 dB coupling loss, while a horizontal misalignment of 0.5 µm resulted in 3 dB coupling loss.

As described in Section 2.8, changes in temperature give rise to a change in the emission wavelength of the QD-LD, which in turn will also affect the coupling loss, depending on the wavelength dependence of the spot size converter.

2.4. *Mach–Zehnder Modulator Design for High-Temperature Operation*

Figure 11a shows the schematic design of the silicon photonic Mach–Zehnder modulator (MZM) with distributed push–pull phase change nodes along each arm. The modulator was designed to provide on-off keying (OOK) over a wide temperature and frequency range. Figure 11b shows the frequency roll-off characteristics of the modulator. IOCore operation at 25 and 32 Gbps NRZ per channel has been reported [3], and the development of 50 Gbps NRZ per channel capability is underway.

2.5. *Optical Output Power*

Figure 12 shows the optical output characteristics (OMA) of the IOCore micro-transceiver as measured from the Tx optical pin. When the drive current is fixed at 75 mA, an average optical output of 0.14 dBm is measured, but by adjusting the drive current, the optical output can be increased to +1 dBm when in use.

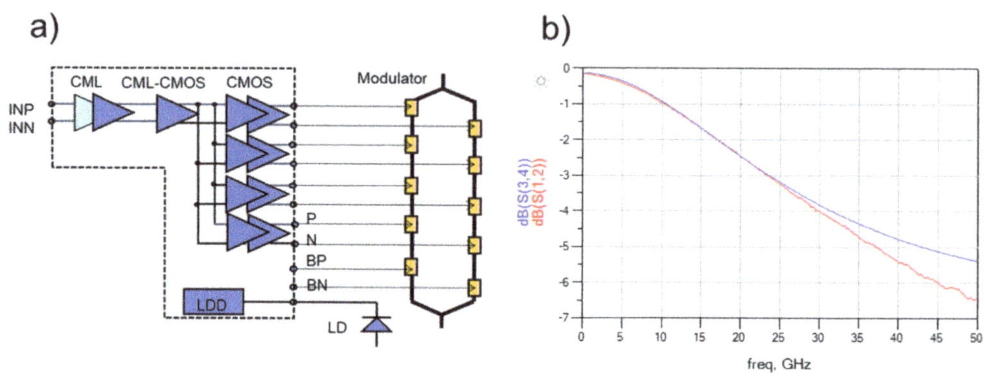

Figure 11. Mach–Zehnder modulator (MZM) design and performance. (**a**) Schematic diagram of MZM driven by CMOS IC; (**b**) frequency characteristics of the MZM.

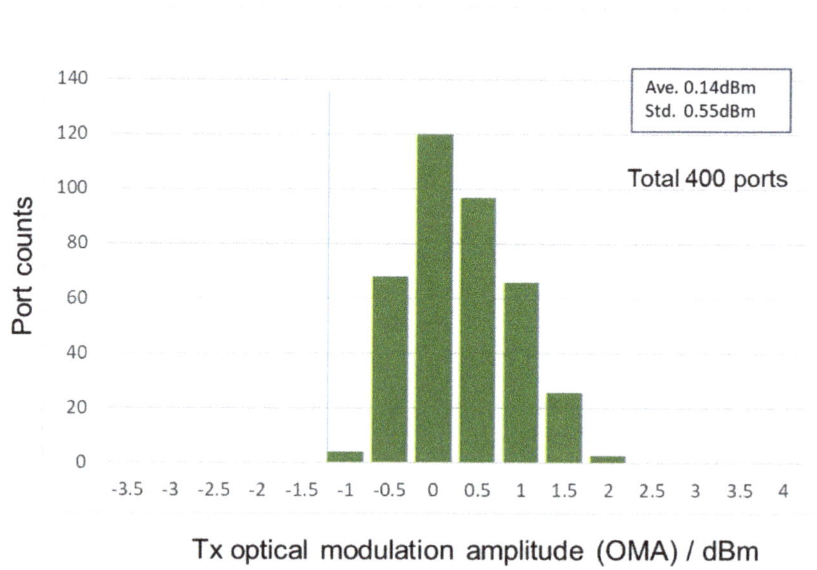

Figure 12. Distribution of optical modulation amplitude as measured on the outputs of the IOCore Tx optical pins.

2.6. Optical Transmitter Signal Integrity Measurements

Figure 13 shows the optical eye diagrams from the four output channels driven at 25.78 Gbps. The eye diagrams in the left column were recorded at an ambient temperature of 25 °C, while the eye diagrams in the right-hand column were recorded at an ambient temperature of 105 °C. The IOCore micro-transceiver does not include a clock and data recovery (CDR) circuit; nonetheless, a negligible difference in signal integrity and low jitter were recorded across the temperature range.

Transceivers with data rates of 25 and 32 Gbps have been reported [3,19], and a 50 Gbps version is currently under development with demonstration planned in 2022.

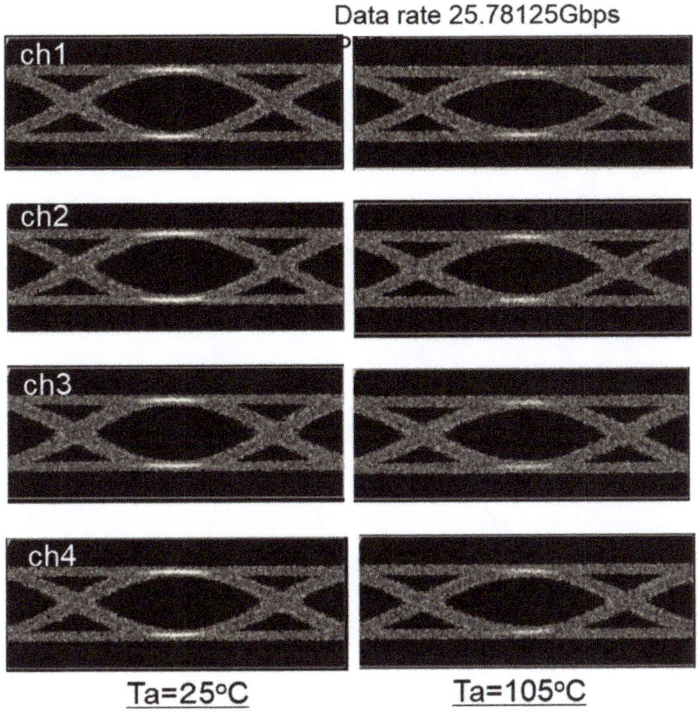

Figure 13. Optical eye diagrams on all four channels at 25 and 105 °C.

2.7. Optical Receiver Performance at 105 °C

Figure 14 shows the optical receiver sensitivity profile. The responsivity of the integrated germanium photodetector is 0.5 A/W with a frequency response of 18 GHz. The receiver IC has a gain adjustment function for high-temperature operation, and the degradation in receiver sensitivity is negligible at high temperatures.

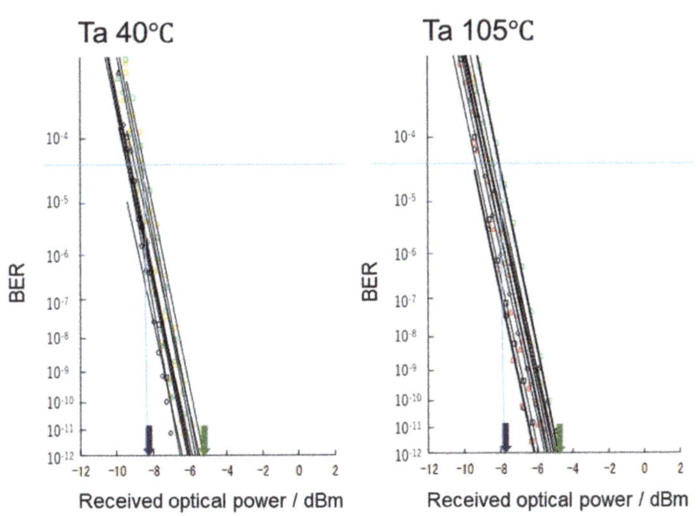

Figure 14. Optical bit error rate vs. received optical power.

Table 1 shows that with IEEE 802.3 bm-compliant forward error correction (FEC), reception sensitivity for a BER of $\geq 5 \times 10^{-5}$ can be achieved with received optical power >−8.2 dBm at an ambient temperature of 40 °C, increasing slightly to −7.8 dBm at an ambient temperature of 105 °C, while without FEC applied, a BER of $\geq 10^{-12}$ can be achieved with a received optical power of −5.6 dBm at an ambient temperature of 40 °C, increasing slightly to −5.0 dBm at an ambient temperature of 105 °C.

Table 1. Reception sensitivity.

	Ta 40 °C (Tj 60 °C)	Ta 105 °C (Tj 125 °C)
w/ FEC	−8.2 dBm	−7.8 dBm
w/o FEC	−5.6 dBm	−5.0 dBm

2.8. Multimode Optical Interface Performance Challenges at High Temperature

In the transmitter section, the modulated light is coupled through a vertical grating coupler into a tilted waveguide called an optical pin (Figure 15a). The emission angle of the grating coupler varies with wavelength (Figure 15b), and quantum dot lasers exhibit a wavelength shift with temperature (Figure 15c); this in turn results in a change in the direction of optical output over the temperature range. In the case of Fabry-Perot lasers, the wavelength has a temperature dependence of ~0.56 nm/°C, so when the temperature rises from 25 to 105 °C, the emission angle of the grating coupler changes by about 5° due to the corresponding variation in wavelength.

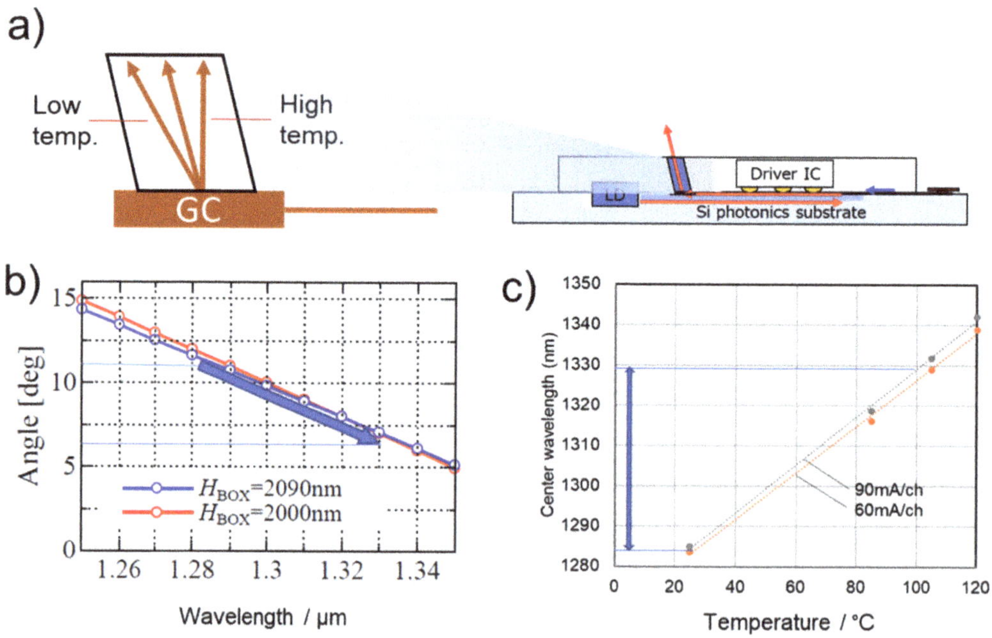

Figure 15. Optical coupling interface performance over temperature. (**a**) Schematic view of optical pin on Tx side accommodating variation in directional angle from the grating coupler over temperature; (**b**) variation of angle with wavelength; (**c**) variation of wavelength with temperature.

If the multimode optical fiber were directly connected to the grating coupler, the coupling loss would change more strongly over the temperature range in response to the change in emission angle. However, the optical pin has a higher NA of 0.4 and a core diameter of 35 μm, which confines the light more strongly in the optical pin core than a multimode fiber would and can thus better accommodate this variation in emission angle.

This intermediary waveguide then couples light into a multimode fiber with little variation in coupling loss over the temperature range.

2.9. Electro-Optical Module for High-Temperature Operation

The IOCore micro-transceiver is a building block that can be integrated into different types of electro-optical modules or co-packaged configurations. Figure 16b shows a low-form factor electro-optically pluggable module called the EOB (Electro-Optical Blade), which includes advanced heat dissipation structures on the upper and lower surfaces of the IOCore micro-transceiver, enabling operation at 105 °C. When used at an ambient temperature of 105 °C, the junction temperature, Tj, is designed to be 120 °C or less, which is within the temperature specifications for the ICs and laser.

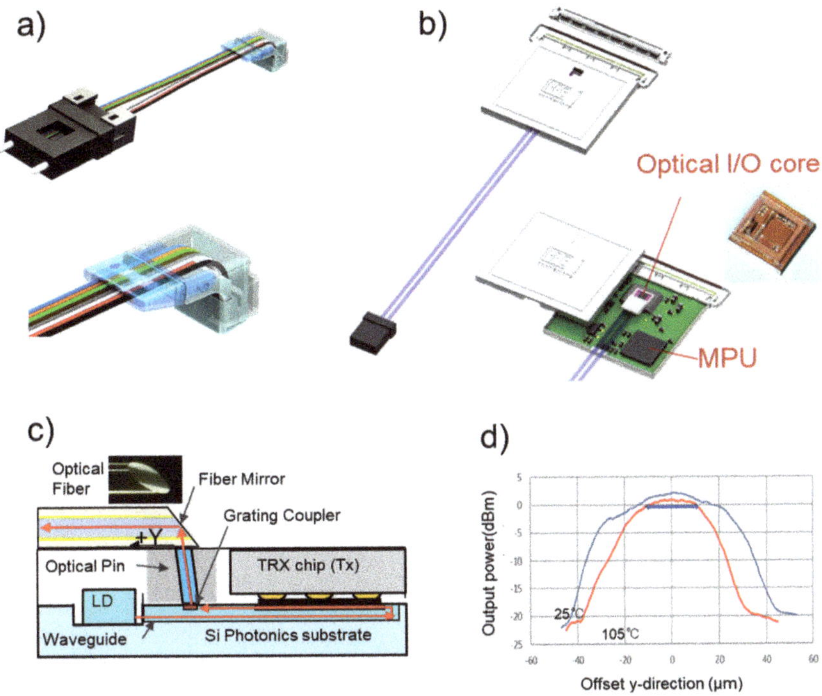

Figure 16. (**a**) Commercial vertical fiber array unit (courtesy of Senko); (**b**) electro-optically pluggable micro-transceiver module "EOB"; (**c**) cross-sectional schematic showing low-form factor coupling based on cleaved fiber; (**d**) misalignment tolerance of cleaved fiber at 25 to 105 °C.

2.10. Low-Form Factor Optical Coupling Interface

Commercial vertical multimode fiber array units, as shown in Figure 16a, can be used to couple to the optical interface; however, such units take up an additional height of at least 3.8 mm today and would therefore be unsuitable for many space-constrained applications such as co-packaged optics in a 5G+ environment or data centers.

The EOB instead uses a fiber with a cleaved angled end facet (Figure 16c) allowing light to be coupled directly from the optical pin to a horizontally arranged fiber, thus enabling very low profile interfaces. The 1 dB misalignment tolerance of the cleaved fiber along the axis of the horizontal fiber (denoted as offset y-direction) is about ±10 μm (Figure 16d) at 25 °C and is slightly tighter at around ±8 μm at 105 °C.

3. Discussion

In this paper we have primarily addressed the performance of the micro-transceiver in high-temperature environments, including, in particular, the channel speed, total throughput and low power consumption, where increasing the data rate per channel reduces the power consumption per bit of optical interconnect. It should be noted that the micro-transceiver is particularly very well suited to integration in board-mountable and co-packaged optical modules, which are currently of great importance to hyperscale data centers, and this has been reported by Kurata et al. [3].

In addition to these important performance metrics, we also focused on the required operating conditions. Here, in particular, there is a need for high-temperature operation combined with high reliability which exceeds that of pluggable components by an order of magnitude and enables a substantially lower cost solution than conventional single-mode silicon photonic devices. In order to meet these requirements, we have proposed a silicon photonic micro-transceiver with a multimode interface. Stable operation, without the need for temperature control systems, is possible at temperatures up to 105 °C. Furthermore, the introduction of a redundant laser circuit ensures over an order of magnitude higher reliability than a system with a non-redundant laser circuit. Thus, the reported micro-transceiver meets two of the key requirements outlined above.

To achieve the all-important low cost point, we need to consider both the device and mounting/assembly costs. On the device side, a reduction in the chip size is crucial as it enables a lower cost than conventional devices. This is of particular importance for high-volume applications such as optical interconnects, where small cost differences will quickly scale. On the packaging side, we can achieve low-cost assembly through the use of a multimode interface, which eliminates the high costs associated with high-precision single-mode fiber-to-chip assembly of conventional silicon photonics transceivers. The final important factor affecting the device cost is the fabrication and assembly yield. We have presented a micro-transceiver with very robust processing and assembly schemes, enabling the requisite high yield.

Although it is possible to implement Tbps with more channels in a single micro-transceiver chip by applying this technology, we are actively working towards transceivers with aggregate data rates above 1 Tbps whereby multiple "standard" four-channel micro-transceivers are mounted into different module form factors. Since the four-channel micro-transceiver chip size is small, many chips can be incorporated into standard module packages, allowing bandwidths to be scaled. This allows modules with n x four-channel micro-transceiver building blocks to address a wide diversity of applications beyond 5G and 6G, where n can be anything from '1' (e.g., for internal optical communication in automotive applications) to '64' (and beyond) for co-packaged optical ASICs in hyperscale data centers.

The development of 128 Gbps (4 × 32 Gbps/ch) transceivers with the same configuration as the 4 × 25 Gbps/ch transceivers reported in this paper is planned to be completed before 2022, and the development of 50 Gbps NRZ/ch 200 Gbps chips and 2.4 Tbps modules is in progress.

In the future, optimization of the optical wiring system in housing will become an important consideration in the realization of micro-transceivers. Here, we anticipate the use of multimode optical wiring because of its ease of connection, low cost and increased reliability, but the transmission medium and optical connector present areas of further research and development.

4. Conclusions

In this paper, we have reported on the design and performance of the IOCore multimode micro-transceiver and its viability for deployment in large-scale integration or co-packaged assemblies in 5G+ and other high-temperature, space-constrained environments. One important feature of the second-generation IOCore micro-transceiver is the integrated quantum dot laser and laser redundancy circuit, which enable operation in high-temperature environments with substantially improved reliability. The relaxed assembly constraints of a multimode interface give rise to a reduced-cost solution at the expense of restricting the device-to-device optical link distance to under 500 m. This trade-off allows multimode silicon photonics micro-transceivers to satisfy very short-reach, low-cost, temperature-resilient requirements in diverse applications. In addition, the IOCore micro-transceiver has been evaluated under single- and dual-phase immersion cooling environments, with a negligible change in performance compared to operation in a standard air-cooled environment as reported in [19].

Author Contributions: Conceptualization, K.K., K.N., F.C. and L.G.; methodology, K.K., K.N., L.G. and F.C.; validation, K.N., Y.H., K.Y., T.M., S.K. and M.K.; formal analysis, K.K., M.K. and S.K.; investigation, K.N., Y.H., K.Y., T.M., S.K., M.K., L.G. and F.C.; writing—original draft preparation, R.P., K.K., L.G. and F.C.; writing—review and editing, L.O., S.A.S. and R.P.; project administration, K.K.; supervision, K.K.; funding acquisition, K.K. All authors have read and agreed to the published version of the manuscript.

Funding: This paper incorporates results obtained from the project JPNP20017, commissioned by the New Energy and Industrial Technology Development Organization (NEDO) in Japan.

Institutional Review Board Statement: Not applicable.

Informed Consent Statement: Not applicable.

Data Availability Statement: The data presented in this study are available on request from the corresponding author. The data are not publicly available due to confidentiality restrictions.

Conflicts of Interest: The authors declare no conflict of interest.

References

1. Kurata, K.; Pitwon, R. Short reach, low cost silicon photonic micro-transceivers for embedded and co-packaged system integration. *Proc. SPIE* **2020**, *11286*, 112860R. [CrossRef]
2. CEA-LETI. Horizon 2020 COSMICC Project. Available online: http://www.h2020-cosmicc.com (accessed on 5 November 2021).
3. Kurata, K.; Hagihara, Y.; Kurihara, M.; Yashiki, K.; Kinoshita, K. Short reach, high temperature operation and high reliability silicon photonic micro-transceivers for embedded and co-packaged system integration. *Proc. SPIE* **2021**, *11692*, 1169204. [CrossRef]
4. Pitwon, R. *IEC 62150-6. Fibre Optic Active Components and Devices. Basic Test and Measurement Procedures Part 6. Universal Mezzanine Boards for Test and Measurement of Photonic Devices*; Standard in Progress; International Electrotechnical Commission: Geneva, Switzerland, 2021.
5. Pitwon, R.; O'Faolain, L.; Kurata, K.; Lee, B.; Ninomyia, T. Hyperscale Integrated Optical and Photonic Interconnect Platform. In Proceedings of the 2020 IEEE Photonics Conference (IPC), Vancouver, BC, Canada, 28 September–1 October 2020; pp. 1–2. [CrossRef]
6. Pitwon, R.; Wang, K.; Worrall, A. Converged photonic data storage and switch platform for exascale disaggregated data centers. In Proceedings of the SPIE—The International Society for Optical Engineering, SPIE OPTO, San Francisco, CA, USA, 28 January–2 February 2017; Volume 10109. [CrossRef]
7. Pitwon, R.; Wang, K.; Immonen, M.; Schröder, H.; Neitz, M. Universal test system for system embedded optical interconnect. *Proc. SPIE* **2018**, *10538*, 1053804–1053810. [CrossRef]
8. Co-Packaged Optics Collaboration. 2021. Available online: http://www.copackagedoptics.com (accessed on 5 November 2021).
9. 3.2 Tb/s Copackaged Optics Optical Module Product Requirements Document. 2021. Specification document from Copackaged Optics Collaboration. Available online: http://www.copackagedoptics.com/wp-content/uploads/2021/02/JDF-3.2-Tb_s-Copackaged-Optics-Module-PRD-1.0.pdf (accessed on 5 November 2021).
10. ITU-T. G.8300 Characteristics of Transport Networks to Support IMT-2020/5G International Standard. 2020. Available online: https://www.itu.int/rec/T-REC-G.8300-202005-I (accessed on 12 November 2021).
11. von Butovitsch, P.; Astely, D.; Furuskär, A.; Göransson, B.; Hogan, B.; Karlsson, J.; Larsson, E. Ericsson White Paper—Advanced Antenna Systems for 5G Networks. 2020. Available online: https://www.ericsson.com/en/reports-and-papers/white-papers/advanced-antenna-systems-for-5g-networks (accessed on 12 November 2021).
12. Ericsson. Leveraging the Potential of 5G Millimeter Wave. Available online: https://www.ericsson.com/en/reports-and-papers/further-insights/leveraging-the-potential-of-5g-millimeter-wave (accessed on 12 November 2021).
13. Stone, R.; Chen, R.; Rahn, J.; Venkataraman, S.; Wang, X.; Schmidtke, K.; Stewart, J. Co-packaged Optics for Data Center Switching. In Proceedings of the 2020 European Conference on Optical Communications (ECOC), Brussels, Belgium, 6–10 December 2020; pp. 1–3. [CrossRef]

14. Testa, F.; Giorgi, L.; Bigongiari, A.; Bianchi, A. Experimental evaluation of silicon photonics transceiver operating at 120 °C for 5G antenna array systems. *Electron. Lett.* **2018**, *54*, 1391–1393. [CrossRef]
15. Klopf, F.; Krebs, R.; Reithmaier, J.P.; Forchel, A. High-temperature operating 1.3-μm quantum-dot lasers for telecommunication applications. *IEEE Photonics Technol. Lett.* **2001**, *13*, 764–766. [CrossRef]
16. Zimmermann, L.; Tekin, T.; Schroeder, H.; Dumon, P.; Bogaerts, W. How to bring nanophotonics to application—Silicon photonics packaging. *IEEE Leos Newsl.* **2008**, *22*, 4–14.
17. H2020 EU Project 'Teraboard'. Available online: http://www.teraboard.eu (accessed on 5 November 2021).
18. IPSR-I. 2020 IPSR-I Integrated Photonic Systems Roadmap. 2020. Available online: https://photonicsmanufacturing.org (accessed on 5 November 2021).
19. Kurata, K.; Hagihara, Y.; Kuwata, M.; Muto, T.; Kobayashi, S.; Pitwon, R. Silicon photonics transceiver evaluation for immersion cooled data center and HPC environments. In Proceedings of the IEEE Group Four Photonics Conference, Málaga, Spain; 2021.

applied sciences

MDPI

Article

Packet Optical Transport Network Slicing with Hard and Soft Isolation

Samier Barguil [1,*], Victor Lopez Alvarez [2], Luis Miguel Contreras Murillo [2], Oscar Gonzalez de Dios [2], Alejandro Alcala Alvarez [1], Carlos Manso [3], Pol Alemany [3], Ramon Casellas [3], Ricardo Martinez [3], David Gonzalez-Perez [4], Xufeng Liu [4], Jose-Miguel Pulido [4], Juan Pedro Fernandez-Palacios [2], Raul Muñoz [3] and Ricard Vilalta [3]

1 Escuela Politécnica Superior, Universidad Autónoma de Madrid, 28049 Madrid, Spain; alejandro.alcala@estudiante.uam.es
2 Telefónica I+D, 28033 Madrid, Spain; victor.lopez@telefonica.com (V.L.A.); luismiguel.contrerasmurillo@telefonica.com (L.M.C.M.); oscar.gonzalezdedios@telefonica.com (O.G.d.D.); juanpedro.fernandez-palaciosgimenez@telefonica.com (J.P.F.-P.)
3 Centre Tecnològic de Telecomunicacions de Catalunya (CTTC/CERCA), 08860 Castelldefels, Spain; carlos.manso@cttc.es (C.M.); pol.alemany@cttc.es (P.A.); ramon.casellas@cttc.es (R.C.); ricardo.martinez@cttc.cat (R.M.); raul.munoz@cttc.es (R.M.); ricard.vilalta@cttc.es (R.V.)
4 Volta Networks, 08006 Barcelona, Spain; david@voltanet.io (D.G.-P.); xufeng@voltanet.io (X.L.); jm@voltanet.io (J.-M.P.)
* Correspondence: samier.barguil@estudiante.uam.es

Abstract: Network operators have been dealing with the necessity of a dynamic network resources allocation to provide a new generation of customer-tailored applications. In that sense, Telecom providers have to migrate their BSS/OSS systems and network infrastructure to more modern solutions to introduce end-to-end automation and support the new use cases derived from the 5G adoption and transport network slices. In general, there is a joint agreement on making this transition to an architecture defined by programmable interfaces and standard protocols. Hence, this paper uses the iFusion architecture to control and program the network infrastructure. The work presents an experimental validation of the network slicing instantiation in an IP/Optical environment using a set of standard protocols and interfaces. The work provides results of the creation, modification and deletion of the network slices. Furthermore, it demonstrates the usage of standard communication protocols (Netconf and Restconf) in combination with standard YANG data models.

Keywords: Software Defined Networking; network slicing; 5G; network automation; L3VPN

Citation: Barguil, S.; Lopez Alvarez, V.; Contreras Murillo, L.M.; Gonzalez de Dios, O.; Alcala Alvarez, A.; Manso, C.; Alemany, P.; Casellas, R.; Martinez, R.; Gonzalez-Perez, D.; et al. Packet Optical Transport Network Slicing with Hard and Soft Isolation. *Appl. Sci.* **2021**, *11*, 6219. https://doi.org/10.3390/app11136219

Academic Editors: Fabio Cavaliere and Luca Potì

Received: 6 June 2021
Accepted: 1 July 2021
Published: 5 July 2021

1. Introduction

Network slicing is positioned as the next paradigm for service delivery in telecom operator networks. the slicing concept departs from the idea of allocating network resources to customer services on demand, leverages on the paradigms of Software Defined Networking (SDN) and Network Function Virtualization (NFV) [1], those resources being dedicated to them during service lifetime. Those assigned resources (including compute, storage and transport ones) can be either physical or virtual, tailored to the customer need expressed through the slice request. Since distinct customers could have different requirements to be satisfied, different slice types can be considered from the point of view of management and control, as defined in [2].

The way of slicing can be seen as an alternative form of consuming network capabilities, permitting us to address the particular needs of new industries and markets [1] as was not possible before. Specific service characteristics like deterministic extreme low latency or guaranteed high bandwidth, different degrees of isolation (with respect other customers' services), scalability adapted to the actual demand, or resource management and control can be now enabled, allowing more sophisticated services to emerge and co-exist in a commonly shared infrastructure.

In order to enable this flexible consumption of network capabilities, it is necessary to develop new forms of network control to make possible a dynamic partitioning and assignment of resources end-to-end. Therefore, several initiatives have been discussed across the industry, including the IP network resources management (L2 [3] and L3VPNs [4]), the optical layer management (Transport API [5]) or the network devices configuration (Openconfig [6]).

This paper extends the work presented in the latest OFC conference [7]. In addition, this work adds a set of tests; Including creating two isolated network slices, prefixes addition and modification to each slice and slices deletion. A last set of tests included the service continuity evaluation after a physical device reset. For the whole set of experiments, a hierarchical control architecture over multi-layer IP over DWDM networks was used.

This paper is organized as follows: The authors describe those concepts in Section 2. Section 3 described the whole control architecture used for the network slice instantiation in a service provider environment. Section 3.5 describes the proposed architecture's implementation choices. The authors selected one of those choices and made a proof-of-concept in Telefonica and CTTC Laboratories, Section 4 includes the results.

2. Network Slicing Concepts

Network slicing refers to the partitioning of one physical network into multiple virtual networks, each network slice is architected and optimized for a specific application/service. In this context, the Next Generation Mobile Networks (NGMN) defines two main concepts [8]:

- The Service Instance (SI) is an end-user service or a business service realized in a Network Slice.
- The Network Slice Instance (NSI) as the complete, instantiated logical network that meets specific characteristics required by a Service instance.

Hence, network slicing is sharing network infrastructure across different Service Instance(s) to meet network-specific requirements. Depending on the communication layer where a network operator implements network slicing, resource management, the network characteristics or the toolbox used to implement the network slice can change. Some examples of the network characteristics requested by a Service Instance include ultra-low-latency or ultra-reliability, among others [8].

The network slicing concept provides a framework for broad applicability across various industries. The majority of these scenarios envisioned to suit emerging and diverse business models are based on the Network As A Service (NASS) approach, creating opportunities for intelligent services and a new business ecosystem [9–11]. One of the main enablers to drive the network slicing is the Fifth-generation (5G) networks realization. Due to the necessity to integrate multiple services with various performance requirements—such as high throughput, low latency, high reliability, high mobility, and high security—into a single physical network infrastructure, and provide each service with a customized logical network. the Network Slicing being the key technology to achieve the aforementioned goals.

5G expects to support a new generation of customer-tailored applications with diverse requirements regarding capacity, latency, level of mobility, number of users, and user density. Hence, 5G sets the stage for innovation and transformation in customer services and vertical industries, such as the ones described in above. Table 1 [12].

Table 1. Possible Applications derived from the 5G implementations.

Applications	Definition	Example
Ultra Reliable Low Latency Communications (uRLLC)	Requires support for 1 ms latencies, 0.001% packet loss, user mobility up to 100 km/h	Autonomous driving or Industrial automation
Massive Machine Type Communications (mMTC)	Requires support for 1 million devices per square kilometer, tens of bps bandwidth, and latency minimization for battery life optimization	Massive IoT
Enhanced Mobile Broadband (eMBB)	Requires Gbps bandwidth, real time or not	Immersive UIs based Augmented Reality Virtual Reality. Streaming of High Quality Video

The services described in Table 1 demonstrate highly diversified traffic characteristics and differentiated quality-of-service (QoS) requirements. For example, the mMTC is based on its application in machine-to-machine communication [13]. The actual realization of those services in telecom networks implies the migration of the current heterogeneous IP/MPLS + DWDM transmission networks to more modern and 5G-ready designs, using the Network programmability, Software-Defined Networking (SDN) and Machine Learning (ML) as their main pillars.

Implementing the network slices has two main alternatives: Hard and Soft. The way the network resources are shared between the services in each alternative is described in the following sections.

2.1. Soft Network Slicing

Soft slicing corresponds to a lower level of isolation between the services a network is transporting. Soft slicing implies sharing the physical infrastructure but creates logical segmentations between the customers. According to this definition, soft slicing is not a new concept from the IP/MPLS networks perspective [4,14,15].

Traditional L3VPNs are samples of soft-slicing implementations in an MPLS network, because a VPN can be thought of as a series of tunnels connecting customer sites, each site can potentially have different QoS treatment, and all traffic to and from each site is internal to the customer. In the VPN service, the provider ensures that each customer's traffic is logically discriminated over shared physical infrastructure based on routing policies configuration across the network.

In that sense, Network slicing at the MPLS level can be implemented using:

- **Virtual Routing and Forwarding** (VRF), that enables multiple routing environments over a shared MPLS transport network.
- **Virtual Interfaces** (VSI), that enables multiple switching environments over the same shared infrastructure.

Each physical router is able to host multiple VRF(s) and multiple VSI(s) (along with their attached logical interfaces), effectively slicing it into multiple routing and switching environments that can be assigned to different tenants/customers/services.

2.2. Hard Network Slicing

Despite the apparent advantages that an overlay MPLS tunnels could provide, there are some disadvantages. Overlay tunnels built by data encapsulation have neither visibility nor control of the underlying physical network. With more and more tunnels deployed on shared physical infrastructure, network congestion necessarily becomes an attention point. Thus, VRFs, VSIs or Optical Data units (ODus) cannot be managed by their corresponding tenants directly because they are part of the same administrative domain represented by the physical device, and they need to be operated by the network administrator. Due to the massive number of tunnels and services in a complex network, the QoS and Traffic Engineering policies management becomes a crucial task to guarantee the right SLAs to its customers [13,16].

The solution to this limitation and the way to move the networking industry to more automated scenarios is Hard slicing. Hard slicing refers to the provision of dedicated resources to a specific Network Slice Instance. For example, Data-plane resources are provided by allocating time-domain multiplexed resources such as a Flex Ethernet channel or as a service such as an MPLS hard-pipe, route diversity (disjoint paths), wavelength selection, among others.

Disaggregated routers is an example of hard-slicing. In the disaggregated case, the data plane runs in the physical device, and the control plane runs outside the device in a remote cloud or server. In that case, the 1:1 relationship between physical device and routing logic is disassociated, enabling the support of multiple virtual routers over a single physical network device. Each of these virtual routers is a router in its full sense, being able to host multiple VRFs and VSIs, or to be managed independently of the other virtual routers running in the same device. Virtual routers are thus administratively separated from each other, which means separate virtual routers can be assigned to different tenants, and each tenant can manage the virtual routers directly, without the need from the operator that owns the physical network to intervene intervene (e.g., following an approach similar to the one described in [17]).

Network slicing at the IP transport level can then be accomplished by grouping multiple virtual routers running over a shared physical network infrastructure into a common virtual infrastructure under its separated administrative domain. Virtual routers under the same administrative domain are then known as hard network slice, so different hard slices can be assigned to different tenants.

2.3. 5G Network Slicing

5G must enable network operators to ensure the same network can fulfil the heterogeneous demands of diverse types of applications. To efficiently achieve those requirements and determining how network resources are assigned, a service provider must integrate technologies like Software-Defined Networking (SDN), Network Functions Virtualization (NFV), and Machine Learning (ML) with a hierarchical transport architecture.

According to the 5G definition the architecture has three main components [18,19]:

- **Radio Access Network (RAN)**: It covers everything related to the air interface between the user element and the base station. The RAN interfaces and interconnections is specified within 3G PPP Architecture Working Group.
- **Mobile Core (MC)**: Its central role is to act as a gateway for user traffic to and from the internet. The mobile core is composed of a set of network functions (NF) responsible for managing user mobility, access authentication, access authorization, location service management, registration and establish per-user tunnels between base stations for each different traffic type.
- **Backhaul Network** is the network that interconnects the RAN with the MC. It is not part of the 5G specification, so it is up to each network operator to decide how to implement it. It requires functionalities such as QoS management, synchronization and a stack of protocols like IP/MPLS or segment routing.

Network slicing capabilities need to be available across all components of the 5G cellular network (RAN, MC, Backhaul network) in order to ensure a differentiated treatment of the packets in the network. Thus, the 5G working groups have specified a standard set of network slices, denoted Standardized Slice Type (SST), to determine how resources should be assigned at the RAN and MC [20].

5G specifies two mechanisms for network slicing. The first one is based on QoS techniques, by applying a dynamic allocation of available network resources to different classes of traffic, and it is denoted as soft network slicing. The second one takes advantage of the software-based, cloud-based architecture of 5G, as well as component disaggregation, and achieves slicing through 5G component virtualization and replication. This second approach is denoted hard network slicing.

3. Proposed Architecture

The iFUSION architecture is an architecture defined by Telefonica to strengthen the network automation and programmability in a service provider environment, as depicted in Figure 1 [21]. iFUSION is a two-layer control architecture, with specific domain controllers per technological domain (IP/MPLS, microwave and optical) in the bottom and a Software-Defined Transport Network controller (SDTN controller) to handle the multi-layer and multi-domain transport network resources. The domain controllers directly communicate with the network elements and the SDTN controller with the OSS/BSS systems. More than the functional block definition, iFUSION includes the usage of: (1) Standard interfaces based on RESTCONF/YANG [22] for the communication between control components and NETCONF/YANG [23] to configure the network elements; (2) YANG data models based on latest releases in the standards-development organizations (SDOs): IETF, ONF and OpenConfig.

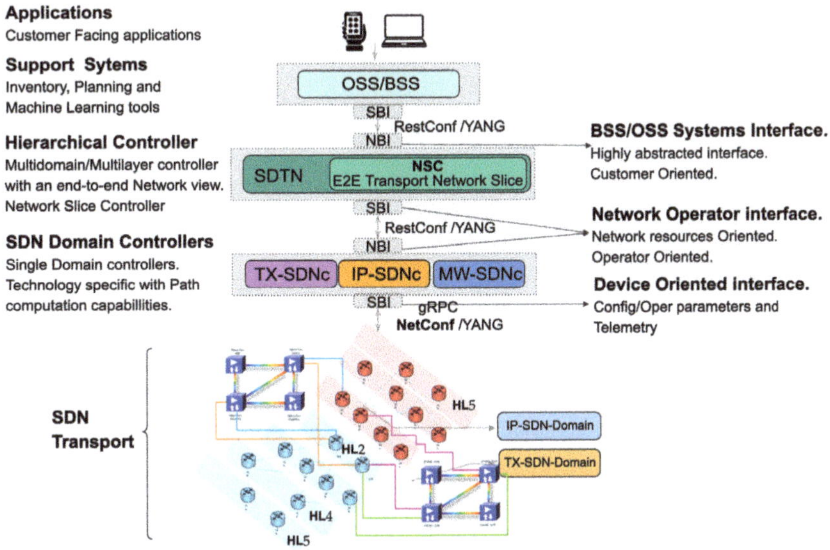

Figure 1. Overall proposed architecture.

Figure 1 shows the network scheme of the iFUSION architecture in terms of components and relationships among them. The following sections define each of the structural pillars in the architecture, including its role in network slicing.

3.1. Software Defined Transport Network Controller

The Software-Defined Transport Network Controller is a functional block with the following purposes [21]; It is The main entry point from the OSS/BSS systems to the network. It is in charge of coordinating/providing services through several domains and layers. It has the multi-layer/multi-domain topological view of the network. The SDTN controller can split requirements based on the technological requirements. During this process, the SDTN controller can add/assign logical resources to be used by the network at the service implementation. The SDTN controller have two RESTCONF interfaces, one to process the OSS/BSS systems' requirements and one to send the specific requests to the domain controllers. TeraFlow project [24] is proposing the development of a cloud-native SDN controller that will serve as SDTN.

The SDTN controller can have incorporated the Network Slice Controller as a working piece of its implementation.

3.2. Network Slice Controller

The Network Slice Controller (NSC) effectuates a transport network slice in the underlying transport infrastructure, manages and control the state of resources and topologies associated with it. The NSC receives a transport network slice request from the Operation Support System and Business Support System (OSS/BSS). The NSC runs an internal workflow for transport network slice life-cycle management and interacts with underlying IP and Optical Domain controllers via a RESTCONF client. As described in the following sections, depending on the NSC location in the architecture, the NSC will delegate to SDN Domain controllers to configure the network (Section 3.5 (a) and (b)) or make it directly through a NETCONF SBI (Section 3.5 (c)).

The network slice controller is the key building block for control and management of network slice. It provides the creation/modification/deletion, monitoring an optimization of network slices in a multi-domain, a multi-technology and multi-vendor environment. It has two main functionalities, defined in [25,26]:

- *Map*: The NSC must map the Network Slice requests to the underlying technology-specific infrastructure. Accordingly, It maintains a record of the mapping from user requests to slice instantiations, as needed to allow for subsequent control functions like modification or deletion.
- *Realize*: The NSC should realize the network slice request using its SBI interface against the domain controllers in either physical or logical connectivity through VPNs or various tunnelling technologies such as Segment Routing, MPLS, etc.

3.3. Network Domain Controller

The network domain controller (SDN Controller) is in charge of network elements (network domain). It has standard southbound interfaces to communicate with the network elements. The Domain controller SBI relies on using the Network configuration protocol (NETCONF) to interact with the underlying technology's network elements. The SDN controller also has a northbound interface to communicate with the SDTN controller or the OSS/BSS Systems using RESTCONF.

3.4. Yang Models for Network Controllers

As described until now, the three control elements: SDTN controller, Network Slicing Controller and Network Domain controllers, has standard SBI and NBI interfaces to communicate between them and against the network or the OSS/BSS systems. The standard interfaces are composed of selecting a protocol to transfer the data and YANG data models to define how the message is formed. For that sense, the YANG modelling activities have acquired significant relevance across the standardization entities. This is to such an extent that by 2019, the number of correctly extracted YANG models from IETF drafts was 283, in the Broadband forum was 214, and in Openconfig was 137 [27]; similarly, other organizations like the MEF, 3GPP or ONF produced YANG data models to describe technologies, protocols or connectivity services as well. Thus, navigating through the massive set of YANGs available and selecting the suitable pack of data models to define each functional block's interfaces becomes an essential task from the architectural definitions point of view.

To request and after instantiate the network slices a set of these data models are described as the experimental base of this work are described in the Table 2.

Table 2. Set of these data models are described as the experimental base of this work.

Models	Description	Example
LxVPN	These models describe a VPN service from the customer or network operator point of view.	L3SM: [15] L2SM: [28] L3NM: [4] L2SM: [3]
Traffic Engineering	These models allow to manipulate Traffic Engineering tunnels within the network segment. Technology-specific extensions allow to work with a desired technology (e.g., MPLS RSVP-TE tunnels, Segment Routing paths, OTN tunnels, etc.)	TE: [29] TE Topology: [30,31]
TE Service Mapping extensions	These extensions allow to specify for LxVPN the details of an underlay based on Traffic Engineering	Service Mapping: [32]
ACLs and Routing policies	Even though ACLs and routing policies are device models, It's exposure in the NBI of a domain controller allows to provide an additional granularity that the network domain controller is not able to infer on its own.	ACL: [33] Routing Policy: [34,35]
OTN	As a part of the transport network, OTN can provide hard pipes with guaranteed data isolation and deterministic low latency, which are highly demanded in the Service Level Agreement (SLA).	OTN Slice: [36]
Slicing	Set of data models available to map and realize the network Slices.	Network Slice NBI: [25,26]

3.5. Instantiating of Network Slices in SDN Transport Networks

As described before in the Section 3.1, the OSS/BSS systems may request the deployment of a new network slices with certain transport characteristics. Each network slice must be isolated from any other network slices or different services delivered to particular customers and naturally, other network slices or services must not negatively impact the requested transport network slice's delivery.

To provide this isolation and instantiate the slice in the network there are certain implementation options ranging from softer to hardest grades of isolation, as follows:

- No-isolation, meaning that slices are not separated.
- Logical-isolation, where slices are logically separated, only a certain degree of isolation is performed through QoS mechanisms.
- Service-isolation, where virtual resources and NFs are shared.
- Process-isolation, where slices include process and threads isolation.
- Virtual-resource-isolation, where slices have dedicated virtual resources.
- Network-functions-isolation, where Network Function (NF) are dedicated to a single network slice.
- Physical-isolation, where slices are completely physically separated, for example, in different locations.
- Physical-network-isolation, where slices contain physically separated links.

As the isolation grade is a significant constraint to consider for the network slice implementation, the selected network infrastructure and the control elements selected would generate different sets of capabilities in the Network Slice Controller. Figure 2 describes three of the possibilities analyzed from the point of view of the mapping and realization tasks of the Network Slice Controller:

(a) **Network Slice Controller as part of the hierarchical controller:** When the Network Slice Controller is a Hierarchical SDN controller module, the NSC's and the Hierarchical Network Controller should share the same internal data and the same NBI. Thus, to process the customer, view the H-SDN module must be able to:

- *Map*: The customer request received using the must be processed by the NCS. The mapping process takes the network-slice SLAs selected by the customer to available Routing Policies and Forwarding policies.
- *Realize*: Create necessary network requests. The slice's realization can be translated into one or several LXNM Network requests, depending on the number of underlay controllers. Thus, the NCS must have a complete view of the network to map the orders and distribute them across domains. The realization should include the expansion/selection of Forwarding Policies, Routing Policies, VPN policies and Underlay transport preference.

To maintain the data coherence between the control layers, the `network-slice-id` used must be directly mapped to the `transport-instance-id` at the VPN-Node level.

(b) **Network Slice Controller as an stand-alone controller:**
When the Network Slice Controller is a stand-alone controller module, the NSC's should perform the same two tasks described before:

- *Map*: Process the customer request. The customer request can be sent using the [draft-liu-teas-transport-network-slice-yang-01]. This draft allows the topology mapping of the Slice request.
- *Realize*: Create necessary network requests. The slice's realization will be translated into one LXNM Network request. As the NCS has a topological view of the network, the realization can include the customer's traffic engineering transport preferences and policies.

(c) **Network Slice Controller at the domain controller level:**
The Network Slice Controller can at the same level as the network domain controllers. The SDTN controller should handle the slices request, and the realization can be done by the NSC directly communicating with the Network Elements. The SDTN controller should create unified network views, including each transport domain and the network slices.

- *Map*: The SDTN would Process the customer request. The customer request can be sent using the [draft-liu-teas-transport-network-slice-yang-01]. This draft allows the topology mapping of the Slice request.
- *Realize*: The realization can be done by the NCS controller applying the service logic to create policies directly on the Network elements. The SDTN should handle the shared resources management between domains.

(d) **Network Slice Controller as part of the domain controller:**
When the Network Slice Controller is part of the domain controller, the OSS/BSS systems process the Slices requests and introduce the network abstraction layer. At the network level, the same device data model would be used in the NBI and SBI of the SDN controller. The direct translation would reduce the service logic implemented at the SDN controller level, grouping the mapping and translation into a single task. :

- *Map & Realize*:The mapping and realization can be done by the Domain controller applying the service logic to create policies directly on the Network elements.

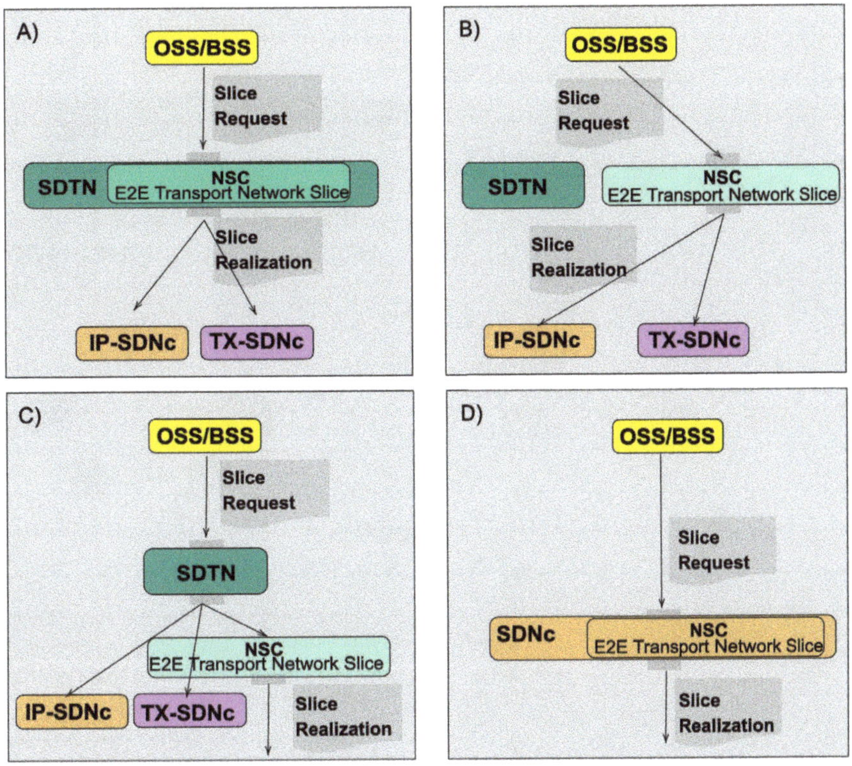

Figure 2. Possible Network Slice instantation in a service provider network. (**A**) Network Slice Controller as part of the hierarchical controller, (**B**) Network Slice Controller as an independent hierarchical controller, (**C**) Network Slice Controller as an independent domain controller and (**D**) Network Slice Controller as part of the domain controller.

4. Experimental Validation

The experimental testbed has been distributed between Telefónica and CTTC laboratory premises, as depicted in Figure 3. The testbed includes two layers, a control layer and a transport layer.

Telefonica and CTTC deployed a control layer composed of three relevant items. First, to receive all the service requests, a Network Slice Controller was developed as part of the SDTN controller. Second, to interact with the transport domains, the testbed included Two (2) controllers, one for IP and one for Optical. To configure the routers, the IP SDN Controller used gRPC, while the Optical SDN controller used Netconf with T-API for the configuration tasks.

The physical transport layer was composed of:

- Two (2) 7316 Edgecore switches running ONIE.
- Two (2) Spirent traffic generator.
- Four (4) Flexi-grid DWDM nodes.

The Edgecore used a 10 Gigabit Ethernet (10G) interface for the connection against the Spirent testers and 1G interfaces to the Optical devices. Furthermore, on each Edgecore, two (2) separate virtual routers were configured to test redundancy and route filtering between the ends. The Network Operating system (NOS) running on the virtual-routers were volta-stack deployment version 20.4-2-36-g0ba8807.

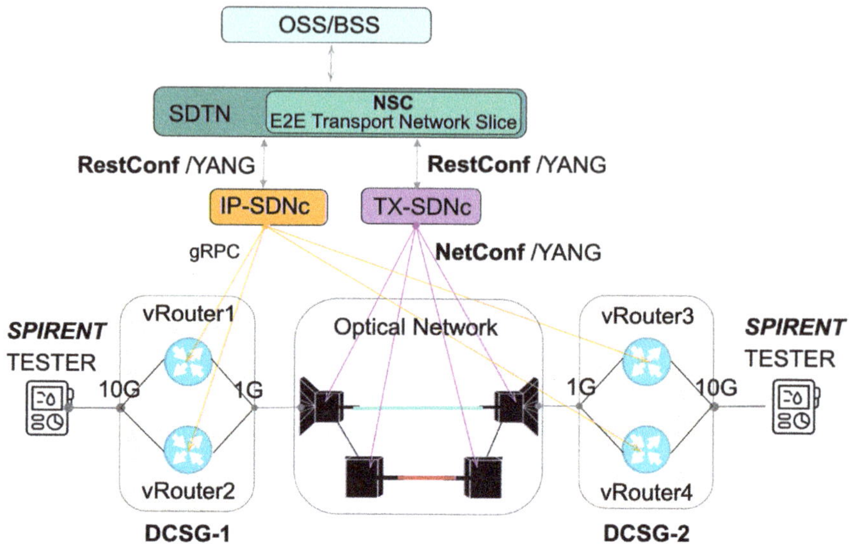

Figure 3. Network Design, including the Logical and Physical Infrastructure to support the test cases.

As we have described previously, there are several requirements for network automation for network slicing. However, a service provider can not all treat all the needs in the same way. Thus, a set of use cases were designed and executed as part of the whole testing process, the use cases were the following:

1. Slice Creation: To request a Standards-Based and Model-Driven isolated Network Slices using. The SDTN controller would receive and translate the network slice service into specific per domain requests as follows:
   (a) L3VPN service creation: In this use case, each network slice request requires an L3VPN service creation; thus, each L3VPN service would map to a single network slice. The endpoints defined in the slice request would map to Virtual Router and Forwarding (VRFs) instances on the virtual routers. The Yang data model used for the SDTN requests the L3VPN services to the IP domain controller was the L3NM [4].
   (b) DWDM connectivity: The T-API [5] was used to create a new connectivity service in the transport network to enable the L2-L1 communication between the network slice endpoints.
2. Add Prefixes and destroy: The second use case expects to validate the IP connectivity between the network slices. Hence the Spirent testers announced a set of 5k IP prefixes through the prior created network slices. Afterwards, the testers make an automatic IP reachability test and a CLI route redistribution validation. Once the route propagation was validated, the testing team stopped all the Spirent prefixes' announcement, and a new automatic connectivity test was done.
3. Device recovery test: The third use case was to verify the network slices service continuity so. The testing team manually rebooted all the DCSGs. Once the devices got online again, we have checked the service status with a measurement of the service restoration time.

4.1. Slice Creation

The iFusion architecture enhancement proposed for this paper performs the management of services and resources through the use of information models that capture the definitions of managed entities in terms of attributes and supported operations. Hence, a set of Yang Data models has been defined to render and realize a network slice between each of the control entities. The workflow proposed used postman to simulate, the network slice creation requests. The YANG data-model used for this request was defined in [37]. It includes the endpoints, the customer information and the service level agreement of the slice requests. The PE-CE and the end-to-end connectivity protocols for the network slice realization was automatically expanded by the SDTN controller and received by the IP SDN controller to properly configure the network elements. Each network slice, The workflow can be seen in the Figure 4.

A capture of the messages exchanged between the control items is depicted in Figure 5. The Figure illustrates in Orange the messages exchanged for the Network Slice Creation using the IETF network slices data model. The subsequent statements show how the SDTN controller unwrapped the creation request message to the optical and IP domains. It illustrates in red the IP network massages using gRPC, including the access ports, routing protocols, and PE-CE connectivity parameters. In purple, the T-API messages have the connectivity and optical service requirements.

The time consumed in the Slices creation was 10.34 s on average. The left part of the Figure 6 depicts a cumulative histogram of the time consumed for the slice creation in the controller layer using ten samples. Additionally, the rigth side Figure 6 shows the percentage of time spent for the VPN-instantiation in the Virtual routers. The most time-consuming task is the BGP Neighbours creation (13 s), followed by the BGP instance redistribution (10 s). In contrast, the device implemented the primary VPN configuration parameters (Rute-Distinghuiser, Route Target and Router ID) in less than a second.

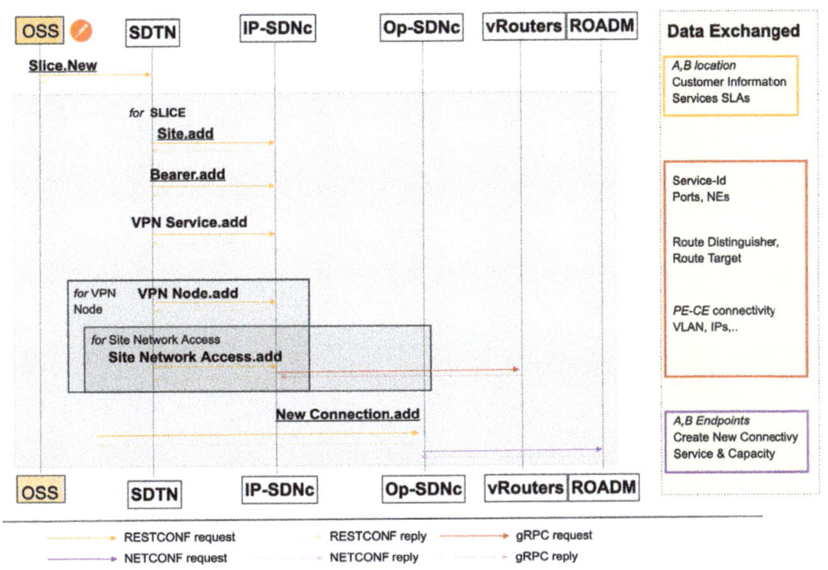

Figure 4. Workflow used to instantiate the network slices in the network. From Postman, a user sends a slice request to the SDTN controller. The SDTN controller automatically split the request based on technological requirements. The L3NM and the T-API are used for the IP and Optical domains, respectively.

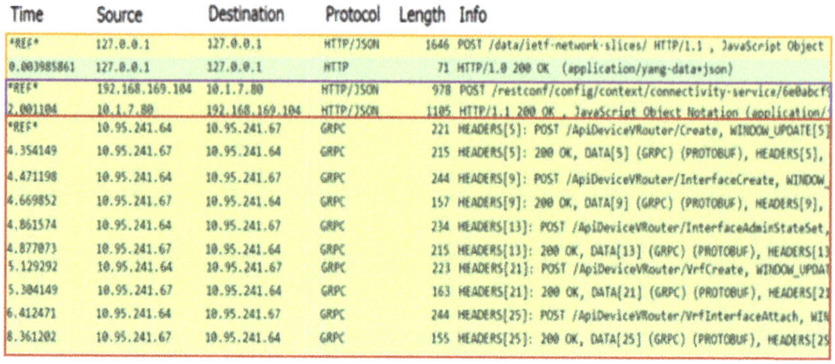

Time	Source	Destination	Protocol	Length	Info
REF	127.0.0.1	127.0.0.1	HTTP/JSON	1646	POST /data/ietf-network-slices/ HTTP/1.1 , JavaScript Object
0.003985861	127.0.0.1	127.0.0.1	HTTP	71	HTTP/1.0 200 OK (application/yang-data+json)
REF	192.168.169.104	10.1.7.80	HTTP/JSON	978	POST /restconf/config/context/connectivity-service/6e0abcf
2.001104	10.1.7.80	192.168.169.104	HTTP/JSON	1105	HTTP/1.1 200 OK , JavaScript Object Notation (application/
REF	10.95.241.64	10.95.241.67	GRPC	221	HEADERS[5]: POST /ApiDeviceVRouter/Create, WINDOW_UPDATE[5]
4.354149	10.95.241.67	10.95.241.64	GRPC	215	HEADERS[5]: 200 OK, DATA[5] (GRPC) (PROTOBUF), HEADERS[5],
4.471198	10.95.241.64	10.95.241.67	GRPC	244	HEADERS[9]: POST /ApiDeviceVRouter/InterfaceCreate, WINDOW_
4.669852	10.95.241.67	10.95.241.64	GRPC	157	HEADERS[9]: 200 OK, DATA[9] (GRPC) (PROTOBUF), HEADERS[9],
4.861574	10.95.241.64	10.95.241.67	GRPC	234	HEADERS[13]: POST /ApiDeviceVRouter/InterfaceAdminStateSet,
4.877073	10.95.241.67	10.95.241.64	GRPC	215	HEADERS[13]: 200 OK, DATA[13] (GRPC) (PROTOBUF), HEADERS[1
5.129292	10.95.241.64	10.95.241.67	GRPC	223	HEADERS[21]: POST /ApiDeviceVRouter/VrfCreate, WINDOW_UPDA
5.304149	10.95.241.67	10.95.241.64	GRPC	163	HEADERS[21]: 200 OK, DATA[21] (GRPC) (PROTOBUF), HEADERS[2
6.412471	10.95.241.64	10.95.241.67	GRPC	244	HEADERS[25]: POST /ApiDeviceVRouter/VrfInterfaceAttach, WI
8.361202	10.95.241.67	10.95.241.64	GRPC	155	HEADERS[25]: 200 OK, DATA[25] (GRPC) (PROTOBUF), HEADERS[2

Figure 5. Capture of the workflow used to instantiate the network slices in the network (Orange). The capture shows the IP network configuration (Red) and the T-API Optical domains (Purple).

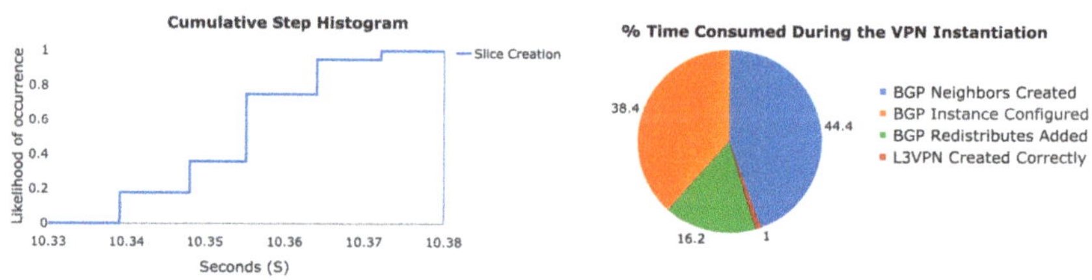

Figure 6. A cumulative step histogram of the time consumed for the slice creation, and the time consumed in during the VPN instantiation in the vRoutes.

4.2. Add Prefixes and Destroy

Once the domain controller realized the network slices, and we have validated the service status. The next step vas to verify the correct establishment of the data-plane sessions. Thus, the Spirent testers added 5k prefixes to each of the services(VRF-Blue and VRF-Red). To confirm the status of each service, we have captured the virtual routers BGP information, as depicted in Figure 7, where each VRF has 5002 prefixes received (PfxRcd).

After the end-to-end service creation and control and data plane validation, we have rebooted the devices to validate the service continuity after a simulated power failure. We validate that after the recovery was complete, all the traffic flows again between the network slices. To confirm the status of each device, we have captured the counters before and after the reboot process Figure 8. Each instance recovered sequentially, and it took up to 4.6 min for the latest service to completely restore the traffic flow.

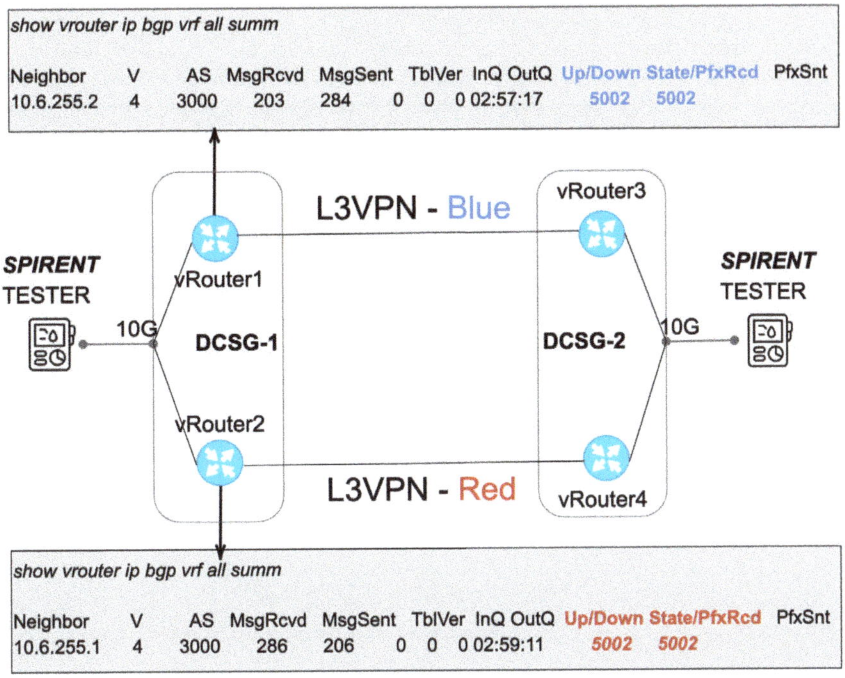

Figure 7. Comparison of the Prefixes received and Sent in each of the network slices deployed in the network. Each Network slice (Blue and Red) has 5K prefixes announced.

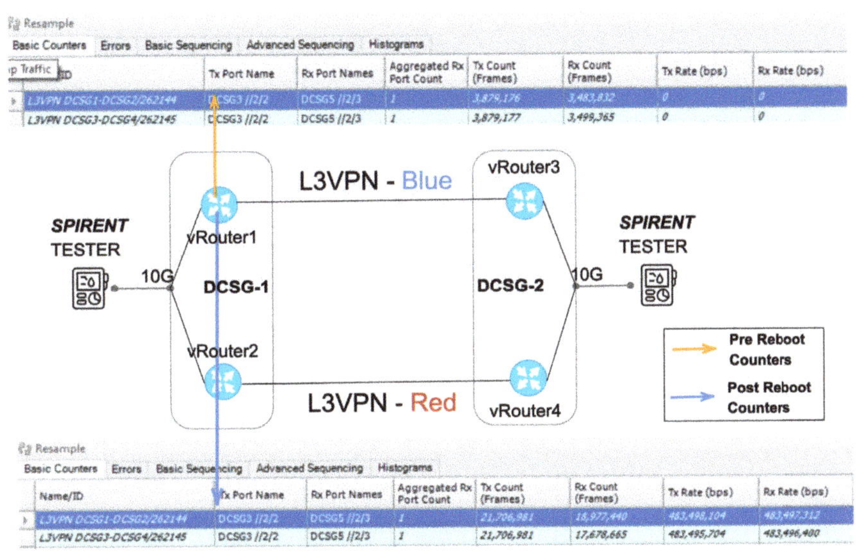

Figure 8. Depicts comparing the counters in the network slice Blue, prior (Up) and after (Down) the device reboot. Once the devices are down, the counters go down to zero; after 4.6 min and once the device is ready, the counters move up again.

5. Conclusions

Dealing with the necessity of a dynamic network resources allocation to provide a new generation of customer-tailored applications is a primary concern nowadays. In that sense, Telecom providers have to prepare their whole set of systems and network infrastructure to allow the introduction of end-to-end network automation. In that sense, this paper uses and validates the iFusion architecture defined by Telefonica, which is ready to support new use cases derived from the 5G adoption and transport network slices. Additionally, this work validates an end-to-end creation, modification and deletion of transport network slices with several degrees of isolation. Furthermore, the results indicate the feasibility of deploying multi-layer IP over DWDM transport network slices based on virtual routers and disjoint optical paths. Future work testing the map and realization of network slices in the different NSC controller positions is required. This testing would allow us to fully understand the information exchanged/stored in each layer to make feasible the deployments in real networks.

Author Contributions: Conceptualization, V.L.A., L.M.C.M., O.G.d.D. and R.V.; Data curation, J.-M.P.; Investigation, S.B. and R.V.; Software, D.G.-P. and X.L. Supervision, O.G.d.D., C.M., P.A., R.C., R.M. (Ricardo Martinez), J.P.F.-P. and R.M. (Raul Muñoz); Writing–original draft, S.B. and A.A.A.; Writing–review & editing, S.B., V.L.A., L.M.C.M., J.-M.P. and R.V. All authors have read and agreed to the published version of the manuscript.

Funding: This research was partially supported by the EC H2020 TeraFlow (101015857) and SpanishAURORAS (RTI2018-099178-I00).

Institutional Review Board Statement: Not applicable.

Informed Consent Statement: Not applicable.

Data Availability Statement: Not applicable.

Acknowledgments: Work partially supported by the EC H2020 TeraFlow (101015857) and Spanish AURORAS (RTI2018-099178-I00).

Conflicts of Interest: The authors declare no conflict of interest.

Abbreviations

3GPP	3rd Generation Partnership Project
API	Application programming interface
BSS	Business Support System
BGP	Border Gateway Protocol
CE	Customer Edge
IETF	Internet Engineering Task Force
eMBB	Enhanced Mobile Broadband
L2SM	L2VPN Service Model
L2NM	L2VPN Network Model
L3SM	L3VPN Service Model
L3NM	L3VPN Network Model
L3VPN	Layer Three Virtual Private Network
MC	Mobile Core
mMTC	Massive Machine Type Communication
MPLS	Multiprotocol Label Switching
NASS	Network As A Service
NGMN	Next Generation Mobile Networks
NSI	Network Slice Instance
ODU	Optical Distribution Unit
ONF	Open Networking Foundation
OSS	Operation Support Systems
PE	Provider Edge
QoS	Quality of service

RAN	Radio Access Network
SDN	Software Defined Network
SI	Service Instance
URLLC	Ultra Reliable Low Latency Communications
VSI	Virtual switching Interfaces
VPN	Virtual Private Network
VRF	Virtual Routing and Forwarding
YANG	Yet Another Next Generation

References

1. Ordonez-Lucena, J.; Ameigeiras, P.; Lopez, D.; Ramos-Munoz, J.J.; Lorca, J.; Folgueira, J. Network slicing for 5G with SDN/NFV: Concepts, architectures, and challenges. *IEEE Commun. Mag.* **2017**, *55*, 80–87. [CrossRef]
2. Contreras, L.M.; López, D.R. A Network Service Provider Perspective on Network Slicing. IEEE Softwarization: A Collection of Short Technical Articles. 2018. Available online: https://sdn.ieee.org/newsletter/january-2018/a-network-service-provider-perspective-on-network-slicing?highlight=WyJhIiwiJ2EiLCJuZXR3b3JrIiwibmV0d29yaydzIiwiJ25ldHdvcmsiLCJzZXJ2aWNlIiwic2VydmljZScsIiwiJ3NlcnZpY2UiLCJwcm92aWRlciIsInByb3ZpZGVyJ3MiLCJhIG5ldHdvcmsiLCJhIG5ldHdvcmsgc2VydmljZSIsIm5ldHdvcmsgc2VydmljZSIsIm5ldHdvcmsgc2VydmljZSBwcm92aWRlciIsInNlcnZpY2UgcHJvdmlkZXIiXQ== (accessed on 1 July 2021).
3. Barguil, S.; de Dios, O.G.; Boucadair, M.; Munoz, L.; Jalil, L.; Ma, J. A Layer 2 VPN Network Yang Model. draft-ietf-opsawg-l2nm-02. Available online: https://datatracker.ietf.org/doc/html/draft-ietf-opsawg-l2nm-02 (accessed on 1 July 2021).
4. Barguil, S.; de Dios, O.G.; Boucadair, M.; Munoz, L.; Aguado, A. A Layer 3 VPN Network YANG Model. draft-ietf-opsawg-l3sm-l3nm-05. Available online: https://datatracker.ietf.org/doc/html/draft-ietf-opsawg-l3sm-l3nm (accessed on 1 July 2021).
5. Vilalta, R.; Muñoz, R.; Landi, G.; Rodriguez, L.; Capitani, M.; Casellas, R.; Martínez, R. Experimental Demonstration of the BlueSPACE's NFV MANO Framework for the Control of SDM/WDM-enabled Fronthaul and Packet-based Transport Networks by Extending the TAPI. In Proceedings of the IEEE 2018 European Conference on Optical Communication (ECOC), Rome, Italy, 23–27 September 2018; pp. 1–3.
6. Shaikh, A. OpenConfig-progress toward vendor-neutral network management. In Proceedings of the North American Network Operators' Group (NANOG) 71, San Jose, CA, USA, 2–4 October 2017.
7. Alcalá, A.; Barguil, S.; López, V.; Contreras, L.M.; Manso, C.; Alemany, P.; Casellas, R.; Martínez, R.; Gonzalez-Perez, D.; Liu, X.; et al. Multi-layer Transport Network Slicing with Hard and Soft Isolation. In Proceedings of the 2021 Optical Fiber Conference (OFC), Virtual Event, 6–11 June 2021.
8. Alliance, N. Description of network slicing concept. *NGMN 5G P* **2016**. Available online: https://www.ngmn.org/publications/description-of-network-slicing-concept.html (accessed on 1 July 2021).
9. Kukliński, S.; Tomaszewski, L. Business models of network slicing. In Proceedings of 2018 9th International Conference on the Network of the Future (NOF), Poznan, Poland, 19–21 November 2018; pp. 19–21.
10. Zhang, N.; Liu, Y.F.; Farmanbar, H.; Chang, T.H.; Hong, M.; Luo, Z.Q. Network slicing for service-oriented networks under resource constraints. *IEEE J. Sel. Areas Commun.* **2017**, *35*, 2512–2521. [CrossRef]
11. Khan, L.U.; Yaqoob, I.; Tran, N.H.; Han, Z.; Hong, C.S. Network slicing: Recent advances, taxonomy, requirements, and open research challenges. *IEEE Access* **2020**, *8*, 36009–36028. [CrossRef]
12. Foukas, X.; Patounas, G.; Elmokashfi, A.; Marina, M.K. Network slicing in 5G: Survey and challenges. *IEEE Commun. Mag.* **2017**, *55*, 94–100. [CrossRef]
13. Huang, S.; Guo, B.; Liu, Y. 5G-Oriented optical underlay network slicing technology and challenges. *IEEE Commun. Mag.* **2020**, *58*, 13–19. [CrossRef]
14. Rosen, E.; Rekhter, Y. BGP/MPLS IP Virtual Private Networks (VPNs). RFC 4364. Available online: https://www.rfc-editor.org/info/rfc8049 (accessed on 1 July 2021). doi:10.17487/RFC4364. [CrossRef]
15. Litkowski, S.; Tomotaki, L.; Ogaki, K. YANG Data Model for L3VPN Service Delivery. RFC 8049. Available online: https://www.rfc-editor.org/info/rfc4364 (accessed on 1 July 2021). doi:10.17487/RFC8049. [CrossRef]
16. Marotta, A.; Cassioli, D.; Tornatore, M.; Hirota, Y.; Awaji, Y.; Mukherjee, B. Reliable slicing with isolation in optical metro-aggregation networks. In Proceedings of the IEEE 2020 Optical Fiber Communications Conference and Exhibition (OFC), San Diego, CA, USA, 8–12 March 2020; pp. 1–3.
17. Contreras, L.M.; Barguil, S.; Vilalta, R.; López, V. Architecture for integrating vertical customer's programmability control of network functions and connectivity in a slice-as-a-service schema. *EURASIP J. Wirel. Commun. Netw.* **2021**, *2021*, 1–16. [CrossRef]
18. 3GPP TS 23.501. System Architecture for the 5G System (3GPP TS 23.501 Version 15.3. 0 Release 15), September 2018. Available online: https://www.etsi.org/deliver/etsi_ts/123500_123599/123501/15.03.00_60/ts_123501v150300p.pdf (accessed on 1 July 2021).
19. Redana, S.; Bulakci, Ö.; Zafeiropoulos, A.; Gavras, A.; Tzanakaki, A.; Albanese, A.; Kousaridas, A.; Weit, A.; Sayadi, B.; Jou, B.T.; et al. 5G PPP Architecture Working Group: View on 5G Architecture. 2019. Available online: https://bura.brunel.ac.uk/bitstream/2438/18546/1/FullText.pdf (accessed on 1 July 2021).

20. Ferrús, R.; Sallent, O.; Pérez-Romero, J.; Agusti, R. On the automation of RAN slicing provisioning and cell planning in NG-RAN. In Proceedings of the IEEE 2018 European Conference on Networks and Communications (EuCNC), Ljubljana, Slovenia, 18–21 June 2018; pp. 37–42.
21. Contreras, L.; González, Ó.; López, V.; Fernández-Palacios, J.; Folgueira, J. iFUSION: Standards-based SDN Architecture for Carrier Transport Network. In Proceedings of the 2019 IEEE Conference on Standards for Communications and Networking (CSCN), Granada, Spain, 28–30 October 2019; pp. 1–7.
22. Bierman, A.; Bjorklund, M.; Watsen, K. RESTCONF protocol. RFC 8040. Available online: https://www.rfc-editor.org/info/rfc8040 (accessed on 1 July 2021). doi: 10.17487/RFC8040. [CrossRef]
23. Enns, R.; Bjorklund, M.; Schoenwaelder, J.; Bierman, A. Network configuration protocol (NETCONF). RFC 6241. Available online: https://www.rfc-editor.org/info/rfc6241 (accessed on 1 July 2021). doi:10.17487/RFC6241. [CrossRef]
24. Vilalta, R.; Muñoz, R.; Casellas, R.; Martínez, R.; López, V.; González-de-Dios, O.; Pastor, A.; Katsikas, G.; Klaedtke, F.; Monti, P.; et al. TeraFlow: Secured Autonomic Traffic Management for a Tera of SDN Flows. In Proceedings of the IEEE 2021 European Conference on Networks and Communications (EuCNC), Virtual Event, 8–11 June 2021.
25. Rokui, R.; Homma, S.; Makhijani, K.; Contreras, L.; Tantsura, J. Definition of IETF Network Slices. draft-ietf-teas-ietf-network-slice-definition-01. Available online: https://datatracker.ietf.org/doc/html/draft-ietf-teas-ietf-network-slice-definition (accessed on 1 July 2021).
26. Gray, E.; Drake, J. Framework for IETF Network Slices. draft-ietf-teas-ietf-network-slice-framework-00. Available online: https://datatracker.ietf.org/doc/html/draft-ietf-teas-ietf-network-slice-framework (accessed on 1 July 2021).
27. Available online: http://www.claise.be/YANGPageMain.html (accessed on 3 April 2021).
28. Wen, B.; Fioccola, G.; Xie, C.; Jalil, L. A YANG data model for layer 2 virtual private network (L2VPN) service delivery. RFC 8466. Available online: https://www.rfc-editor.org/info/rfc8466 (accessed on 1 July 2021). doi:10.17487/RFC8466. [CrossRef]
29. Saad, T.; Gandhi, R.; Liu, X.; Beeram, V.; Bryskin, I. A YANG Data Model for Traffic Engineering Tunnels, Label Switched Paths and Interfaces. draft-ietf-teas-yang-te-25. Available online: https://datatracker.ietf.org/doc/html/draft-ietf-teas-yang-te-25 (accessed on 1 July 2021).
30. Liu, X.; Bryskin, I.; Beeram, V.; Saad, T.; Shah, H.; de Dios, O.G. YANG Data Model for Traffic Engineering (TE) Topologies. RFC 8795. Available online: https://www.rfc-editor.org/info/rfc8795 (accessed on 1 July 2021). doi:10.17487/RFC8795. [CrossRef]
31. Liu, X.; Bryskin, I.; Beeram, V.; Saad, T.; Shah, H.; de Dios, O.G. YANG Data Model for Traffic Engineering (TE) Topologies. draft-ietf-teas-yang-l3-te-topo-10. Available online: https://datatracker.ietf.org/doc/html/draft-ietf-teas-yang-l3-te-topo (accessed on 1 July 2021).
32. Lee, Y.; Dhody, D.; Fioccola, G.; Wu, Q. Traffic Engineering (TE) and Service Mapping Yang Model. draft-ietf-teas-te-service-mapping-yang-07. Available online: https://datatracker.ietf.org/doc/html/draft-ietf-teas-te-service-mapping-yang-07 (accessed on 1 July 2021).
33. Jethanandani, M.; Agarwal, S. YANG Data Model for Network Access Control Lists (ACLs). RFC 8519. Available online: https://www.rfc-editor.org/info/rfc8519 (accessed on 1 July 2021). doi:10.17487/RFC8519. [CrossRef]
34. Qu, Y.; Tantsura, J.; Lindem, A.; Liu, X. A YANG Data Model for Routing Policy. draft-ietf-rtgwg-policy-model-27. Available online: https://datatracker.ietf.org/doc/html/draft-ietf-rtgwg-policy-model-27 (accessed on 1 July 2021).
35. Jethanandani, M.; Patel, K.; Hares, S.; Haas, J. BGP YANG Model for Service Provider Networks. draft-ietf-idr-bgp-model-10. Available online: https://datatracker.ietf.org/doc/html/draft-ietf-idr-bgp-model-10 (accessed on 1 July 2021).
36. Zheng, H.; Busi, I.; Guo, A.; Lopez, V. Framework and Data Model for OTN Network Slicing. draft-zheng-ccamp-yang-otn-slicing-01. Available online: https://datatracker.ietf.org/doc/html/draft-zheng-ccamp-yang-otn-slicing-01 (accessed on 1 July 2021).
37. Liu, X.; Tantsura, J.; Bryskin, I.; Contreras, L.; Wu, Q.; Belotti, S.; Rokui, R. IETF Network Slice YANG Data Model. draft-liu-teas-transport-network-slice-yang-02. Available online: https://datatracker.ietf.org/doc/html/draft-liu-teas-transport-network-slice-yang-02 (accessed on 1 July 2021).

applied sciences

MDPI

Article

Integrating Optical and Wireless Techniques towards Novel Fronthaul and Access Architectures in a 5G NR Framework

Ramon Maia Borges [1,2], Celso Henrique de Souza Lopes [1], Eduardo Saia Lima [1], Marco Aurélio de Oliveira [1], Matheus Sêda Borsato Cunha [1,2], Luciano Camilo Alexandre [1], Luis Gustavo da Silva [1], Luiz Augusto Melo Pereira [1], Danilo Henrique Spadoti [2], Murilo Araujo Romero [3] and Arismar Cerqueira Sodré Junior [1,*]

1 Laboratory WOCA, National Institute of Telecommunications (Inatel), Santa Rita do Sapucaí 37540-000, MG, Brazil; ramonmb@inatel.br (R.M.B.); celso.lopes@mtel.inatel.br (C.H.d.S.L.); elima@get.inatel.br (E.S.L.); marcoaurelio@mtel.inatel.br (M.A.d.O.); matheusseda@gee.inatel.br (M.S.B.C.); luciano.camilo@mtel.inatel.br (L.C.A.); luis.gustavo@inatel.br (L.G.d.S.); luiz_augusto@get.inatel.br (L.A.M.P.)

2 Department of Electrical Engineering, Federal University of Itajubá (UNIFEI), Itajubá 37500-903, MG, Brazil; spadoti@unifei.edu.br

3 Department of Electrical and Computer Engineering, University of São Paulo (USP), EESC/USP, São Carlos 13566-590, SP, Brazil; murilo.romero@usp.br

* Correspondence: arismar@inatel.br; Tel.: +55-35-3471-9200

Abstract: The fifth-generation of mobile network (5G) and beyond requires a radio access network (RAN) update in order to cope with the incoming increase of wireless data traffic and new applications. In this context, we propose an efficient optical-wireless architecture applied to the non-standalone (NSA) 5G new radio (NR) framework. Several distinct electrical- and optical-based fronthaul configurations combining free-space optical (FSO), wireless links, and radio over fiber (RoF) techniques were implemented and properly analyzed for selection according to network operator deployment requirements. In addition, visible light communication (VLC) was investigated as a future access network technology when immunity to electromagnetic interference is paramount. Experimental results demonstrated fourth-generation of mobile network (4G) and 5G coexistence at Gbit/s throughput and error vector magnitude (EVM) in accordance with 5G NR Release 15.

Keywords: 4G/5G; fronthaul; optical-wireless systems; VLC

Citation: Borges, R.M.; de Souza Lopes, C.H.; Lima, E.S.; de Oliveira, M.A.; Cunha, M.S.B.; Alexandre, L.C.; da Silva, L.G.; Pereira, L.A.M.; Spadoti, D.H.; Romero, M.A.; et al. Integrating Optical and Wireless Techniques towards Novel Fronthaul and Access Architectures in a 5G NR Framework. *Appl. Sci.* **2021**, *11*, 5048. https://doi.org/10.3390/app11115048

Academic Editor: Fabio Cavaliere

Received: 1 May 2021
Accepted: 24 May 2021
Published: 29 May 2021

1. Introduction

The fifth-generation of mobile network (5G) has already started to be commercially implemented around the world. At this initial implementation stage, 5G typically operates in the non-standalone (NSA) mode, aiming to take advantage of the fourth-generation (4G) infrastructure to deploy 5G new radio (5G NR) systems. Therefore, 5G and 4G share the network infrastructure and must coexist seamlessly [1]. Phase 1 of 5G systems is based on the 3rd Generation Partnership Project (3GPP) Release 15, which defines the 5G NR standard by focusing on the enhanced mobile broadband (eMBB) scenario. Technical solutions explored at this stage include radio equipment design, use of additional spectral bands encompassing millimeter-waves (mm-waves), flexible resource allocation, and multiple input multiple output (MIMO) schemes [2,3]. In addition, 5G deployment requires a radio access network (RAN) update in order to cope with the planned increase in wireless data traffic. Trends in network planning point to optical-wireless convergence, mobile dense heterogeneous networks (HetNet), centralized RAN (C-RAN), and the so-called "Xhaul", which integrates backhaul (BH), midhaul (MH), and fronthaul (FH) transport networks [4–6].

Recently, 3GPP completed Release 16 for phase 2 of 5G systems, which considers ultra-reliable low latency communication (URLLC) and massive machine-type communication (mMTC) applications. According to the release, the technical solutions to be explored include integrated access and backhaul (IAB), industrial Internet of Things (IoT), satellite access, and NR-based access to unlicensed spectrum [7]. Releases 17 and 18 for 5G are currently in production and scheduled to end in 2021, with the goal of further enhancing mobile systems [8]. By the same token, planning of sixth-generation (6G) communications has already begun, focused on providing performance superior to 5G and satisfying future demands as far as 2030. Among the potential technologies for 6G, optical-wireless convergence continues to play an important role for future RANs, whereas terahertz and visible light communications (VLC) have also emerged [9,10].

In this context, a key area of study is microwave photonics (MWP), which takes advantage of optical and mobile communications for signal distribution in a heterogeneous architecture [11,12]. MWP techniques have been strongly applied to radiofrequency (RF) generation, processing, detection, transport, and distribution. Notably, radio over fiber (RoF) technology enables transportation and distribution of digital and/or analog signals among central office (CO) and remote radio units (RRUs) via fiber-optic links, giving rise to the fiber-wireless (FiWi) systems [11,13]. From a C-RAN point of view, HetNet in conjunction with picocell and femtocell implementations allows system coverage and data rate to be enhanced. In C-RAN architecture, the backhaul and fronthaul links connect CO to the core network, typically carrying digital user data and CO to RRUs, respectively.

Fifth-generation systems are mostly based on digital radio over fiber (D-RoF) using the common radio public interface (CPRI) protocol, in which radio signals are sampled and directly digitalized into baseband data at the remote radio head (RRH) from the cell site. Millimeter-wave operation requires high-speed A/D (analog-to-digital) and D/A (digital-to-analog) converters, creating a bandwidth bottleneck in uncompressed CPRI. To address this issue, two main network alternatives have been actively investigated. One possibility is functional split (FS), widespread by 3GPP under e-CPRI. The FS establishes that some network functionalities be carried out at RRHs, thereby alleviating the transmission rate requirements for the 5G fronthaul. Specifically, 3GPP has defined a series of options for functional split between BBU and RRHs. In short, functional splits decentralize control functionalities in such way as to establish a trade-off between reducing fronthaul throughput and increasing latency. The other option is analog fronthaul (A-RoF), in which the RF wireless signal modulates an optical carrier that is optically distributed. Because there is no high-speed A/D conversion, the system is far less complex and the overall latency is much smaller, dictated essentially by the optical fiber length.

Our research group has been intensely investigating A-RoF solutions for multiband and Gbit/s 5G systems in recent years [14–17]. The solution presented in [14] takes advantage of an operating gigabit passive optical network (GPON) from a local Internet service provider to distribute 5G-like signals, thus making use of an existing network infrastructure for 5G fronthaul in accordance with C-RAN. The approaches presented in [15,16] explore optical link not only for data transmission but also as a medium to perform photonics-assisted RF amplification. In [17], a dual-band wireless fronthaul using an FSS-based focal point/Cassegrain antenna assisted by an optical midhaul was demonstrated.

Alternatively, free-space optical (FSO) technology has also shown potential for 5G transport network deployment using the unlicensed terahertz bands [18,19]. For instance, Mufutau et al. successfully reported a hybrid RoF/FSO fronthaul approach in which 5G coexists with 4G by means of coarse wavelength division multiplexing (CWDM) [19]. These previous works considered wireless links operating at a variety of radiofrequencies for providing mobile access, which might be either indoor or outdoor. Overall, RF links correspond to the main technical solution employed for access, although visible light communications (VLC) have also been recognized as an alternative for indoor environments, particularly when RF electromagnetic interference must be avoided [20,21].

Apart from a tiny segment of the worldwide market already using VLC, wireless access networks are almost entirely based on electrical solutions, with approximately 70%

of macrosites in the world being connected by microwave fronthauls. Although there are many published works investigating individual proposals regarding fronthaul solutions, only a few of them present comparisons between different optical-based fronthaul solutions [22,23]. To the best of our knowledge, work addressing and confronting diverse electrical- and optical-based fronthauls is still missing in the specialized literature.

In this context, the current study relies on integrating optical and wireless techniques towards novel fronthaul and access architectures in a NSA 5G NR framework, as proposed in Figure 1. In other words, we report implementations and comparison of diverse MWP techniques towards a non-standalone 5G NR optical-wireless communications (OWC) architecture in such a way that a variety of integrated services may be offered according to the network operator needs.

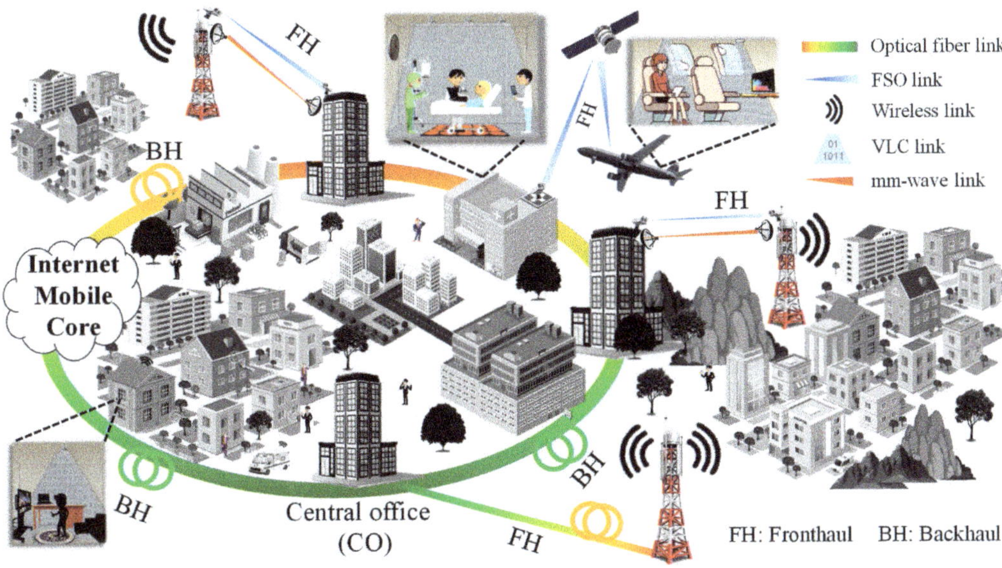

Figure 1. Optical-wireless communication system based on RoF/FSO/wireless fronthaul configurations combined to VLC/wireless access for 5G and beyond.

The main contributions of this study are as follows: proposal and implementation of optical-wireless architectures applied to the NSA 5G NR framework; evaluation of diverse transport and access network solutions, including 5G NR dual wireless and FSO-based fronthaul, 4G/5G NR fiber-wireless fronthaul, combined fiber-optic and FSO-based 5G fronthaul, and 4G/5G NR VLC-based access network; discussion of experimental results regarding 4G and 5G coexistence at Gbit/s throughput and error vector magnitude (EVM) in accordance with 3GPP Release 15.

The focus of our work is data transmission in the physical layer, although Figure 1 provides a unifying view of our applications in a single ecosystem. The rest of the paper is structured as follows. In Section 2, we address different fronthaul configurations. First, in Section 2.1, a dual wireless and FSO-based fronthaul that enables a backup link for high-reliability services is reported. In Section 2.2, a RoF-based fronthaul is demonstrated for simultaneous RF signal distribution at distinct spectral bands. Unlike previous works [7,8], a single wavelength is used to simultaneously cover 4G and 5G NR applications, including millimeter-wave bands. Next, in Section 2.3, we discuss a RoF/FSO-based fronthaul that enables high-capacity fronthaul extension for regions where optical fibers cannot be deployed all the way. A VLC-based access network for indoor applications is presented in Section 3, targeting applications in which immunity to electromagnetic interference is crucial. Section 4 compares the obtained results to related works from the literature. Conclusions, final remarks, and future works are outlined in Section 5.

2. Fronthaul Solutions for 4G/5G

This section reports on the architectures and experimental results for the three fronthaul configurations illustrated in Figure 1, namely 5G NR dual wireless and FSO-based fronthaul, 4G/5G NR FiWi-based fronthaul, and combined fiber-optic and FSO-based 5G fronthaul.

2.1. The 5G NR Dual Wireless and FSO-Based Fronthaul

Figure 2a depicts a block diagram of the dual wireless and FSO-based fronthaul architecture, in which the two parallel data transmission paths enable link dynamic selection according to transmission performance at a given instant, while the other link serves as a backup. The FSO and RF links were implemented at 1550 nm and 38 GHz, respectively. The hybrid FSO/RF system was analyzed under the 5G NR standard operating with 400 MHz bandwidth and 120 kHz between orthogonal frequency division multiplexing (OFDM) subcarriers for quadrature phase shift keying (QPSK) and 16- and 64-quadrature amplitude modulation (QAM).

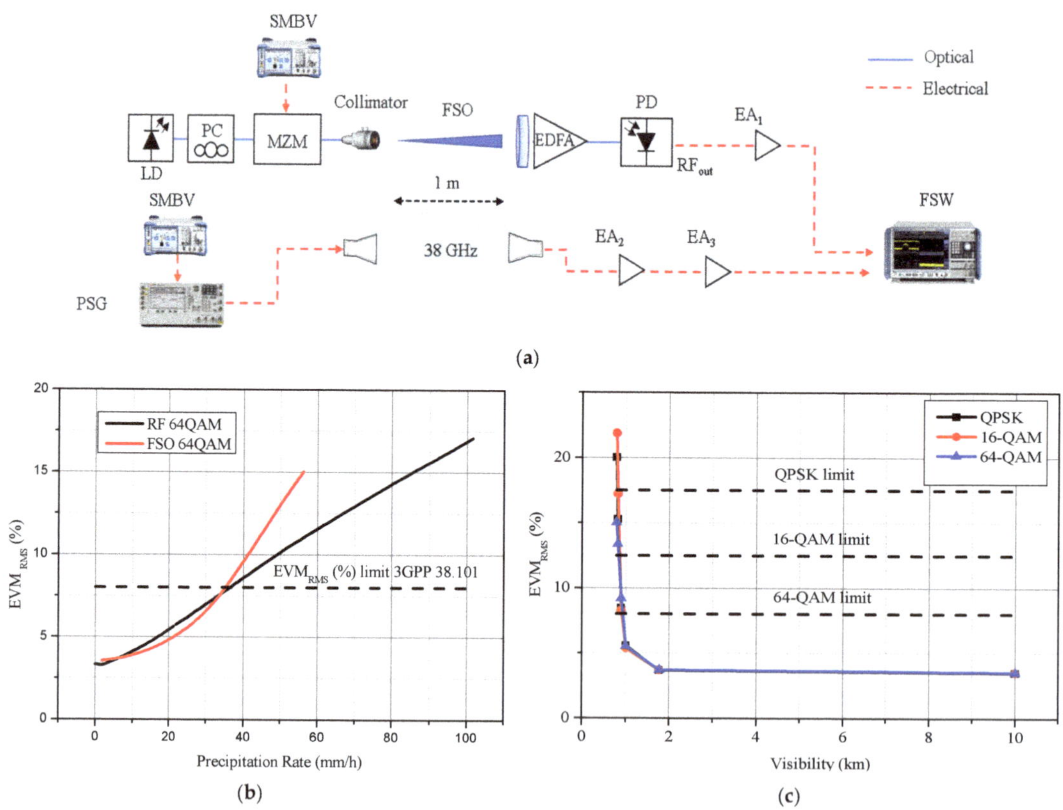

Figure 2. The 5G NR dual wireless and FSO-based fronthaul. (**a**) Block diagram of the proposed hybrid wireless-FSO system. From the left, the signal generated by the vector signal generator (VSG) SMVB100A reaches the FSW signal and spectrum analyzer via the two different links. (**b**) Dependence of hybrid FSO/RF EVM_{RMS} as a function of rain precipitation rate. (**c**) Dependence of hybrid FSO/RF EVM_{RMS} as a function of visibility.

In the experiments, a SMVB100A vector signal generator (VSG) was responsible for generating 5G NR signal at 1 GHz with 400 MHz bandwidth for the QPSK, 16-QAM, and 64-QAM schemes. In the FSO link, the modulated RF signal was inserted into a single-drive Mach–Zehnder modulator (MZM) based at the quadrature point. A distributed feedback (DFB) laser at 1550 nm provided the optical carrier, which fed the MZM after passing through a polarization controller (PC). The MZM output was coupled to a collimator to prepare the optical beam for transmission in free space. The transmitted optical beam was then captured by a M-5X objective lens and coupled to a single-mode fiber (SMF). The received signal was amplified by an Erbium-doped fiber amplifier (EDFA) and then photodetected by a p-intrinsic-n (PIN) broadband photodetector (u^2t XPDV2120RA). Before demodulation in the electrical spectrum analyzer (FSW), the RF signal was amplified by a low-noise amplifier (LNA, EA_1). As in references [24] and [25], an FSO link of a few meters was taken as a proof-of-concept for 5G networks to assure an eye-safe indoor environment. As shown below, real-life performance was emulated by varying the optical power reaching the optical receiver.

Meanwhile, for the RF link, the same 5G NR signal, once again generated by a SMVB100A VSG, was used to drive the PSG E8267D equipment, which upconverted the 5G NR frequency to 38 GHz. Posterior, its output was connected to a 25 dBi horn antenna, and an identical antenna was used at the receiver side with the same height and polarization. A two-stage amplification (EA_2 and EA_3) scheme was employed before demodulation.

The performance of the hybrid FSO/RF is presented in Figure 2b,c through EVM_{RMS} measurements at different power levels at the receiver, emulating equivalent attenuation levels according to visibility and rain conditions. These power levels at the receiver ranged from -41.9 to -7.4 dBm. Regarding the effect of rain attenuation, the attenuation levels used for the FSO link ranged from 0 to 56 mm/h, while the values ranged from 0 to 102 mm/h for the wireless link. For the specific case of 64-QAM, the 3GPP EVM_{RMS} limit was reached at about the same precipitation intensity for both FSO and 38 GHz wireless links (36 mm/h, Figure 2b). However, as shown in Figure 2b, while operating in a rainy environment, the wireless link suffered less significant performance degradation as the rain intensity increased.

Figure 2c depicts the dependance of EVM regarding visibility for the FSO link for three distinct modulation formats. In every case, successful transmission within the 3GPP standards required a visibility above 780 m. In contrast, the visibility issue was much less severe for the wireless link. In fact, our experimental results demonstrated a 1.61 Gbps 64-QAM throughput, which is within the 3GPP EVM requirements, as long as the rain intensity was less than 36.4 mm/h (see Figure 2a) and the visibility exceeded 40 m.

2.2. 4G/5G NR Fiber-Wireless (FiWi) Fronthaul

The long-term evolution-advanced (LTE-A) and 5G NR signals have similar physical channel configurations, including primary (P-SS) and secondary synchronization signal (S-SS), physical transmission channel (PBCH), demodulation reference (DMRS), and shared physical data channel (PDSCH). Different modulation formats are used for the data and control channels. For instance, data signals might be modulated in 64-QAM format, while binary phase shift keying (BPSK) modulation is applied for modulating the P-SS and S-SS channels. In addition, QPSK and 64-QAM are utilized for PBCH-DMRS and PDSCH signals, respectively [19].

The EVM_{RMS} performance of the FiWi system was investigated for the eRAC scenario operating in the 700 MHz frequency range. Figure 3a displays a block diagram of the experimental arrangement of the proposed FiWi system, which employs an A-RoF configuration based on external modulation and direct detection followed by a link access for increased reach.

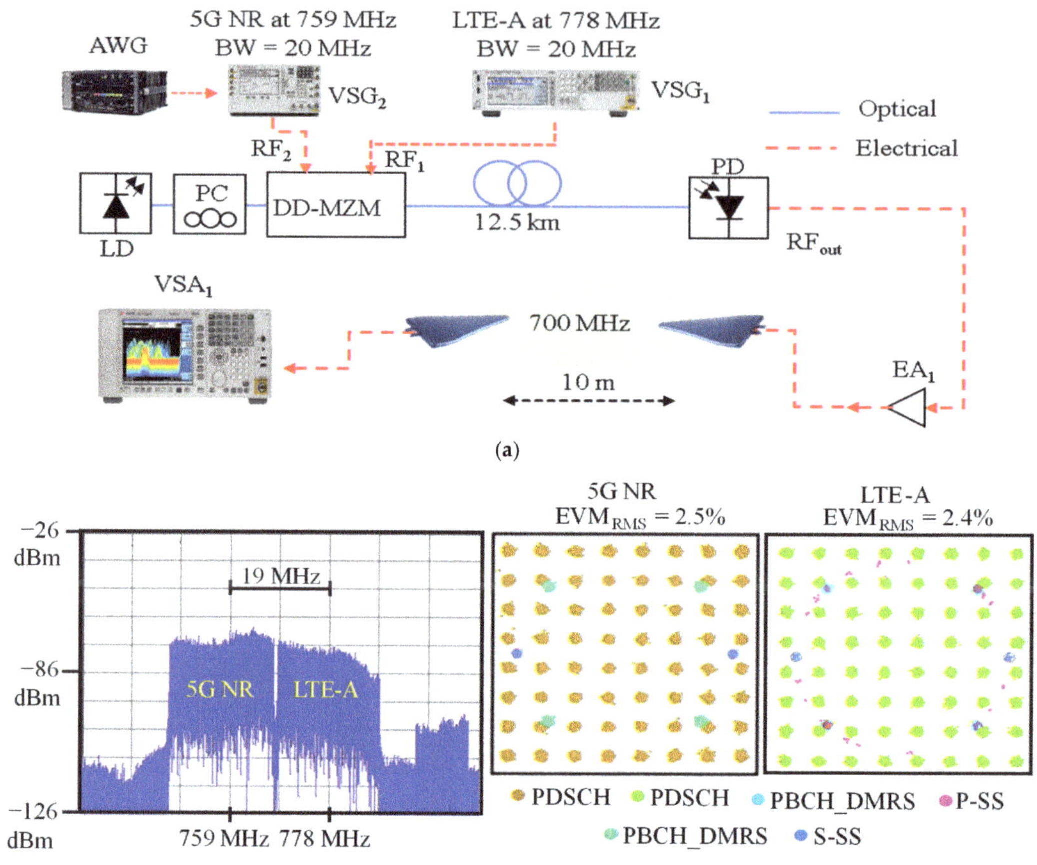

Figure 3. The 4G/5G NR fiber-optic-based fronthaul architecture. (**a**) Block diagram; (**b**) coexistence analysis between 4G and 5G technologies in the optical-wireless system.

An EXG generator (N5171B) from Keysight generated the LTE-A signal (VSG_2) at 778 MHz, which was configured offline using the Keysight Signal Studio software for generating signals with 20 MHz bandwidth and 64-QAM. The 20 MHz baseband 5G NR signal was also configured by means of the Signal Studio software in conjunction with an arbitrary waveform generator (AWG M9505A). Next, the PSG E8267D converted the baseband 5G NR signal to the 759 MHz band (VSG_1). This approach allowed the transmission of the LTE-A and 5G NR signals coexisting in adjacent channels, thereby making use of the non-standalone transmission mode proposed by 3GPP in Release 15 [3].

A dual-drive Mach–Zehnder modulator (DD-MZM) modulated an optical carrier provided by a laser at 1560 nm with two distinct RF signals. The optical beam was then launched into a single-mode fiber with 0 dBm of optical power. Nonlinearities were not observed in our approach as the optical power level in the SMF did not reach the power thresholds leading to nonlinear effects [26]. Regarding the electrical power levels, RF_1 and RF_2 signals were transmitted at −5 dBm, while the PIN photodetector (EOT ET-5000F) received a fixed optical power of −3 dBm. After photodetection, the signals in the electrical domain were amplified by about 44 dB (EA_1) to compensate for the overall system losses. Two log-periodic broadband antennas with 5 dBi gain were used to transmit and receive both RF_1 and RF_2, and a 10 m wireless transmission was performed as a proof-of-concept. At the receiver side, a Keysight MXA N9020A vector signal analyzer (VSA_1) was used to demodulate the received signals and provide the EVM_{RMS} measurement results.

In the experiments, a 19 MHz frequency offset was employed between RF_1 and RF_2. This offset was determined as a preliminary step in order to find the minimum frequency offset allowing seamless coexistence between the RF_1 and RF_2 signals. Specifically, we measured the EVM_{RMS} as a function of the central frequency offset between the LTE-A and 5G NR signals. The LTE-A central frequency was kept at 778 MHz central frequency, whereas the 5G NR central frequency was varied from 760 to 758 MHz. For a 18 MHz frequency offset, there was a spectral overlap of 2 MHz between the signals, resulting in EVM_{RMS} higher than 19%. This exceeds the limits specified by 3GPP [27], making the system unsuitable for 64-QAM transmission. In contrast, the 19 MHz offset was enough to assure an EVM_{RMS} value below 3%. For this offset, there was still a spectral overlap of 1 MHz between the signals. However, this overlap occurred in the lateral lobes and did not significantly affect the quality of the received signal after the optical fronthaul.

Having established an offset of 19 MHz, the optical-wireless system based on non-standalone mode was implemented using a RF power level of −5 dBm for the RF_1 and RF_2 signals and an optical power of −3 dBm at the input of the photodetector. The RF signals were configured with 19 MHz frequency offset, 64-QAM, and 20 MHz bandwidth, thereby achieving a maximum throughput of 182.8 Mbit/s for the 5G NR and LTE-A signals. Figure 3b reports the measured spectrum and detected constellations for the LTE-A and 5G NR signals. One can clearly distinguish each synchronism, control, and data symbols transmitted as the obtained EVM_{RMS} were 2.5% and 2.4% for 5G NR and LTE-A, respectively. Furthermore, there were no noticeable phase rotation distortions in the detected constellations after the fiber-wireless transmission, thus demonstrating successful and seamless coexistence between 5G NR and LTE-A signals.

2.3. *Combined Fiber-Optic and FSO-Based 5G Fronthaul*

The combined fiber-optic and FSO-based 5G fronthaul followed by wireless access extension is described in a block diagram in Figure 4a. The system encompasses RoF, FSO, and wireless technologies in a single architecture that takes advantage of a hybrid RoF/FSO link for fronthaul integration followed by dual-band 5G NR wireless access. The FSO link enables last-mile applications when there are restrictions to the placement of a fiber-optic connection, thereby increasing system flexibility. After photodetection, further reach is achieved by wireless access.

For the experiments, we simultaneously transmitted two RF signals in the hybrid configuration in accordance with the frequency bands standardized by 3GPP Release 15. A 100 MHz bandwidth 5G NR signal at 3.5 GHz was generated using the VSG_1, whereas a 400 MHz bandwidth *M*-QAM signal was generated in baseband using an AWG and upconverted to 26 GHz by means of the VSG_2. Finally, a diplexer combined both signals, resulting in 3 dBm electrical power at the optical modulator. An OS-TL-D-C-50-200-1-S-FA DFB laser from Golight generated a 13 dBm optical carrier at 1550 nm, which was modulated by the RF-driven signals discussed above using a single-drive MZM. The modulated optical carrier was transmitted throughout a 12.5 km fiber-optic fronthaul to the collimator, and then, as a proof-of-concept, it followed a 1 m long FSO fronthaul in such a way as to assure an eye-safe FSO link.

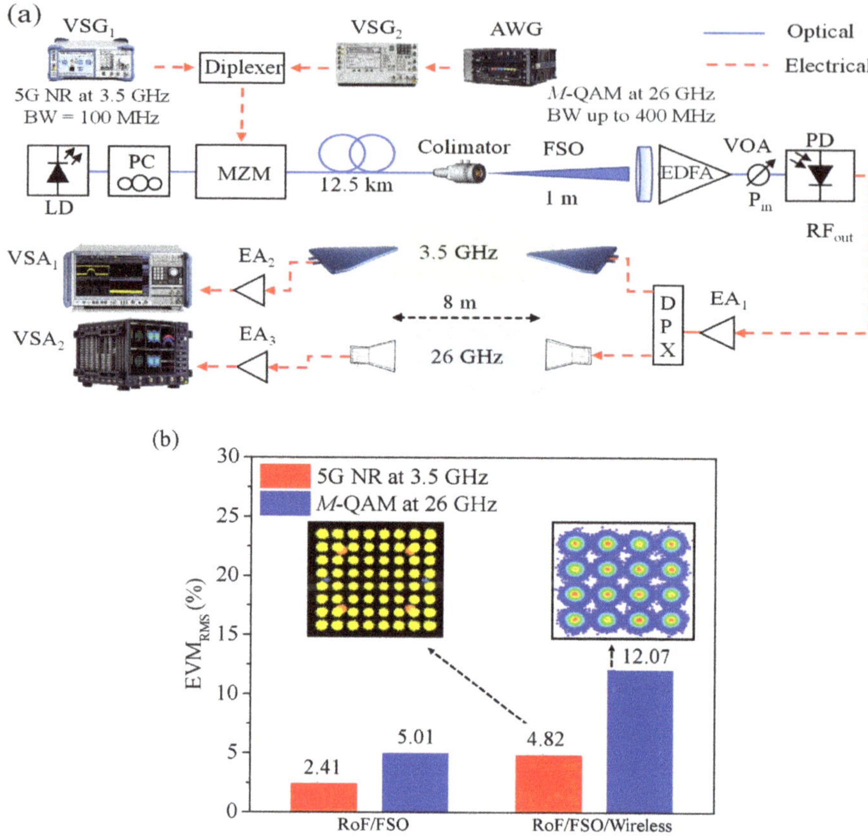

Figure 4. Combined fiber-optic and FSO-based 5G fronthaul architecture. (**a**) Block diagram; (**b**) performance analysis of 5G NR and *M*-QAM signals in the optical-wireless system. The constellations from the RoF/FSO/wireless approach show 16- and 64-QAM received symbols suitable for demodulation.

At the receiver side, an optical lens with a 5-fold collimation factor coupled the free-space optical beam into an optical patch cord. A 3-axis micropositioner was used to minimize vibration and misalignments losses. Next, the optical signal was amplified by an EDFA, whereas a variable optical attenuator (VOA) and an optical power monitor (OPM) were used for controlling and monitoring the optical power at the input of the PIN broadband photodetector from u^{2}t. It is worth highlighting that the EDFA noise was considered in the approach as there was no filter before the photodetector. If the EDFA amplified spontaneous emission (ASE) had been filtered out, an enhanced performance could have been achieved. After photodetection, the resulting RF signals were amplified by the electrical amplifier EA_1 by about 24 dB. A diplexer was used to separate the 3.5 and 26 GHz signals. Two 5 dBi gain log-periodic antennas were used for transmitting and receiving the 5G NR signal at 3.5 GHz. In parallel, two 25 dBi gain horn antennas enabled 8 m of mm-wave wireless link at 26 GHz aimed at indoor applications. At the receiver side, the 5G signal was amplified (EA_2) by 20 dB and evaluated employing an FSW-8351VSA (VSA_1). Meanwhile, the 400 MHz bandwidth *M*-QAM signal was amplified (EA_3) by 35 dB and evaluated using a DSAZ632A high-frequency oscilloscope (VSA_2).

The system performance was investigated by assessing the hybrid architecture in two distinct scenarios in terms of EVM_{RMS} and in accordance with 3GPP Release 15 specifications. The first step was to evaluate the hybrid RoF/FSO fronthaul at the photodetector output, whereas the second step consisted of actually analyzing the dual-band 5G NR wireless system employing the hybrid RoF/FSO fronthaul, as depicted in Figure 4b. It is worth mentioning that the experimental results were achieved using 3 dBm optical power at the photodetector input for both scenarios. In particular, for the 5G NR signal at 3.5 GHz, we selected 64-QAM, which resulted in 578 Mbit/s and measured EVM_{RMS} of around 2.4%, thereby satisfying the 3GPP requirements (8%) by a wide margin. Likewise, the received 400 MHz bandwidth signal at 26 GHz was within the 3GPP recommendations, providing 2.4 Gbit/s throughput. The increased EVM_{RMS} value (5%) for mm-wave transmission was due to the RF cables, SD-MZM, and photodetector frequency responses at the mm-wave band. Overall, the proposed hybrid RoF/FSO fronthaul provided 3 Gbit/s of total throughput, which is within the 3GPP requirements for EVM_{RMS}, thereby enabling further network reach by wireless access.

The second analysis consisted of evaluating the combined fiber-optic and FSO-based 5G fronthaul followed by the 8 m wireless access, as reported in Figure 4b. As expected, the wireless channel impaired the transmitted signal in both phase and magnitude as could be observed by the received symbols in the RoF/FSO/wireless constellations. Nevertheless, the 5G NR signal attained the EVM_{RMS} requirements with 3.2% margin, which may be explored for either increasing the signal data rate or extending the wireless reach. In contrast, the propagation conditions at 26 GHz critically impaired the 400 MHz bandwidth signal, and hence the EVM_{RMS} did not meet the 3GPP recommendations. As an alternative, we employed a 200 MHz bandwidth under 16-QAM to overcome this performance shortcoming. With this bandwidth reduction, the hybrid dual-band 5G NR system achieved an EVM_{RMS} value of around 12%, which is still within the 3GPP limit for 16-QAM (12.5%), due to the severe propagation impairments at mm-waves. In spite of the bandwidth reduction at 26 GHz, the hybrid RoF/FSO/wireless 5G NR system was able to provide 1.4 Gbit/s total throughput, thus fulfilling the Release 15 recommendations and demonstrating feasibility of the system while giving the remarkable additional advantage of FSO fronthaul flexibility.

3. 4G/5G NR VLC-Based Access Network

VLC has emerged as an alternative for transmitting the 5G NR standard or other wireless signals in environments where electromagnetic interferences must be minimal, such as in hospitals, petrochemical industrial plants, and airplanes. Furthermore, LED-based VLC systems can take advantage of environment lighting to provide free-space data transmission using wavelengths in the visible part of the optical spectrum. The LED is directly modulated, and the optical link operates in IM-DD mode at the photodiode [28]. By using an LED, the wide light beam divergence allows the desired area to be fully eliminated, thereby enabling simultaneous access for multiple devices and/or patterns of data traffic [21]. In addition, carrier aggregation strategies can be employed to increase the effective user throughput.

Figure 5a depicts the experimental set-up diagram of the LED-based VLC access point for simultaneous transmission of the 5G NR standard signal and an LTE-A three-band signal. RGBA (red, green, blue, and amber) LEDs were considered as the light source because they provide larger modulation bandwidth in comparison to white commercial LEDs [29]. In particular, the final selection of red LED (623 nm optical carrier) was dictated by the manufacturer's datasheet, which indicates that this wavelength offers the best trade-off between output power and quantum efficiency. In the experiments, the 5G NR signal was generated using a SMBV100B vector signal generator (VSG_1), while a PSGE8267D vector signal generator (VSG_2) and a M9505A arbitrary waveform generator (AWG) were employed to generate the LTE-A three-band signal. The RF signals were electrically combined and directly modulated the red LED by means of a bias tee. The LED light beam directivity was enhanced by a reflector and launched in a free-space link of 0.75 m.

At the receiver side, a plane convex lens with the diameter and focal distance of 25.4 and 35.0 mm, respectively, was used to focus the incident beam at the input of the PIN photodetector (EOT ET-2030). Next, an RLC equalizer similar to the one described in [29] was implemented to compensate for the LED frequency response. A DC blocker ensured that the DC component from the OE conversion did not reach the VSA. Finally, a cascade of four electrical amplifiers (EAs) assured a 50 dB gain in the received signal.

The system performance was evaluated by EVM_{RMS} measurements. The 5G NR signal was generated with a bandwidth of 10 MHz, 64-QAM, centered at 46 MHz. The configuration of the LTE-A signal represented the aggregation of three intra-band carriers in order to triplicate the effective throughput of a single user. Specifically, the LTE-A signal was generated with three sub-bands of 12 MHz, 64-QAM, at 57, 69, and 81 MHz, named as 64-QAM_1, 64-QAM_2, and 64-QAM_3, respectively.

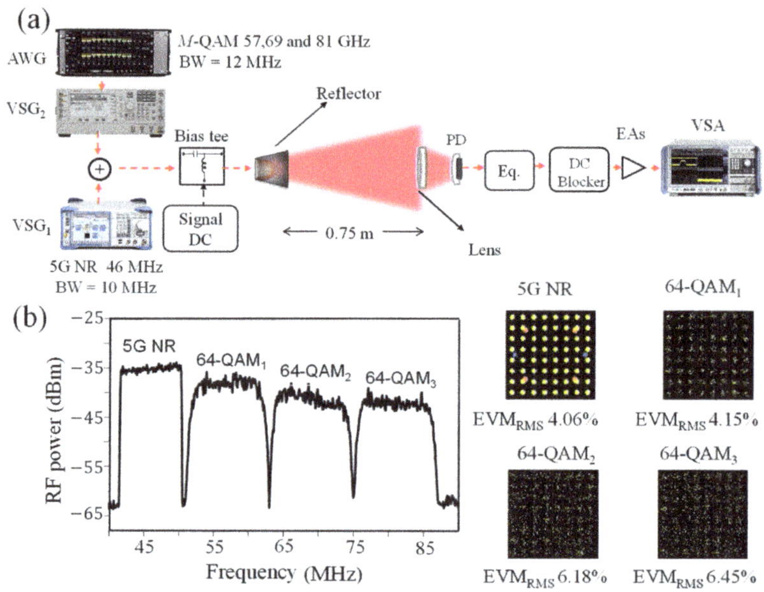

Figure 5. The 4G/5G NR VLC-based access network architecture. (**a**) Block diagram; (**b**) electrical spectrum and EVM_{RMS} at the receiver.

Figure 5b shows the received spectrum as well as the measured constellations and EVM_{RMS} values. The received signal power decreased as the operation frequency increased on account of the LED frequency response, degrading both the signal-noise ratio (SNR) and the EVM_{RMS} value. The constellations in Figure 5b illustrate these performance issues by considering the four transmitted bands. The measured EVM_{RMS} values were 4.06%, 4.15%, 6.18%, and 6.45% for 5G NR, 64-QAM_1, 64-QAM_2, and 64-QAM_3, respectively, which are all within the 3GPP limit of 8% for the 64-QAM format [27]. The effective throughputs measured for the 5G NR signal and the aggregation of the three 64-QAM signals were 38 and 180 Mbit/s, respectively, rates which are attractive for a VLC access point.

4. Related Works and Comparison to the State of the Art

This section reports on a review of the state of the art by comparing optical-wireless techniques for 5G radio access network reported in the literature to the results of our implementation. Five architectures were considered, namely FiWi mm-wave/IFoF links, multiuser FiWi links, parallel FiWi and FSO relay links, RoF links followed by a FSO-section, and VLC access. Table 1 summarizes the aforementioned literature and comparisons to our work with the aim of contributing to and advancing the body of knowledge in this field.

Table 1. State of the art on optical-wireless techniques for 5G radio access network.

Reference	Architecture	Application	RF Bands	Modulation/ Waveform	Transmitter	Throughput
[30]	- FiWi mm-wave/ IFoF link	- RF	- 57 GHz - 91 GHz	- OFDM	- External cavity laser + in-phase and quadrature (IQ)-MZM	- 24 Gbit/s
[31]	- FiWi mm-wave/ IFoF link	- 5G	- 57-64 GHz	- QPSK -16-QAM	- LD + SD-MZM	- 24 Gbit/s
[32]	- FiWi mm-wave/ IFoF link	- 5G - FTTH	- 60 GHz	- 16-QAM	- LD + SD-MZM	- Not specified
[14]	- Multi-user FiWi links	- 5G - FTTH - M2M	- Baseband - 700 MHz - 3.5 GHz - 26 GHz	- OFDM, GFDM, F-OFDM, *M*-QAM and Standard 5G NR	- LD + DD-MZM	- 4.41 Gbit/s
[33]	- Multi-user FiWi links	- 5G	- 60 GHz	- QPSK	- LD + SD-MZM	- 0.6 Gbit/s
[34]	- Multi-user FiWi links	- 5G	- 28 GHz	- 16- and 64-QAM	- RF combiner + LD	- 6 Gbit/s
[35]	- Multi-user FiWi links	- 5G	- 17.6 GHz - 26 GHz	- *M*-QAM	- LD + SD-MZM + optical ring resonators	- 21 Gbit/s
[36]	- mm-wave/FSO links	- Backhaul /fronthaul	- 60 GHz	- BPSK	- Bias tee + LD	- 1 Gbit/s
[37]	- mm-wave/FSO links	- 5G	- 60 GHz	- 16-QAM - OFDM	- DFB laser + MZM	- 4 Gbit/s
[25]	- mm-wave/FSO links	- 5G	- 28 GHz	- 16-QAM - 5G	- LD + MZM	- 1 Gbit/s
[18]	- RoF/FSO/Wireless links	- 5G	- 25 GHz	- Standard LTE	- LD + 2-MZMs	- 120 Mbit/s
[19]	- RoF/FSO/Wireless links	- 5G	- 2.12 GHz - 2.6 GHz - 3.5 GHz	- Standard LTE-A - 5G NR	- SFP module	- Not specified
[24]	- RoF/FSO/Wireless links	- Fronthaul - Access	- 24–26 GHz	- 64-QAM	- Bias tee + LD	- 600 Mbit/s
[38]	- VLC access	- Access point	-	- OOK	- Bias tee + GaN-based LED	- 2.3 Gbit/s
[39]	- VLC access	- 5G	-	- OFDM, GFDM and FBMC	- Bias tee + LED	- 10 Mbit/s
[40]	- VLC access	- Access point	-	- DCO-OFDM	- Bias tee + phosphor LD	- 6.9 Gbit/s
This work	- FiWi multi-user mm-Wave/RoF link - Parallel FiWi and FSO links - RoF/FSO-based link - VLC access	- 4G - 5G	- 700 MHz - 3.5 GHz - 26 GHz - 38 GHz	- *M*-QAM - LTE standard - 5G NR standard	- LD + SD-MZM - LD + DD-MZM - Bias tee + LED	- 182.8 Mbit/s - 1.61 Gbit/s - 1.4 Gbit/s - 218 Mbit/s

FiWi mm-wave/IFoF architectures, which include digital or analog RoF approaches followed by wireless transmission, have benefited from diverse technical solutions aimed at improving capacity and coverage [30–32]. For instance, Li et al. reported a 24 Gbit/s FiWi mm-wave link by exploiting 6 GHz bandwidth [30], whereas Argyris et al. demonstrated a similar bit rate in a mm-wave/IFoF link with electrical subcarrier multiplexing [31]. Mm-wave/IFoF links have even been deployed over legacy PON as an alternative for 5G fronthauling [32].

Advances in the multiuser FiWi links include simultaneous transmission of multiple RF signals at distinct frequency bands [14] as well as the use of phased array antennas for beamforming and beamsteering [33–35]. In [14], we presented a FiWi system over legacy PON based on DD-MZM that was able to cover the 5G eMBB scenario at FR1 and FR2, thereby enabling applications including long reach, fiber-to-the-home, and machine-to-machine communication. In [33], Ruggeri et al. reported a multiuser FiWi mm-wave/IFoF system taking advantage of a 32-element 60 GHz phased array antenna with beamsteering capability. In [34], Huang et al. demonstrated the possibility of integrating self-steering array beamforming to the FiWi architecture. Additionally, a photonic beamforming-based FiWi architecture was experimentally demonstrated in [35].

Multiple works have reported a hybrid mm-wave and FSO fronthaul aimed at increasing system reliability. The FSO and mm-waves links can operate in distinct architectures, namely simultaneous transmission for increased throughput in parallel with switchable transmission and a cascade system with FSO followed by a wireless link. Shakir et al. reported a hybrid system that simultaneously transmitted the same data over both links, namely an FSO at 1550 nm and mm-waves at 60 GHz [36]. The link selection was based on SNR for both approaches in accordance with decision thresholds aimed at evaluating BPSK transmission in a turbulent channel. Zhang et al. presented a hybrid FSO and mm-wave link for 5G backhauls operating at 1553.96 nm and 60 GHz, respectively. The link option was based on maximum-ratio combining (MRC) and was evaluated in terms of EVM_{RMS} for a 16-QAM signal in turbulent, fog, and rain conditions [37]. Additionally, the authors in [25] reported a hybrid 1550 nm FSO link cascaded with a 28 GHz RF link for 5G backhaul and fronthaul applications. The experiments consisted of evaluating the cascaded approach in terms of bit error rate (BER), EVM, and visibility range under the same dusty conditions. The experimental results demonstrated that the dust environment did not impair the RF link at 28 GHz, enabling the RF link to operate as a backup for FSO [25].

The FSO link has also been employed for extending the RoF fronthaul to sites where fiber optics do not reach, thus maintaining the optical communication advantages. In this way, Mufutau et al. demonstrated a hybrid 4G/5G optical fiber-wireless system, including a free-space optics fronthaul [19]. The results showed that a 100 MHz bandwidth 5G signal transmission using A-RoF throughout a 8 km fiber-optic fronthaul could coexist on the same LTE fronthaul infrastructure. Additionally, a hybrid fiber and FSO fronthaul demonstrated the 4G and 5G operability at 2.6 and 3.5 GHz, respectively. Nguyen et al. experimentally implemented a radio-over-fiber followed by an FSO fronthaul and mm-waves wireless transmission aimed at broadband wireless access [18]. The performance was evaluated by transmitting 4-, 16-, and 64-QAM LTE signals over the hybrid link and evaluating the EVM parameter as a function of weak-to-strong atmospheric turbulence regimes. Furthermore, Bohata et al. presented a radio-over-fiber combined with FSO and RF transmission for the 5G networks [24]. The authors used the 24–26 GHz frequency band and a particular direct intensity modulation solution for transmitting a 100 MHz bandwidth 64-QAM signal. The link performance was evaluated by exposing the FSO section to atmospheric turbulence and analyzing the SNR and EVM_{RMS}.

In parallel with the wireless techniques for access point, VLC systems have been emerging as a potential technology to accomplish the 5G requirements and beyond. They are mainly composed of light-emitting diodes, which can be used for simultaneously lighting the environment and transmitting data. In this way, Yeh et al. reported a GaN LED-based visible light communication using a 4 × 4 color polarization multiplexing and on–off keying (OOK) [38]. The system attained 2.3 Gbit/s over 1 m employing MIMO and pre-equalization under the illumination of 6.9–136.1 lux. Recent works have shown the use of VLC for 5G. For instance, Monteiro et al. reported a comparison among 5G waveform candidates, namely filter bank multicarrier (FBMC), generalized frequency division multiplexing (GFDM), and OFDM, with regard to a single-input single-output VLC system [39]. On the other hand, VLC systems may employ lasers as a light source to achieve higher throughput. Wei et al. reported a VLC system employing a white-light phosphor laser for illuminating and providing 6.9 Gbit/s over 1.5 m [40].

In this work, we examined an innovative proposal for optical-wireless architectures applied to the NSA 5G new radio framework. In contrast to previous works, we experimentally demonstrated 4G and 5G NR coexistence in the 700 MHz band taking into account a FiWi system.

Regarding the dual wireless and FSO-based fronthaul architecture, our implementation employed the 5G NR standard. The parallel FSO at 1550 nm and mm-wave link at 38 GHz enabled a dynamic selection according to its transmission performance, whereas the other link operated as a backup. Its performance was investigated by assessing the EVM_{RMS} for different power levels at the receiver to emulate the equivalent attenuation levels for visibility and rain conditions. The attenuation levels used for the FSO link were from 0 to 56 mm/h, whereas the values ranged from 0 to 102 mm/h for the wireless link. Overall, our 5G NR experimental results demonstrated 1.61 Gbit/s total throughput, which is within the 3GPP EVM_{RMS} requirements, for rain intensity less than 36.4 mm/h and visibility above 40 m.

Moreover, we examined a fiber-optic and FSO-based 5G fronthaul followed by a dual-band 5G NR wireless access, including mm-waves. The FSO link enables last-mile applications, thereby increasing system flexibility and providing optical fiber-like throughput. The system performance was investigated by assessing the hybrid architecture in two distinct scenarios in terms of EVM_{RMS} and in accordance with 3GPP, namely the RoF/FSO fronthaul at the photodetector output and the dual-band 5G NR wireless system employing RoF/FSO as fronthaul. Overall, the proposed hybrid RoF/FSO fronthaul provided 3 Gbit/s total throughput with plenty of margin in terms of EVM_{RMS}, thereby enabling wireless access. The dual-band 5G NR wireless system assisted by a hybrid RoF/FSO fronthaul demonstrated the feasibility of the system by providing 1.4 Gbit/s total throughput, thus fulfilling the 3GPP Release 15 specifications.

Finally, the RGB-based VLC system for indoor applications was examined, in which immunity to electromagnetic interference is crucial. We transmitted a standard 5G NR signal and 3 LTE-A bands in order to attend to multiple users employing a single red LED, which offers the best trade-off between output power and quantum efficiency. The system performance was evaluated by assessing each transmitted signal in terms of SNR, constellation symbols dispersion, and consequently the EVM_{RMS}. The 5G NR VLC system provided 218 Mbit/s throughput and EVM_{RMS} as low as 6.45% considering all transmitted signals, thereby demonstrating the viability of VLC for 5G networks.

5. Conclusions

We have successfully proposed and reported on an efficient optical-wireless architecture applied to the non-standalone 5G new radio framework and compared the system to related works in the literature. Several distinct electrical- and optical-based fronthaul approaches combining free-space optical, wireless links, and RoF techniques were deployed for selection according to network operator requirements. First, a dual 38 GHz wireless and FSO-based fronthaul architecture enabled a backup link for high-reliability services. Next, a fiber-wireless configuration based on an analog RoF link followed by a wireless extension was studied for simultaneous RF signal distribution of both LTE-A and 5G NR signals. Finally, fiber optics and FSO techniques were integrated into a single fronthaul to extend the fronthaul reach where fiber optics cannot be deployed. After photodetection, further access extension was achieved by wireless signal transmission of the two 3GPP Release 15 spectral bands of 3.5 and 26 GHz. All three investigated approaches demonstrated performance within the 3GPP EVM_{RMS} requirements prescribed by 5G NR Release 15. In addition, a VLC access network with carrier aggregation was examined to increase user throughput when immunity to EM interference is paramount. The reported solutions may be integrated in a flexible optical-wireless platform for 5G and beyond, allowing their coexistence with 4G over a shared fronthaul infrastructure. Future works should consider implementation of the proposed architecture to either downlink an uplink or to increase the FSO reach up to hundreds of meters in an outdoor scenario.

Author Contributions: Conceptualization, A.C.S.J.; methodology, R.M.B.; validation, D.H.S. and M.A.R.; formal analysis, R.M.B., D.H.S., M.A.R. and A.C.S.J.; investigation, C.H.d.S.L., E.S.L., M.A.d.O., M.S.B.C., L.C.A., L.G.d.S. and L.A.M.P.; resources, A.C.S.J.; writing—original draft preparation, R.M.B., C.H.d.S.L., E.S.L., M.A.d.O., M.S.B.C., L.C.A., L.G.d.S. and L.A.M.P.; writing—review and editing, R.M.B. and M.A.R.; visualization, R.M.B.; supervision, D.H.S., M.A.R. and A.C.S.J.; project administration, A.C.S.J.; funding acquisition, M.A.R. and A.C.S.J. All authors have read and agreed to the published version of the manuscript.

Funding: This work was partially supported by RNP with resources from MCTIC, Grant No. 01245.010604/2020-14, under the 6G Mobile Communications Systems project of the Radiocommunication Reference Center (Centro de Referência em Radiocomunicações-CRR) of the National Institute of Telecommunications (Instituto Nacional de Telecomunicações-Inatel), Brazil. The authors would also like to thank CNPq, CAPES, FINEP, and FAPEMIG for the financial support.

Institutional Review Board Statement: Not applicable.

Informed Consent Statement: Not applicable.

Data Availability Statement: Not applicable.

Conflicts of Interest: The authors declare no conflict of interest.

References

1. Parkvall, S.; Dahlman, E.; Furuskär, A.; Frenne, M. NR: The new 5G radio access technology. *IEEE Commun. Stand. Mag.* **2017**, *1*, 24–30. [CrossRef]
2. 5GPP. View on 5G architecture. *White Paper*. 2017. Available online: https://5g-ppp.eu/wp-content/uploads/2018/01/5G-PPP-5G-Architecture-White-Paper-Jan-2018-v2.0.pdf (accessed on 26 May 2021).
3. 3GPP. 5G.; NR.; Overall description; Stage-2. TS 38.300 version 15.8.0 Release 15. 2020. Available online: https://www.etsi.org/deliver/etsi_ts/138300_138399/138300/15.08.00_60/ts_138300v150800p.pdf (accessed on 26 May 2021).
4. Tzanakaki, A.; Anastasopoulos, M.; Berberana, I.; Syrivelis, D.; Flegkas, P.; Korakis, T.; Mur, D.C.; Demirkol, I.; Gutierrez, J.; Grass, E.; et al. Wireless-optical network convergence: Enabling the 5G architecture to support operational and end-user services. *IEEE Commun. Mag.* **2017**, *55*, 184–192. [CrossRef]
5. Zhang, N.; Cheng, N.; Gamage, A.P.K.T.; Zhang, K.; Mark, J.W.; Shen, X. Cloud assisted HetNets toward 5G wireless networks. *IEEE Commun. Mag.* **2015**, *53*, 59–65. [CrossRef]
6. I, C.-L.; Li, H.; Korhonen, J.; Huang, J.; Han, L. RAN revolution with NGFI (xhaul) for 5G. *J. Lightw. Technol.* **2018**, *36*, 541–550. [CrossRef]
7. 3GPP. Release 16 description; Summary of Rel-16 work items. TR 21.916 v0.5.0. 2020. Available online: https://www.3gpp.org/release-16 (accessed on 26 May 2021).
8. 3GPP. Release17. Available online: 3gpp.org/release-17 (accessed on 26 May 2021).
9. Dang, S.; Amin, O.; Shihada, B.; Alouini, M. What should 6G be? *Nat. Electron.* **2020**, *3*, 20–29. [CrossRef]
10. Zhang, Z.; Xiao, Y.; Ma, Z.; Xiao, M.; Ding, Z.; Lei, X.; Karagiannidis, G.K.; Fan, P. 6G wireless networks: Vision, requirements, architecture and key technologies. *IEEE Veh. Technol. Mag.* **2019**, *14*, 28–41. [CrossRef]
11. Kalfas, G.; Vagionas, C.; Antonopoulos, A.; Kartsakli, E.; Mesodiakaki, A.; Papaioannou, S.; Maniotis, P.; Vardakas, J.S.; Verikoukis, C.; Pleros, N. Next generation fiber-wireless fronthaul for 5G mmWave networks. *IEEE Commun. Mag.* **2019**, *57*, 138–144. [CrossRef]
12. Liu, C.; Wang, J.; Cheng, L.; Zhu, M.; Chang, G.-K. Key microwave-photonics technologies for next generation cloud-based radio access networks. *J. Lightw. Technol.* **2014**, *32*, 3452–3460. [CrossRef]
13. Yao, S.; Chen, Y.-W.; Su, S.-J.; Alfadhli, Y.; Shen, S.; Zhang, R.; Zhou, Q.; Chang, G.-K. Non-orthogonal uplink services through co-transport of D-RoF/A-RoF in mobile fronthaul. *J. Lightw. Technol.* **2020**, *38*, 3637–3643. [CrossRef]
14. Borges, R.M.; Pereira, L.A.M.; Filgueiras, H.R.D.; Ferreira, A.C.; Cunha, M.S.B.; Neto, E.R.; Spadoti, D.H.; Mendes, L.L.; Sodré Jr., A.C. DSP-based flexible-waveform and multi-application 5G fiber-wireless system. *J. Lightw. Technol.* **2020**, *38*, 642–653. [CrossRef]
15. Borges, R.M.; Lima, E.S.; Ferreira, A.C.; Spadoti, D.H.; Abreu, M.; Mendes, L.L.; Sodré Jr., A.C.; Junior, A.C.S. Multiband 5G NR system with photonic-assisted RF amplification. *Opt. Lett.* **2020**, *45*, 1539–1542. [CrossRef]
16. Lima, E.S.; Borges, R.M.; Pereira, L.A.M.; Filgueiras, H.R.D.; Alberti, A.M.; Sodré Jr., A.C. Multiband and photonically amplified fiber-wireless Xhaul. *IEEE Access* **2020**, *8*, 44381–44390. [CrossRef]
17. Filgueiras, H.R.D.; Borges, R.M.; Melo, M.C.; Brandao, T.H.; Sodré Jr., A.C. Dual-band wireless fronthaul using a FSS-based focal-point/Cassegrain antenna assisted by an optical midhaul. *IEEE Access* **2019**, *7*, 112578–112587. [CrossRef]
18. Nguyen, D.-N.; Bohata, J.; Komanec, M.; Zvanovec, S.; Ortega, B.; Ghassemlooy, Z. Seamless 25 GHz transmission of LTE 4/16/64-QAM signals over hybrid SMF/FSO and wireless link. *J. Lightw. Technol.* **2019**, *37*, 6040–6047. [CrossRef]
19. Mufutau, A.O.; Guiomar, F.P.; Fernandes, M.A.; Lorences-Riesgo, A.; Oliveira, A.; Monteiro, P.P. Demonstration of a hybrid optical fiber-wireless 5G fronthaul coexisting with end-to-end 4G networks. *J. Opt. Commun. Netw.* **2020**, *12*, 72–78. [CrossRef]
20. Koonen, T. Indoor optical wireless systems: Technology, trends, and applications. *J. Lightw. Technol.* **2018**, *36*, 1459–1467. [CrossRef]
21. Feng, L.; Hu, R.Q.; Wang, J.; Xu, P.; Qian, Y. Applying VLC in 5G networks: Architectures and key technologies. *IEEE Netw.* **2016**, *30*, 77–83. [CrossRef]
22. Ranaweera, C.; Wong, E.; Nirmalathas, A.; Jayasundara, C.; Lim, C. 5G C-RAN With Optical Fronthaul: An Analysis From a Deployment Perspective. *J. Lightw. Technol.* **2018**, *36*, 2059–2068. [CrossRef]

23. ACG Research. An Economic Comparison of Fronthaul Architectures for 5G Networks. Available online: https://www.acgcc.com (accessed on 17 August 2020).
24. Bohata, J.; Komanec, M.; Spácil, J.; Ghassemlooy, Z.; Zvánovec, S.; Slavık, R. 24–26 GHz radio-over-fiber and free-space optics for fifth- generation systems. *Opt. Lett.* **2018**, *43*, 1035–1038. [CrossRef] [PubMed]
25. Esmail, M.A.; Ragheb, A.M.; Fathallah, H.A.; Altamimi, M.; Alshebeili, S.A. 5G-28 GHz signal transmission over hybrid all-optical FSO/RF link in dusty weather conditions. *IEEE Access* **2019**, *7*, 24404–24410. [CrossRef]
26. Agrawal, G.P. *Applications of Nonlinear Fiber Optics*, 1st ed.; Academic Press: Cambridge, MA, USA, 2001.
27. 3GPP. *Group radio access network; NR.; User equipment (UE) radio transmission and reception; Part 1: Range 1 Standalone. TS 38.101–1 version 15.5.0 Release 15*. 2019. Available online: https://www.etsi.org/deliver/etsi_ts/138100_138199/13810101/15.05.00_60/ts_13810101v150500p.pdf (accessed on 17 August 2020).
28. Haas, H.; Yin, L.; Wang, Y.; Chen, C. What is lifi? *J. Lightw. Technol.* **2016**, *34*, 1533–1544. [CrossRef]
29. Huang, X.; Shi, J.; Li, J.; Wang, Y.; Wang, Y.; Chi, N. 750Mbit/s visible light communications employing 64QAM-OFDM based on amplitude equalization circuit. In Proceedings of the Optical Fiber Communication Conference, Los Angeles, CA, USA, 22–26 March 2015.
30. Li, X.; Xiao, X.; Xu, Y.; Wang, K.; Zhao, L.; Xiao, J.; Yu, J. Real-time demonstration of over 20 Gbit/s V- and W-band wireless transmission capacity in one OFDM-RoF system. In Proceedings of the Optical Fiber Communications Conference (OFC), Los Angeles, CA, USA, 19–23 March 2017.
31. Argyris, N.; Giannoulis, G.; Kanta, K.; Iliadis, N.; Vagionas, C.; Papaioannou, S.; Kalfas, G.; Apostolopoulos, D.; Caillaud, C.; Debrégeas, H.; et al. A 5G mmWave fiber-wireless IFoF analog mobile fronthaul link with up to 24 Gb/s multi-band wireless capacity. *J. Lightw. Technol.* **2019**, *37*, 2883–2891. [CrossRef]
32. Kanta, K.; Pagano, A.; Ruggeri, E.; Agus, M.; Stratakos, I.; Mercinelli, R.; Vagionas, C.; Toumasis, P.; Kalfas, G.; Giannoulis, G.; et al. Analog fiber-wireless downlink transmission of IFoF/mmWave over in-field deployed legacy PON infrastructure for 5G fronthauling. *J. Opt. Commun. Netw.* **2020**, *12*, D57–D65. [CrossRef]
33. Ruggeri, E.; Tsakyridis, A.; Vagionas, C.; Leiba, Y.; Kalfas, G.; Pleros, N.; Miliou, A. Multi-user V-band uplink using a massive MIMO antenna and a fiber-wireless IFoF fronthaul for 5G mmWave small-cells. *J. Lightw. Technol.* **2020**, *38*, 5368–5374. [CrossRef]
34. Huang, M.-Y.; Chen, Y.-W.; Peng, P.-C.; Wang, H.; Chang, G.-K. A full field-of-view self-steering beamformer for 5G mm-wave fiber-wireless mobile fronthaul. *J. Lightw. Technol.* **2020**, *38*, 1221–1229. [CrossRef]
35. Morant, M.; Trinidad, A.; Tangdiongga, E.; Koonen, T.; Llorente, R. Experimental demonstration of mm-Wave 5G NR photonic beamforming based on ORRs and multicore fiber. *IEEE Trans. Microw. Theory Techn.* **2019**, *67*, 2928–2935. [CrossRef]
36. Shakir, W.M.R. Performance evaluation of a selection combining scheme for the hybrid FSO/RF system. *IEEE Photon. J.* **2017**, *10*, 1–10. [CrossRef]
37. Zhang, J.; Wang, J.; Xu, Y.; Xu, M.; Lu, F.; Cheng, L.; Yu, J.; Chang, G.-K. Fiber–wireless integrated mobile backhaul network based on a hybrid millimeter-wave and free-space-optics architecture with an adaptive diversity combining technique. *Opt. Lett.* **2016**, *41*, 1909–1912. [CrossRef]
38. Yeh, C.-H.; Weng, J.-H.; Chow, C.-W.; Luo, C.-M.; Xie, Y.-R.; Chen, C.-J.; Wu, M.-C. 1.7 to 2.3 Gbps OOK LED VLC transmission based on 4× 4 color-polarization-multiplexing at extremely low illumination. *IEEE Photon. J.* **2019**, *11*, 1–6. [CrossRef]
39. Monteiro, F.T.; Costa, W.S.; Neves, J.L.C.; Silva, D.M.I.; Rocha, H.R.O.; Salles, E.O.T.; Silva, J.A.L. Experimental evaluation of pulse shaping based 5G multicarrier modulation formats in visible light communication systems. *Opt. Commun.* **2020**, *457*. [CrossRef]
40. Wei, L.-Y.; Liu, Y.; Chow, C.-W.; Chen, G.-H.; Peng, C.-W.; Guo, P.-C.; Tsai, J.-F.; Yeh, C.-H. 6.915-Gbit/s white-light phosphor laser diode-based DCO-OFDM visible light communication (VLC) system with functional transmission distance. *Electron. Lett.* **2020**, *56*, 945–947. [CrossRef]

applied sciences

MDPI

Article

A Programmable ROADM System for SDM/WDM Networks

Ruizhi Yang, Lida Liu †, Shuangyi Yan * and Dimitra Simeonidou

The HPN Group, Smart Internet Lab, University of Bristol, Bristol BS8 1UB, UK; ruizhi.yang@bristol.ac.uk (R.Y.); lida.liu@segulagrp.com (L.L.); dimitra.simeonidou@bristol.ac.uk (D.S.)
* Correspondence: shuangyi.yan@bristol.ac.uk
† Current address: Segula Technology AB, 41749 Gothenburg, Sweden.

Abstract: This paper proposed and evaluated a programmable ROADM system for MCF-based SDM/WDM networks. The proposed ROADM system employing both bypass connection and Route-and-Select wavelength switching enables adaptable virtual topology in optical networks by dynamically configuring bypass connection cores. The simulation results confirmed this ROADM system could provide acceptable performance with an around 10–20% reduction in the total cost including the number of ports and WSSs by comparing with a fully flexible SDM/WDM ROADM system, which cannot be implemented due to the required extremely high-port-count WSSs.

Keywords: optical network; multicore fibre; programmable ROADM; bypass connection

Citation: Ruizhi, Y.; Lida, L.; Shuangyi, Y.; Dimitra, S. A Programmable ROADM System for SDM/WDM Networks. *Appl. Sci.* **2021**, *11*, 4195. https://doi.org/10.3390/app11094195

Academic Editor: Fabio Cavaliere

Received: 31 March 2021
Accepted: 30 April 2021
Published: 5 May 2021

1. Introduction

Optical networks, as the key Internet infrastructure in the information era, have served the explosive traffic growth successfully in the past decades. Innovations in digital signal processing (DSP) and high-order modulation signals have pushed the spectral efficiency in single-mode optical fibres (SMF) around 10 bits/s/Hz with a total capacity more than 40 Tbit/s per fibre [1]. However, the increased optical signal to noise ratio (OSNR) requirement for optical signals with advanced modulation formats reduces the transmission distance significantly. Mitigating optical nonlinearity helps improve the transmission distance but at the cost of a significant amount of computational power in the receiver. Furthermore, it has been becoming increasingly challenging to squeeze more capacity out of SMF-based wavelength-division multiplexing (WDM) technology.

Space-division multiplexing (SDM) can offer significantly higher capacity by bringing another dimension in optical networks [2]. Several technologies have been demonstrated to enable SDM in optical networks by using single-mode fibre bundles, multicore fibre (MCF), few-mode fibre, or orbital angular momentum multiplexing (OAM). Most of the efforts focus on link capacity improvement without considering the implementation of switching system in SDM networks. Among these technologies, few-mode fibre and OAM technologies present strong coupling between different spatial modes, prohibiting traditional WDM switching without cumbersome mode decoupling either by high-complex and power-hungry digital signal processing or advanced optical decoupling systems. Weakly-coupled MCFs provides a promising SDM solution with more matured fabrication and availability of relevant optical components [3,4]. Besides, MCF-based SDM networks provide full compatibility with current wide-deployed WDM systems. The latest research has demonstrated C+L band transmission of 19-core fibres for trans-oceanic communications [5]. In addition, a field-deployed MCF testbed has been installed in L'Aquila, Italy [6]. It is believed that MCF-based SDM networks provide a feasible solution for potential deployment in future optical networks.

Comparing the SMF, MCFs with a large number of cores incorporated with WDM technology could potentially increase the total capacities by 10 to 20 times. To manage the growing traffic in MCF-based SDM networks, switching strategies should be reconsidered to allow efficient network resource allocation and management. As the key switching element in optical networks, ROADM systems for SDM networks will also be at the heart of the switching solution. Traditional ROADM that conducts band switching in WDM-based optical networks poses a significant challenge for efficient scaling in MCF-based SDM networks [7]. Providing flexible spectrum switching for all fibre cores not only requires a large amount of cost; it may also introduce a significant amount of spectrum fragmentation [8]. Moreover, the implementation of ROADM with full-spectrum switching requires wavelength-selective switches (WSSs) with a considerable number of ports, which poses a limitation to further scaling up. Besides, spectrum flexibility in all fibre cores becomes unnecessary, providing the node degrees are far less than the core numbers. With a sufficient number of cores in one fibre, fibre cores should be treated as a type of switching entity compared to wavelength in the conventional WDW system. Therefore, the ROADM system for SDM-WDM networks providing both wavelength and fibre core switching granularity should be adopted to allow effective and efficient resource management.

In this paper, a programmable SDM/WDM ROADM system is proposed and evaluated for MCF-based SDM/WDM networks, with an extension of our previous work [9]. The proposed ROADM system is featured with programmable node synthesization and dynamic bypass-fibre-core configuration. The bypass core configuration provides direct core-to-core connections from the source node to the destination node without any intermediate wavelength switching, which allows adaptable virtual typology in optical networks for different network traffic. At each node, the total number of fibre cores that require wavelength switching is reduced significantly with the fibre core bypass configuration so that a small-scale traditional WDM ROADM can be setup to provide wavelength switching capability. The proposed SDM/WDM ROADM system minimizes the required scale of WDM-ROADM at each node and reduces the required number of WSSs. The network simulation is carried out to validate the proposed SDM/WDM ROADM system. The simulation results indicate that the proposed SDM/WDM ROADM system provides similar performance as that of a baseline ROADM design, which provides WDM switching in all fibre cores. It is worth mentioning that the baseline SDM/WDM ROADM system is a fully flexible ROADM that treats each core as a single mode fibre. This baseline ROADM system cannot be implemented practically as it requires extremely high-port-count WSSs that is far beyond the current technologies. To evaluate the system cost, we choose the required quantities of WSSs and the total power quantity to reflect the cost. Our proposed SDM/WDM ROADM system requires less number of WSSs with 10–20 % reduction, which depends on the fibre core number and network scale. Most importantly, the port number per WSS in our proposed system is in the practical range. In the simulation network, the maximum port number of WSSs in a fully flexible ROADM scenario is 41 (assuming three degrees in the node, bidirectional transmission in the MCFs and 13 cores in each MCF). Moreover, the maximum total port count of practical WSSs is limited around 30 [10]. In addition, the SDM/WDM ROADM design is fully compatible with the current SMF-based optical networks and allows smooth migration to multidimensional optical networks. With the increase in core number in MCFs, the ratio of the saved hardware can be further increased for better hardware utilization.

The remainder of this paper unfolds as follows. The latest development of the SDM ROADM system is reviewed in Section 2. Section 3 presents the design of the programmable SDM/WDM ROADM. Simulation results to validate the proposed ROADM architecture are presented and discussed in Sections 4 and 5 concludes the paper.

2. Relevant Works

MCF-based SDM networks are designed to integrate with the traditional WDM technologies by adding another dimension. Multicore fibre is a typical approach to introduce multiple spatial paths into a fibre from an SDM fibre point of view [3]. The transmission distance and maximum cores number in MCFs are significantly limited by both cross talk and attenuation. Several compensating methods have been proposed. For example, crosstalk-aware spectrum defragmentation [11] was proposed to remedy the issue of spectrum fragmentation in the SDM optical network, and the EDFAs in switching nodes can compensate for the attenuation. In SDM/WDM networks, fully flexible spectrum switching would be a straightforward solution to use network resources efficiently [7,12]. The fully flexible ROADM will require WSSs, the key switching device in ROADMs, to provide switching ports more than the total fibre core numbers in all ingress fibres. Sato [13] explained how to create very large-scale ROADM, but the requirement of WSS port number is very high. With the increased core numbers offered by MCFs, WSSs with such high port counts become unpractical. Therefore, traffic aggregation emerged as an option to reduce switching complexity. Besides, the work in [14] confirmed that traffic aggregation in space and spatial can reduce the routing complexity and save a considerable amount of switching components.

One approach for traffic aggregation is to provide network resources as superchannel signals either in space or spectral [15]. Regarding space superchannel signals, several cores can be grouped together to perform core switching as shown in [16]. However, such coarse granularity in multiple core levels prohibits deployment in optical networks, which require flexibility in the channel level according to the client interface speed. By combining with spectra and cores, spatial superchannels can be set up over multiple cores [12]. Compared with the traditional spectral superchannel signals, the spatial superchannel signals require specially designed WSSs for switching operations. By taking account of the recent standardisation of 400GbE, the spatial superchannels over multiple cores will not be foreseen for any commercial deployment in a few years.

Another approach to handle SDM switching is to offer both core switching and wavelength switching in a collaborative approach. In 2017, we proposed the concept of core bypassing in [17], to reduce the scale of the ROADM for better hardware utilization. The followed work evaluated the design at the network level and further reduced the required port count of the WSSs in the ROADM by cascading two-level WSSs [9]. Jinno [18] recognized core bypassing as a key enabling technology for the massive SDM era. However, the work separates the whole network into a pure SDM network and a traditional WDM network. It is believed that integrating SDM and WDM networks with fibre core bypassing will achieve better switching performance. Therefore, several key enabling technologies are listed here to take full advantage of combined fibre core bypassing and traditional WDM in SDM/WDM networks.

1. The bypass cores should be integrated with the WDM system that consists of the residual cores with traditional ROADM switching. The integration allows the bypass core to serve as a part of optical links. Without the integration, bypass cores will impact network flexibility, especially when core numbers are low. The integration requires connecting bypass cores back to the ROADM system at the end of the bypass cores.
2. The bypass cores should be reconfigurable at each node. The configurability of bypass cores enables adaptability of networks to accommodate different network traffic and support the migration from SMF-based networks to MCF-based optical networks. Firstly, variations of network traffic during the life cycle of optical links require fibre cores to be reconfigured as either bypass cores or WDM cores to optimize network performance. The programmable bypass core configuration will provide compatibility to the current wide-deployed SMF-based networks.
3. Intelligent node management is required to handle the bypass core reconfiguration. For example, intelligent node power management is needed to optimize the optical power in the node. More advanced power management can be developed to support the dynamic network reconfiguration [19].

3. Design of Programmable SDM/WDM ROADM System

The proposed programmable SDM/WDM ROADM system creates two integrated networks over the MCF-based fibre networks. Generally, each core in MCFs is treated as a typical SMF with fan-in and fan-out devices for each MCF. Inside of each MCF, some fibre cores are configured as bypass cores and the residual fibre cores are configured as WDM fibre links. The number of bypass cores and WDM cores can be reconfigured by a centralized controller. Therefore, these two integrated networks can be optimized according to the traffic request. Bypass core links, combined with WDM cores, generate programmable virtual topologies in the MCF-based optical networks.

Figure 1 shows the concept of programmable bypass core configuration based on an exemplary SDM network. Nodes are interconnected with 7-core MCF links. Fibre cores in different MCF links that are directly connected without any wavelength switching in the intermediate nodes are defined as bypass cores. As shown in Figure 1, a bypass link is set up between node A to node F by connecting bypass cores directly to node C and node D. Similarly, other bypass links can be established between different nodes. As an example, to further illustrate, several cores in node C are configured as bypass cores, including node A to node E and node B to node D. The residual cores as shown in node D are used to achieve the Route-and-Select ROADM, such as the bypass link from node B bypassed node C and connected to RS ROADM in node D.

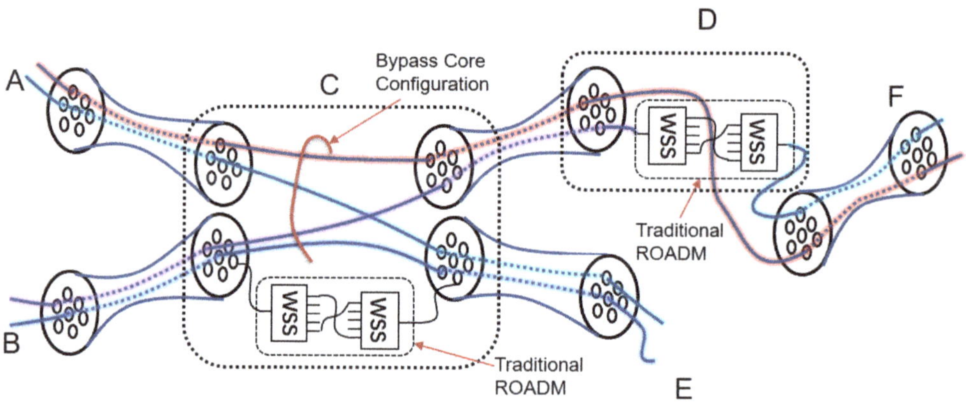

Figure 1. Concept of bypass-core configuration in an exemplary optical network with 7-core fibres. (**A–F**) are the node names. The color lines are the bypass links.

The following sections will illustrate the detailed node structure including bypass connection and traditional WDM fibre links, programmable virtual topology, and hardware requirements.

3.1. Programmable ROADM Node Structure with Bypass Connection

The programmable ROADM node with bypass connection is shown in Figure 2. The node includes bypass configuration through core switching and WDM routing/selection based on Route-and-Select (RS) ROADM. To achieve the required configuration, all fibre cores are connected to a large-port-count fibre switch (LPFS) based on the architecture-in-demand (AoD) concept [20]. The WSSs and other components that are used for RS ROADM are also connected to the fibre switch to synthesize the RS ROADM, shown in Figure 2 separately on the top and the bottom two parts. A fixed degree ROADM can be used to replace the synthesized ROADM by connecting the ROADM as one subsystem for RS ROADM and the other subsystem for bypass connection. Figure 2 shows a 2-degree node which is a proposed node structure with four 13-core MCFs. Inside the node, the 13 cores in MCFs are regarded as 13 single mode fibres. Every core of fibre-in and fibre-out is connected to the AoD-based optical fibre Switch. The LPSF can reconfigure connections between different ports to achieve different selections of the proposed ROADM. As shown in Figure 2, the bypass connection part is on the top part of the node, which can directly connect every two cores of fibre-in and fibre-out. The other components are used to synthesise Route-and-Select ROADM for the residual fibre cores. Route-and-Select ROADM in the programmable node deploys 2-layer WSSs in the Figure 2. The first-layer WSSs demultiplex channels from every core of the input MCFs to select the destination cores in that node. WSSs in the second layer is used for multiplexing channels from different cores in the first-layer WSSs to the destination MCFs. The implementation of ROADM requires low port-count WSSs, which can bring the benefits of reducing hardware devices and saving the cost. Several functions are attached in the traditional Route ROADM part, it includes EDFAs, Couplers, Rx/Tx, and Functions/Subsytems. Based on the AoD concept, the RS ROADM can be reconfigured in terms of input fibre cores and the corresponding add/drop configurations.

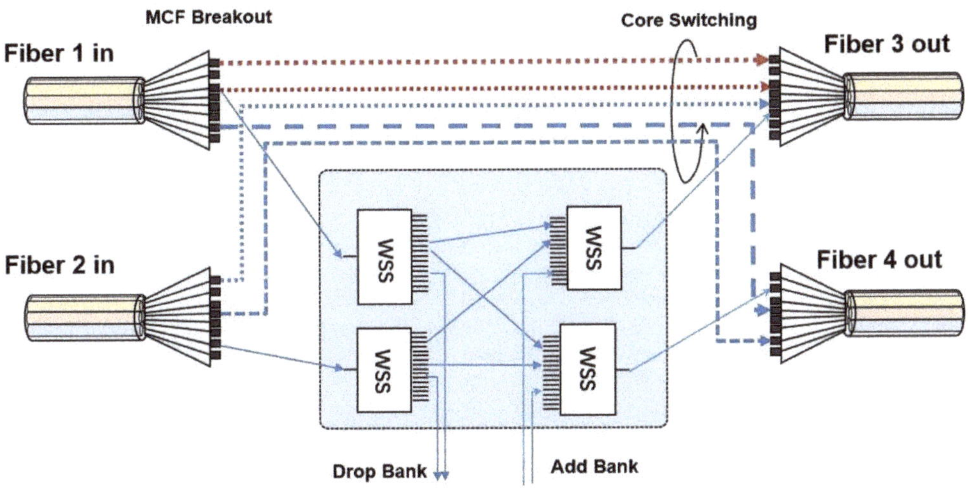

Figure 2. Detailed node structure including bypass connection and Route-and-Select ROADM.

3.2. *Programmable Virtual topology*

To demonstrate the promotion of optical network performance of core switch, 6-node topology optical network which is shown in Figure 3 has been used for analysis. There are four three-degree nodes and two two-degree nodes with 8 multicore fibres used in the whole network. The number of cores in every MCF was assumed as 13 and the fibre-core switch can select different numbers which can be controlled by a centralized controller. Due to there being only six nodes in this optical network, the pairs of transmitting and receiving nodes are including three types: neighbor pairs, one-hop pairs, and two-hop pairs. As mentioned above, the programmable ROADM can dynamically distribute the number of bypass connection cores and WDM RS cores, which can be seen as different virtual topologies. Three different topologies with different bypass connections are used to illustrate the virtual topology. The difference between these topologies is the number of bypass links and where to add them. One-hop core switch was proposed to explain the nondirect two-node whose shortest physical path needs two MCFs and goes through one internode, such as node 0 and node 2. Furthermore, two-hop core switch can be seen as the nondirect two-node whose shortest physical path needs three MCFs and goes through two internodes. Due to there being only six nodes in the proposed optical network, the maximum hop is two.

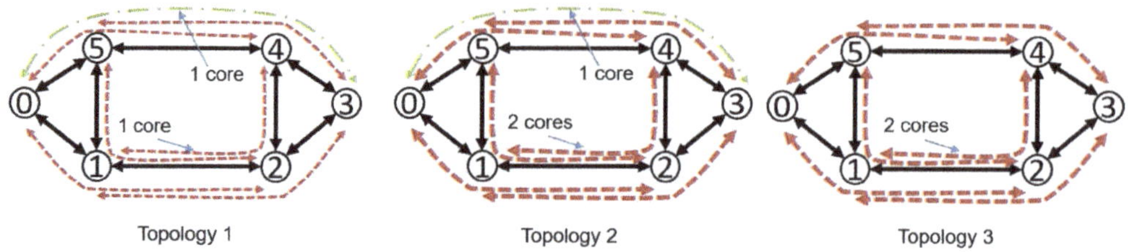

Figure 3. Three different Virtual topologies, named topology 1, 2, 3. The number of one-hop and two-hop bypass connection cores is separately for 1:1, 2:1, 2:0.

In Figure 3, the three different virtual topologies come from using different cores to set up the bypass connection. Topology 1 shows that one core is used for one-hop pair nodes with one core for two-hop pair node bypass connection, while topology 2 presents two cores that are used for one-hop pair-nodes and one core is used for two-hop pair nodes. As for topology 3, only two cores are used for one-hop pair nodes bypass connection.

3.3. Hardware Requirements in Different Topology

In ROADM systems, switching devices, such as WSSs, contribute to most parts of the cost. The cost depends on the number of WSSs as well as the total number of port count in WSSs. Based on the 6-node topology, the total system cost is estimated by calculating the required number and the total port count of WSSs. With the bypass connection, the required WSSs and total WSSs ports are reduced dramatically. To roughly get the hardware saving, the WSS number and total port number in different topologies should be calculated as shown in Table 1. Firstly, the number of cores used for bypass connection in three different virtual topologies has been calculated. After that, the WSS number and the total ports number are presented. As shown in Table 1, comparing with the fully flexible/basic topology, topology 1 could save 7.7% WSSs while the total port number saves 12.6%. Meanwhile, topology 2 and 3 save a little more than topology 1 which has been given the exact percentage in the table. Furthermore, as shown in the last column, the maximum port number in WSSs was calculated in each topology. Comparing with the other three proposed topologies, the fully flexible topology needs an unpractical large-scale WSSs. As mentioned above, the maximum total port count of practical WSSs is limited around 30 so that under the existing conditions, the fully flexible topology cannot be achieved. Besides, the cost of WSSs will be rapidly increased with the port number increasing. In practical networks with more nodes and large core number MCFs, more hardware can be saved based on the proposed SDM/WDM ROADM system.

Table 1. Comparison of hardware request between four topologies.

Topology	Bypass Core/WDM Core	WSS Number	Total Ports Number	Maximum Port Number
Fully flexible topo	0/104	416	10296	41
Topo 1	15/89	38–7.7%	9000–12.6%	33
Topo 2	27/77	360–13.5%	8028–22.1%	25
Topo 3	24/80	368–11.5%	8352–18.9%	25

4. Simulation Results

Simulation analysis was used to demonstrate that an optimal virtual topology with bypass connection can be found to replace the fully flexible topology in optical network and saving a lot of hardware devices. To assess the optical network performance, Blocking Probability is used to evaluate the performance, including bandwidth blocking probability and traffic blocking probability. Both blocking probabilities are simulated and analyzed. Bypass connection would reduce the wavelength switch in the node so that the bandwidth continuity request will be reduced. To begin with this simulation, the assumed configuration should be set. The number of cores in one multi-core fibre (MCF) is 13 and the supposed topology is a typical 6-node core optical network as mentioned above. Assuming the whole bandwidth in one core is 2 THz and every slot is 12.5 GHz, there are 160 slots in one core based on every traffic can be randomly generated from 12.5 GHz, 25 GHz, 37.5 GHz, 50 GHz, 100 GHz, and 200 GHz corresponding with 1, 2, 3, 4, 8, and 16 slots. In addition, bidirection has been considered in this simulation, and it appears that the 2 THz bandwidth is only for the unidirection in one core and the 4 THz assumed used to present bidirection bandwidth. Meanwhile, the cross talk and attenuation could not be considered as mentioned above there are several methods to compensate them. Furthermore, the distance between every two nodes is assumed 40 km so that EDFAs should not be used in the whole optical network.

To evaluate the performance of the proposed network with programmable ROADM combined fibre/core switch with WDM routing/selection node, the traffic blocking probability and bandwidth blocking probability of the same traffic request based on different topology should be given. The blocking probability can be seen as the ratio of the amount of rejected traffic and traffic bandwidth which cannot be transmitted network to the total amount of traffic and the traffic bandwidth in the whole optical network.

4.1. Traffic and Bandwidth Blocking Probability Results in 6-Node Topology

In this section, the traffic request of 6-node core optical network is the same and randomly generated with the ratios of direct-link pair nodes, one-hop pair nodes, and two-hop pair nodes are 1:1:1, 3:2:1, and 6:3:2; three scenarios are separately named traffic request 1, 2, and 3. Regarding every traffic request, four topologies can be simulated to deal with the data. Four topologies are including the basic fully flexible physical 6-node optical network topology and three different virtual topologies with bypass connection, including topology 1, topology 2, and topology 3 as mentioned above.

In Figure 4a,d, they show that the traffic blocking probability and bandwidth blocking probability of four topologies were based on traffic requests 1. The unit of the horizontal axis is Erlang, which is a unit of traffic load and equals the traffic arrival rate times traffic duration. Here we assume all traffic arrives at the same second and only lasts for 1 s [21]. Traffic request 1, which can be seen as a random traffic request, means that the probability of different traffic requests including direct-link pair-nodes, one-hop pair-nodes, and two-hop pair-nodes is equal or 1:1:1. It is obvious that the basic topology is the best one to deal with the data, but the other three topologies appear acceptable performance. Focusing on traffic blocking probability in Figure 4a, virtual topology 2 shows similar performance as the other two topologies while saving much more cost than others so that practically this cost-saving and similar performance topology can be used. As for Figure 4b,e, they show the simulation data on blocking probability based on four different topologies with traffic request 2, as well as Figure 4c,f show the results of four different topologies dealing with traffic request 3. Similar performance with less cost is what all researchers try to get, which can demonstrate that the bypass connection added in the programmable ROADM can improve the hardware efficiency and save hardware cost. To explain this benefit in detail, the wavelength fragmentation which can be reduced by adding the fibre-core switch should be considered as the reason. To illustrate the wavelength fragmentation, the RS ROADM part should be stated. An accepted traffic through RS WDM part needs allocating the same bandwidth in every passed core of MCFs which called bandwidth continuity so that fibre-core switch path does not have this limitation, which can reduce the wavelength fragmentation. As bandwidth blocking probability shown in Figure 4, it shows much better performance than traffic blocking probability, especially in traffic request 2 and 3. Topology 1 and 2 have almost the same bandwidth blocking probability as fully flexible topology in Figure 4e,f.

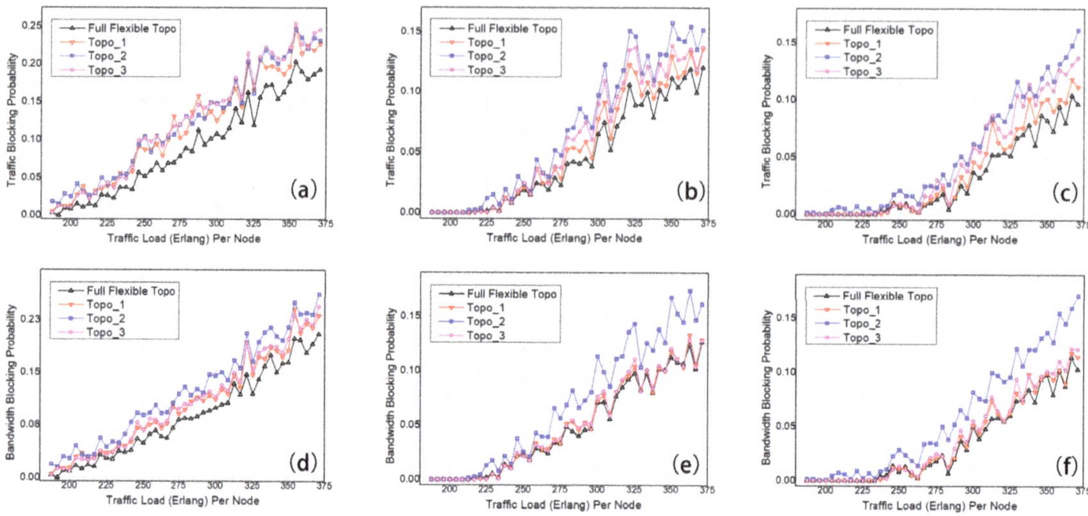

Figure 4. Traffic blocking probabilities and bandwidth blocking probabilities. The traffic request ratio between Neighbor pair-nodes:one-hop pair-nodes:two-hop pair-nodes is 1:1:1, 3:2:1, 6:3:2, respectively shown in (**a**,**d**), (**b**,**e**), (**c**,**f**) named traffic request 1, 2, 3. Three bypass connection topologies are corresponding with Figure 3 Topology 1, 2, 3.

4.2. Traffic and Bandwidth Blocking Probability Results in NSF Topology

The proposed SDM/WDM ROADM system is evaluated in the National Science Foundation (NSF) network. In the NSF network, 7-core MCFs are used for the simulation. As shown in Figure 5a, bypass links are set up based on numbers of neighbor nodes, to ensure direct connections are available between two neighbor nodes of the current node.

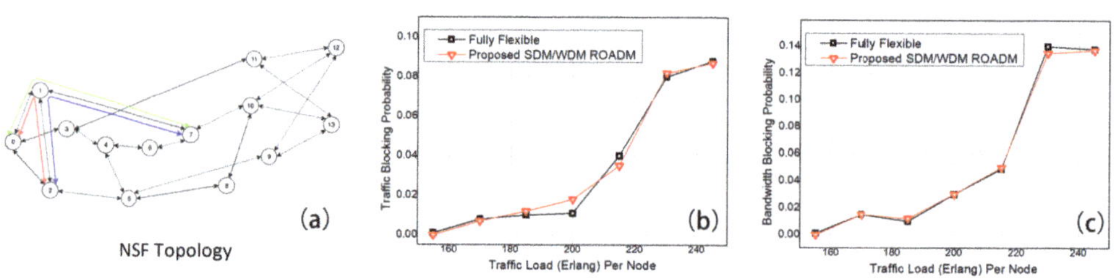

Figure 5. (**a**) NSF topology. (**b**) Traffic and (**c**) Bandwidth blocking probabilities in NSF topology.

Random traffic is used for the simulation. Heuristic routing wavelength/core assignment (RWCA) algorithm is used to evaluate the blocking probability.

Figure 5 shows the NSF topology and the results of blocking probability in NSF topology network by comparing between fully-flexible setup and the proposed setup with bypass configurations. In both traffic and bandwidth blocking probability as shown in Figure 5b,c, the proposed solution shows similar performance as the fully flexible setup.

The simulation results in Figures 4 and 5 suggest the proposed SDM/WDM ROADM system has an acceptable performance by comparing with the fully flexible topology in terms of traffic and bandwidth blocking probabilities. In addition, the proposed system reduces the total cost by roughly calculating the required quantity of WSSs and the total port number.

5. Conclusions

In this paper, a programmable SDM/WDM ROADM system is proposed for MCF-based multidimensional optical networks. The ROADM system introduces reconfigurable fibre-core bypass connections to generate different virtual topologies based on network traffic. The design reduced the required number and port count by using spectrum switching devices significantly by comparing with traditional ROADMs. The design takes account of the vast resources provided by the MCFs and provides a practical solution for wavelength switching. Our simulation analysis shows that the proposed SDM/WDM ROADM system provides acceptable performance as the fully flexible ROADM that treats all fibre cores as individual SMFs while saving considerable cost. The SDM/WDM ROADM design also allows dynamic reconfiguration based on network traffic and provides full compatibility with the current SMF-based networks. It is believed that the proposed SDM/WDM ROADM system provides a potential solution for future multidimensional optical networks.

Author Contributions: Conceptualization, Ruizhi.Yang, Lida.Liu and Shuangyi.Yan; methodology, R.Y. and S.Y.; software, R.Y. and L.L; validation, R.Y., L.L. and S.Y.; formal analysis, R.Y. and S.Y.; data curation, R.Y., L.L. and S.Y.; writing—original draft preparation, R.Y.; writing—review and editing, R.Y. and S.Y.; project administration, S.Y. and D.S.; funding acquisition, S.Y. and D.S. All authors have read and agreed to the published version of the manuscript.

Funding: The authors acknowledge funding support from the EU H2020 (5G-Clarity, No.: 871428).

Institutional Review Board Statement: Not applicable.

Informed Consent Statement: Not applicable.

Data Availability Statement: Not applicable.

Conflicts of Interest: The authors declare no conflict of interest.

References

1. Schuh, K.; Buchali, F.; Dischler, R.; Chagnon, M.; Aref, V.; Buelow, H.; Hu, Q.; Pulka, F.; Frascolla, M.; Alhammadi, E.; et al. 49.2-Tbit/s WDM Transmission over 2×93-Km Field-Deployed Fiber. In Proceedings of the 2020 Optical Fiber Communications Conference and Exhibition (OFC), San Diego, CA, USA, 8–12 March 2020; pp. 1–3.
2. Saridis, G.M.; Alexandropoulos, D.; Zervas, G.; Simeonidou, D. Survey and Evaluation of Space Division Multiplexing: From Technologies to Optical Networks. *IEEE Commun. Surv. Tutor.* **2015**, *17*, 2136–2156. [CrossRef]
3. Saitoh, K.; Matsuo, S. Multicore Fiber Technology. *J. Light. Technol.* **2016**, *34*, 55–66. [CrossRef]
4. Jain, S.; Castro, C.; Jung, Y.; Hayes, J.; Sandoghchi, R.; Mizuno, T.; Sasaki, Y.; Amma, Y.; Miyamoto, Y.; Bohn, M.; et al. 32-Core Erbium/Ytterbium-Doped Multicore Fiber Amplifier for next Generation Space-Division Multiplexed Transmission System. *Opt. Express* **2017**, *25*, 32887–32896. [CrossRef]
5. Puttnam, B.J.; Rademacher, G.; LuÃs, R.S.; Eriksson, T.A.; Klaus, W.; Awaji, Y.; Wada, N.; Maeda, K.; Takasaka, S.; Sugizaki, R. High Data-Rate and Long Distance MCF Transmission With 19-Core C+L Band Cladding-Pumped EDFA. *J. Light. Technol.* **2020**, *38*, 123–130. [CrossRef]
6. Hayashi, T.; Nagashima, T.; Nakanishi, T.; Morishima, T.; Kawawada, R.; Mecozzi, A.; Antonelli, C. Field-deployed multi-core fiber testbed. In Proceedings of the 2019 24th OptoElectronics and Communications Conference (OECC) and 2019 International Conference on Photonics in Switching and Computing (PSC), Fukuoka, Japan, 7–11 July 2019; pp. 1–3.
7. Marom, D.M.; Colbourne, P.D.; D'Errico, A.; Fontaine, N.K.; Ikuma, Y.; Proietti, R.; Zong, L.; Rivas-Moscoso, J.M.; Tomkos, I. Survey of Photonic Switching Architectures and Technologies in Support of Spatially and Spectrally Flexible Optical Networking [Invited]. *J. Opt. Commun. Netw.* **2017**, *9*, 1–26. [CrossRef]
8. Yousefi, F.; Rahbar, A.G. Novel Fragmentation-Aware Algorithms for Multipath Routing and Spectrum Assignment in Elastic Optical Networks-Space Division Multiplexing (EON-SDM). *Opt. Fiber Technol.* **2018**, *46*, 287–296. [CrossRef]
9. Liu, L.; Yan, S.; Maguire, G.Q.; Li, Y.; Simeonidou, D. Hardware-Efficient ROADM Design with Fiber-Core Bypassing for WDM/SDM Networks. In Proceedings of the 2020 Optical Fiber Communications Conference and Exhibition (OFC), San Diego, CA, USA, 8–12 March 2020.
10. Niwa, M.; Mori, Y.; Hasegawa, H.; Sato, K.i. Tipping point for the future scalable OXC: What size MxM WSS is needed? *IEEE/OSA J. Opt. Commun. Netw.* **2017**, *9*, A18–A25. [CrossRef]
11. Zhao, Y.; Hu, L.; Zhu, R.; Yu, X.; Wang, X.; Zhang, J. Crosstalk-aware spectrum defragmentation based on spectrum compactness in space division multiplexing enabled elastic optical networks with multicore fiber. *IEEE Access* **2018**, *6*, 15346–15355. [CrossRef]
12. Jatoba-Neto, A.C.; Mello, D.A.; Rothenberg, C.E.; Arık, S.Ö.; Kahn, J.M. Scaling SDM Optical Networks Using Full-Spectrum Spatial Switching. *IEEE/OSA J. Opt. Commun. Netw.* **2018**, *10*, 991–1004. [CrossRef]

13. Sato, K. Impact of node/fiber/WSS degrees in creating cost effective OXCs. In Proceedings of the 2016 18th International Conference on Transparent Optical Networks (ICTON), Trento, Italy, 10–14 July 2016; pp. 1–4.
14. Arık, S.Ö.; Ho, K.P.; Kahn, J.M. Optical Network Scaling: Roles of Spectral and Spatial Aggregation. *Opt. Express* **2014**, *22*, 29868–29887. [CrossRef] [PubMed]
15. Khodashenas, P.S.; Rivas-Moscoso, J.M.; Siracusa, D.; Pederzolli, F.; Shariati, B.; Klonidis, D.; Salvadori, E.; Tomkos, I. Comparison of Spectral and Spatial Super-Channel Allocation Schemes for SDM Networks. *J. Light. Technol.* **2016**, *34*, 2710–2716. [CrossRef]
16. Feuer, M.D.; Nelson, L.E.; Abedin, K.S.; Zhou, X.; Taunay, T.F.; Fini, J.F.; Zhu, B.; Isaac, R.; Harel, R.; Cohen, G. ROADM System for Space Division Multiplexing with Spatial Superchannels. In Proceedings of the 2013 Optical Fiber Communication Conference and Exposition and the National Fiber Optic Engineers Conference (OFC/NFOEC), Anaheim, CA, USA, 17–21 March 2013.
17. Li, Y.; Yan, S.; Hua, N.; Ou, Y.; Qian, F.; Nejabati, R.; Simeonidou, D.; Zheng, X. Hardware Programmable SDM/WDM ROADM. In Proceedings of the 2017 Optical Fiber Communications Conference and Exhibition (OFC), Los Angeles, CA, USA, 19–23 March 2017.
18. Jinno, M. Spatial Channel Network (SCN): Opportunities and Challenges of Introducing Spatial Bypass Toward the Massive SDM Era [Invited]. *J. Opt. Commun. Netw.* **2019**, *11*, 1–14. [CrossRef]
19. Yan, S.; Aguado, A.; Ou, Y.; Wang, R.; Nejabati, R.; Simeonidou, D. Multi-Layer Network Analytics with SDN-Based Monitoring Framework. *J. Opt. Commun. Netw.* **2017**, *9*, A271–A279. [CrossRef]
20. Amaya, N.; Zervas, G.S.; Simeonidou, D. Architecture on demand for transparent optical networks. In Proceedings of the 2011 13th International Conference on Transparent Optical Networks, Stockholm, Sweden, 26–30 June 2011; pp. 1–4.
21. Simmons, J.M. *Optical Network Design and Planning*; Springer: Cham, Switzerland, 2014.

applied sciences

MDPI

Article

Multi-Tier Heterogeneous Beam Management for Future Indoor FSO Networks

Michael B. Rahaim [1,*], Thomas D.C. Little [2] and Mona Hella [3]

1 Engineering Department, University of Massachusetts, Boston, MA 02125, USA
2 ECE Department, Boston University, Boston, MA 02215, USA; tdcl@bu.edu
3 ECE Department, Rensselaer Polytechnic Institute, Troy, NY 12180, USA; hellam@rpi.edu
* Correspondence: Michael.Rahaim@umb.edu

Featured Application: Wireless Communications and Networking.

Abstract: To meet the growing demand for wireless capacity, communications in the Terahertz (THz) and optical bands are being broadly explored. Communications within these bands provide massive bandwidth potential along with highly directional beam steering capabilities. While the available bandwidth offers incredible link capacity, the directionality of these technologies offers an even more significant potential for spatial capacity or area spectral efficiency. However, this directionality also implies a challenge related to the network's ability to quickly establish a connection. In this paper, we introduce a multi-tier heterogeneous (MTH) beamform management strategy that utilizes various wireless technologies in order to quickly acquire a highly directional indoor free space optical communication (FSO) link. The multi-tier design offers the high resolution of indoor FSO while the millimeter-wave (mmWave) system narrows the FSO search space. By narrowing the search space, the system relaxes the requirements of the FSO network in order to assure a practical search time. This paper introduces the necessary components of the proposed beam management strategy and provides a foundational analysis framework to demonstrate the relative impact of coverage, resolution, and steering velocity across tiers. Furthermore, an optimization analysis is used to define the top tier resolution that minimizes worst-case search time as a function of lower tier resolution and top tier range.

Keywords: beamforming; ultra-dense wireless networks; heterogeneous networks (HetNets); optical wireless communication (OWC); indoor free-space optics (FSO)

Citation: Rahaim, M.B.; Little, T.D.C.; Hella, M. Multi-Tier Heterogeneous Beam Management for Future Indoor FSO Networks. *Appl. Sci.* **2021**, *11*, 3627. https://doi.org/10.3390/app11083627

Academic Editor: Fabio Cavaliere

Received: 18 March 2021
Accepted: 13 April 2021
Published: 17 April 2021

1. Introduction

The continuous growth in demand for wireless capacity projects the use of spectrum into the sub-mm, Terahertz (THz), and optical bands. Extreme directionality is a unique characteristic of THz and optical wireless communications that can be exploited in future ultra-dense networks [1–5]. Component technologies exist for short range (1–10 m) indoor free space optical communications (FSO) with steering capabilities, instantaneous coverage on the order of 1 cm^2, and rates exceeding 400 Gb/s [6]. Research in THz communications is driven by similar potential; however, this research area is still building momentum and is mostly dominated by photonic approaches at this time [7,8]. In general, the challenge is that directional systems require beam management to establish and maintain link connectivity for quasi-static (i.e., portable) and mobile devices. In this paper, we introduce a multi-tier heterogeneous beam management strategy that reconciles the nature of highly directional indoor FSO links and the dynamic beam management needed to maximize network performance for multiple mobile users. While this paper highlights the use of indoor FSO as the highest resolution tier, it should be noted that the modular design of the multi-tier approach could be applied to other highly directional technologies (e.g., THz communications).

We propose ***a multi-tier heterogeneous (MTH) beam management strategy*** that benefits from the resolution and steering velocity of various component technologies (Figure 1). The *multi-tier design* offers the high resolution of indoor FSO links while the millimeter-wave (mmWave) system narrows the FSO search space and relaxes the requirements of the FSO network in order to assure a practical time-frame to search through the device space. *Heterogeneous integration* allows for optimization of traffic distribution based on device context-of-use. This heterogeneous multi-tier design enables high resolution indoor FSO links as components of a high-performance access network serving mobile users and overcoming anomalies such as occlusions, handover, device rotation, and steering/scanning latencies.

The proposed design does not intend to compete with mmWave architectures that exist today; rather, it seeks to establish how dense networks of highly directional indoor FSO links can coexist and work collaboratively with forthcoming mmWave technologies. In addition, the MTH architecture is envisioned under the overarching coverage of a broadly available RF small cell (RFSC). This highlights the design's potential to add supplemental capacity to conventional microwave systems (e.g., WiFi). Furthermore, we envision a modular integration of the tiers within the MTH architecture. This modular design can benefit from beam management improvements at each tier and, accordingly, this manuscript does not attempt to optimize each tier individually. Rather, we consider the implicit benefit of a heterogeneous multi-tier approach to indoor FSO beam management and highlight the relative performance gains related to the coverage and resolution at each tier. In particular, we provide the following primary contributions within this manuscript:

1. A detailed description of the MTH beam management strategy including key requirements for pointing/acquisition/tracking, multi-tier beam refinement, heterogeneous beam management, and configuration/deployment of the system
2. A foundational analysis framework that offers a qualitative visualization of relative performance in terms of coverage, resolution, and steering velocity across tiers
3. An optimization analysis to define the top tier resolution that minimizes worst-case search time as a function of lower tier resolution and top tier range

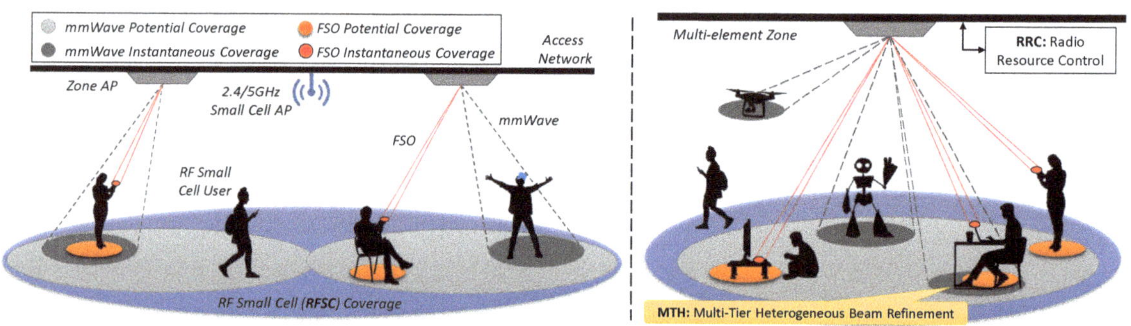

Figure 1. System-level view (**left**); and depiction of a single zone (**right**). *Instantaneous coverage* is defined as the coverage for a specific orientation/emission configuration (beam width, angle, etc.). *Potential coverage* is defined as the achievable coverage over all configurations.

The remainder of the paper is organized as follows. In the following section, we provide background and motivation for Indoor FSO and multi-tier heterogeneous beamforming. Section 3 then introduces the requirements of a system implementing MTH beam management. Sections 4 and 5, respectively, introduce our analysis framework and preliminary results related to optimization of the link acquisition process. Section 6 concludes the paper.

2. Background and Motivation

The number of mobile devices and the performance requirements of applications requiring network access continue to drive an unabated growth in wireless traffic [9]. The increasing device density and per-device data demand continues to drive requirements for area spectral efficiency [b/s/m^2] or volumetric efficiency [b/s/m^3]. The directionality and dynamic control of steerable wireless technologies in the mmWave, THz, and optical bands offer massive potential for improvement of both area spectral efficiency and realizable system capacity [10–13]. In particular, exploiting narrow optical beams for local end-user wireless access can provide data densities and user-experienced data rates that far surpass 5G specifications. This is technically feasible based on the demonstrated potential of steered beams and the inherent signal density of focused light [6,14–20]; however, many research challenges exist at the network and system levels. These challenges must be addressed in order to realize the potential of such indoor FSO networks in practice.

Figure 2 demonstrates the breadth of characteristics for various wireless technologies and use-cases for current and future wireless devices. The left image represents the tradeoffs in directionality and coverage of various wireless technologies. Generally, the high resolution of directional communications offers the potential for massive gains in area spectral efficiency; however, the reduced instantaneous coverage also implies unique challenges when considering network provisioning for multiple dynamic users [21–23]. The right image of Figure 2 depicts the potential alignment of technologies with wireless devices that have various data demand and mobility characteristics. Applications such as interactive cloud-based services, video sourcing, augmented reality, and virtual reality have unique usage characteristics that vary greatly in terms of mobility and demand. These novel applications are changing the way that we interact with wireless networks. Furthermore, we expect unforeseen applications to contribute to the device and traffic dynamics in the coming years. This device heterogeneity motivates our vision of a heterogeneous multi-tier system, depicted in Figure 1, where the unique characteristics of various wireless technologies are optimally aligned with the characteristics of the wireless devices that they are connecting [24–26].

	Coverage Range	Coverage Resolution	Steering Velocity
RF Small Cells (e.g., 2.4/5GHz WiFi)	++	--	
Visible Light Communication (VLC)	+	+	
mmWave Communications	+	+	++
Fixed Free Space Optics (FSO)	--	++	
Mechanically Steered Indoor FSO	-	++	+
Solid State Steerable Indoor FSO	-	++	++
Terahertz Communications	-	++	++

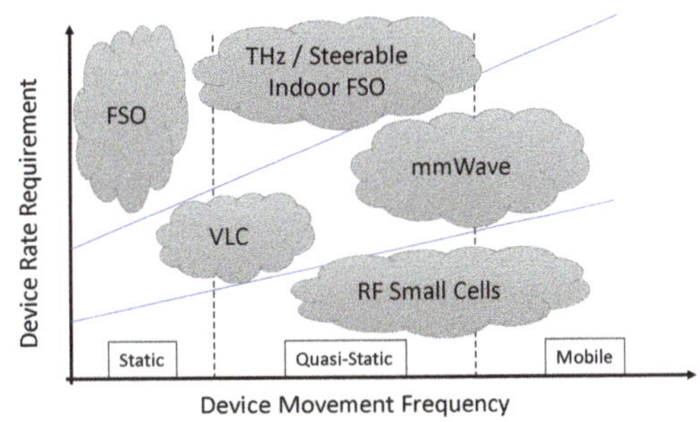

Figure 2. Generalized evaluation of link technology characteristics (**left**) and a context-of-use based analysis of preferred link technologies (**right**). We consider VLC as LED-based systems for both lighting and communications. FSO is broadly categorized as fixed building-to-building FSO or short range steerable FSO for indoor communications. *Range* directly related to potential coverage. *Resolution* is inversely related to instantaneous coverage. Absence of + or − indicates that the technology is typically non-steerable.

2.1. Background and Existing Literature

Fiber optic and FSO communications are two proven high-data-rate technologies, each with successful commercial adoption; however, both are typically used with persistent point-to-point links. Indoor optical wireless communications technologies (e.g., visible light communications or LiFi) are also beginning to see commercial acceptance; however, these systems are typically static emission systems without beam steering due to lighting requirements in the common dual-use paradigm (i.e., systems that provide both data communications and indoor illumination). Furthermore, these systems do not approach the extreme directionality of pencil-beam FSO links. Thus, novel architectures and protocols are needed at the system level in order to address the challenge of connecting mobile and quasi-static users in dynamic environments using highly directional optical beams. At the scale of cellular networks, heterogeneous beam management techniques have been introduced in the literature in order to provide a method of coordination between macrocell and microcell base stations [27–31]. However, this heterogeneity is typically considered in the context of scale rather than wireless technology. In other words, the work referenced in [27–31] primarily considers spatial tiers of similar RF mmWave technologies that are collectively managed to avoid interference. Our vision considers the heterogeneous integration of microwave, mmWave, and Indoor FSO technologies at the scale of indoor wireless networks. Considering optimal tier alignment, mobility-aware networks have also been presented at the cellular level in order to assign resources based on mobility context [32,33]; however the scale of indoor FSO networks implies the need for novel methods that account for translational and rotational motion at higher resolution. Work in the area of heterogeneous radio and optical wireless networks has similarly introduced the idea of mobility aware or context aware resource allocation [26,34,35]; however, this has typically applied to the distribution of traffic across static RF small cells and static directional optical cells.

2.2. System View

Figure 1 shows our system vision where multiple user devices (UDs) receive and transmit data to overhead multi-element access points (APs) using a combination of low GHz/microwave transmission from RFSCs, mmWave directional communications, and narrow electronically steered laser sources. The mmWave APs create multiple coverage zones within the broader coverage of the RFSC. The indoor FSO links may be co-located with mmWave APs or configured as sub-zones in the potential coverage range of a mmWave AP. Each wireless technology has unique resolution and range that impacts the corresponding instantaneous and potential coverage (defined in Figures 1 and 2). Along with steering velocity, these characteristics define the time needed to scan a coverage zone. High resolution is ideal in terms of area spectral efficiency; however, resolution is inversely related to the scan time. The multi-tiered design allows for multiple levels of coverage—addressing the indoor FSO signal acquisition through beam refinement as well as providing various access tiers for UDs with different mobility traits. The radio resource control (RRC) manages the various beam configurations and resource allocation in order to bridge between the network and medium access control (MAC) layers.

3. MTH Beam Management Protocol

The MTH beam management protocol aims to allow multiple mobile UDs to establish and maintain a connection with fast initial access and seamless connectivity. Accordingly, the protocol must account for pointing, acquisition, and tracking (PAT) at each directional tier; heterogeneous beam refinement and handover across tiers; allocation of beams/resources in multi-UD environments; and zone configuration/deployment strategies. The protocol design must also account for physical parameters including: beam width variation, resolution, range, and transition rates at each tier; number of elements and layout of elements in each tier; and control plane feedback latency.

3.1. Pointing, Acquisition, and Tracking (PAT)

Each directional tier must account for the well known PAT requirements of directional communication technologies [36]. Pointing addresses the the need to direct an element's emission towards the desired UD. Pointing protocols may require the AP to scan for available UDs or use an alternative communication medium to make the system aware of the UD's presence [37–39]. After the pointing process has determined the UD's general direction, the acquisition process iteratively adapts the beam emission profile (i.e., width and direction) in order to concentrate emitted signal on the receiver. With an established connection, tracking monitors the UD's movement and potentially adapts the emission profile to maintain the link [40]. There are obvious tradeoffs between narrow and broad emission. Wider emission simplifies tracking for small scale variations in the UD's location. Narrow emission can increase system capacity, but it requires more frequent adaptation and high speed feedback to accommodate the small scale movement.

3.2. Multi-Tier Beam Refinement

A high-level depiction of the beam refinement and link selection protocol for a single UD is shown in Figure 3. The three-tier beam refinement begins with the UD connecting to the RFSC and progresses through the course resolution directional tier (e.g., mmWave) and the fine-resolution indoor FSO tier. Prior to beginning the acquisition process for a higher resolution tier, the UD assesses its current mobility traits and application requirements (e.g., reliability, security, etc.) in order to make a decision about moving to the next tier. For example, a quasi-static UD that is currently in motion is likely to postpone transferring to a higher resolution tier. Similarly, a UD running a latency-constrained application may opt to remain in Tier 2 if the Tier 3 connection has intermittent outages—even if the link capacity is higher in Tier 3. Outages are more likely in Tier 3 due to the resolution of the FSO link and the fact that optical wireless communications have a higher potential for occlusions.

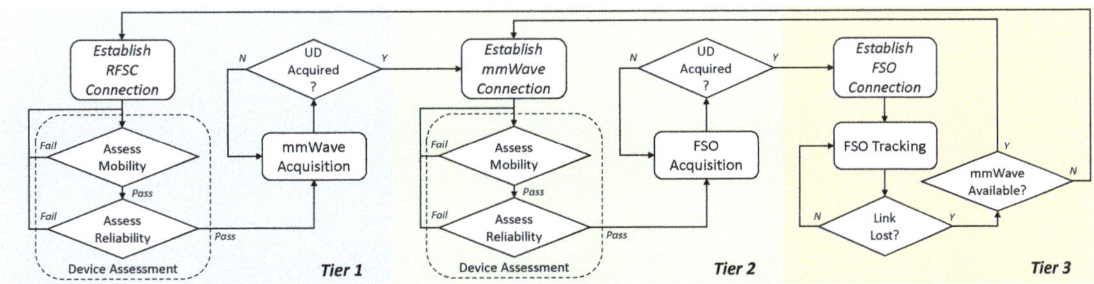

Figure 3. Multi-Tier Heterogeneous (MTH) beam refinement protocol. Device assessment compares UD mobility characteristics and application reliability requirements with next tier characteristics prior to acquiring a higher resolution link.

3.3. Beam Management

A basic instantiation of the MTH protocol can assume that individual UDs operate with a greedy approach and attempt to connect to the tier with highest resolution and link capacity; however, UD mobility, reliability/security requirements, and other device characteristics can also be considered when determining the tier that maximizes average throughput. An extension of the single-UD protocol would account for aggregate system capacity and resource allocation across multiple UDs. To couple the beam management across tiers for both single-UD beam refinement and multi-UD network configuration, the MTH beam management protocol should sit within the radio resource control (RRC) between the Network/Internet layer and the Link layer (Figure 4). This abstract view of the protocol implementation demonstrates the importance of tight integration with both the Data link/Physical layers and the Network layer. The protocol must integrate with the Data link/Physical layers in order to manage the PAT process across tiers. It should also integrate with the Network layer in order to manage routing of data traffic across tiers during beam refinement and/or accommodate resource allocation decisions in scenarios where the optimal tier is related to the UD context-of-use.

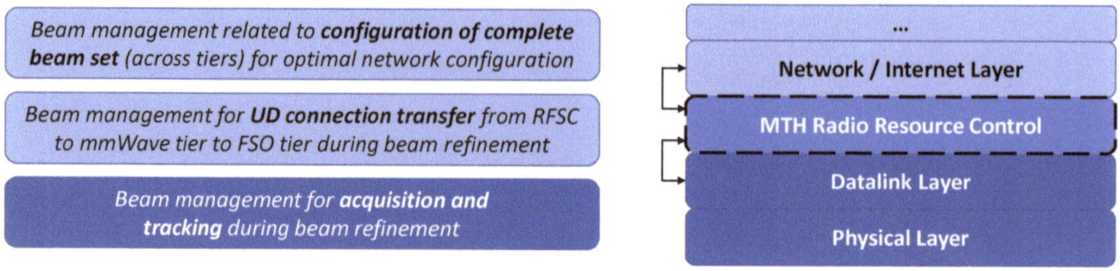

Figure 4. MTH radio resource control for beam management in the TCP/IP 5-layer model.

3.4. Zone Configuration

Optimal system deployment should account for zone configuration—including the number of elements per tier and the coverage associated with each element. The obvious motivation for multiple elements is to provide an opportunity for multiple simultaneous links; however, multiple elements also impact coverage and redundancy. Providing Tier 3 access throughout the Tier 2 zone requires a single element to steer to any location within the zone. To relax this requirement, the indoor FSO network may be deployed to only cover a portion of the zone or, alternatively, be configured with sectors or subzones assigned to different elements (Figure 5). The challenge with sectors is that the protocols must account for additional transitions as UDs move between sectors (i.e., from subzone to subzone). Distributed elements also add redundancy and mitigate outage concerns. When elements are distributed, potential coverage regions can overlap without leading to contention due to the small instantaneous coverage. This provides redundancy in the potential links for a UD to connect to, as we demonstrated in [19].

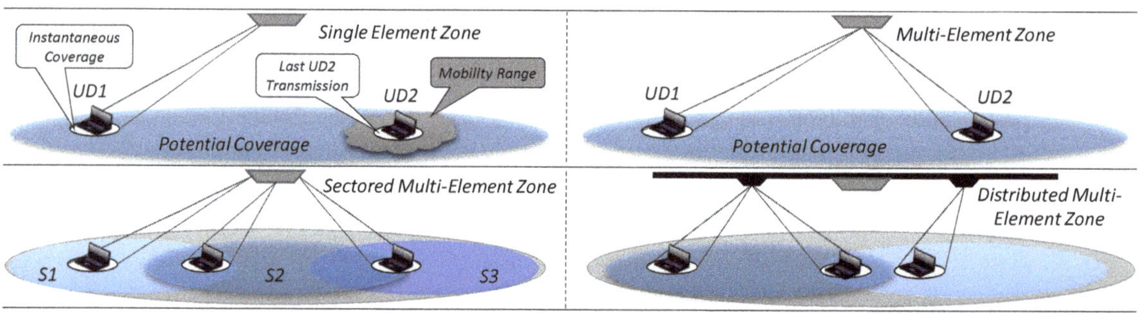

Figure 5. Potential coverage, instantaneous coverage, and associated multiple access strategies for single element (**top left**), multi-element (**top right**), and sectored multi-element (**bottom left**) steerable FSO links as well as a sectored multi-element zone with distributed elements (**bottom right**).

4. Analysis Framework

In this paper, we highlight a specific focus on the optimization of the beam refinement process. We aim to allow UDs to connect to the highest resolution/maximum throughput tier while minimizing access delay — both during initial access and upon occurrence of an outage. The heterogeneous beam refinement allows the Tier 2 search to occur while other UDs use Tier 3 for communication. Furthermore, a variety of search techniques (RSSI, ToF, AoA, etc.) may be used, and the protocol may avoid connecting to the middle tiers during beam refinement in order to speed up access to the higher tier.

Our preliminary analysis focuses on the ideal relationship between range (i.e., potential coverage), resolution, and steering velocity at each tier. To demonstrate the optimization problem, consider a three-tier beam refinement strategy where each tier narrows the search space to the area covered by a single configuration, and the next tier searches over the space covered by a single configuration of the previous tier. Assuming non-overlapping configurations, the worst case search time is

$$T_s = T_d + t_2 n_2 + t_3 n_3 = T_d + t_2 \alpha_2 \beta_2 + t_3 \alpha_3 \beta 3 \tag{1}$$

where T_d is the RFSC's device discovery time, t_i is the time to switch between configurations for tier i, and n_i is the number of configurations in the ith tier. With the assumption of non-overlapping configurations, we can define $n_i = \alpha_i \beta_i$ where α_i and β_i are the resolution and range of tier i, respectively. We define the tier resolution as the number of configurations per unit area and the range as the total area covered by the tier. These parameters are summarized in Table 1.

Table 1. MTH analysis parameters.

Parameter	Description
T_s	Maximum Search Time
T_d	Discovery Time (Tier 1)
t_i	Configuration Switching Time (Tier i)
n_i	Number of Configurations (Tier i)
α_i	Tier i Resolution [configurations/m^2]
β_i	Tier i Range [m^2]

Assuming that the search space of tier $i+1$ is equal to the instantaneous coverage (i.e., area covered by a single configuration) of tier i, then $\beta_{i+1} = 1/\alpha_i$. In other words, tier i narrows the search space to an area covered by a single configuration and tier $i+1$ searches this space. Accordingly, Equation (1) can be rewritten as either of the following:

$$T_s = T_d + t_2 \alpha_2 \beta_2 + t_3 \frac{\alpha_3}{\alpha_2} \tag{2}$$

$$T_s = T_d + t_2 \frac{\beta_2}{\beta_3} + t_3 \alpha_3 \beta_3 \tag{3}$$

The key observation in this analysis is the impact of α_2 (or, similarly, β_3). Specifically, there is an ideal resolution of Tier 2 that minimizes T_s and this resolution is dependent on the range of Tier 2 and resolution of Tier 3. Furthermore, the relative importance of β_2 and α_3 is dependent on the relative switching speeds of Tiers 2 and 3. To determine the optimal Tier 2 resolution, we must determine α_2 that minimizes T_s for given T_d, t_2, t_3, β_2 and α_3. Observing Equations (2) and (3), we can find the rate of change of T_s with respect to α_2 and β_3, respectively.

$$\frac{d}{d\alpha_2} T_s = t_2 \beta_2 - t_3 \frac{\alpha_3}{\alpha_2^2} \tag{4}$$

$$\frac{d}{d\beta_3} T_s = -t_2 \frac{\beta_2}{\beta_3^2} + t_3 \alpha_3 \tag{5}$$

Given that T_s is a convex function for $\alpha_2 \in \mathbb{R}_{>0}$, we can set the derivative from Equation (4) to zero and solve for the positive value of α_2 (i.e., Tier 2 resolution) in order to find the value of α_2 that minimizes T_s. We can similarly observe Equation (5) in order to view the impact in terms of β_3 (i.e., Tier 3 range or, similarly, Tier 2 instantaneous coverage). Following this process, we find the optimal Tier 2 resolution and Tier 3 range in Equations (6) and (7), respectively. Given that $\beta_{i+1} = 1/\alpha_i$ for this specific problem formulation, we recognize the direct relationship between the optimal Tier 2 resolution and Tier 3 range. As such, the optimal value for β_3 could be directly derived from knowledge of α_2 (or vice versa). However, the inverse relationship of the derived results in Equations (6) and (7) demonstrates that the general relationship holds and offers a sense of validation. Furthermore, the direct relationship implies that all relevant information can be defined by either

α_2 or β_3; however, each can be used to better visualize the relative performance impact under different operating conditions (as seen in Section 5).

$$\underset{\alpha_2 \in \mathbb{R}_{>0}}{\arg\min}(T_s) = \sqrt{\frac{t_3\alpha_3}{t_2\beta_2}} \tag{6}$$

$$\underset{\beta_3 \in \mathbb{R}_{>0}}{\arg\min}(T_s) = \sqrt{\frac{t_2\beta_2}{t_3\alpha_3}} \tag{7}$$

Under the given assumptions, these results could be used to define the ideal mmWave (Tier 2) emission profile for a system with known FSO emission profile and FSO/mmWave configuration switching times along with a specified coverage area. However, these simplifying assumptions are obviously limited in terms of practical application since a more realistic system would account for configuration overlap and non-uniform coverage, feedback and pipelined scanning, and other protocol characteristics. Accordingly, these results are intended primarily for qualitative analysis in order to observe the relative impact of various system parameters.

5. Results and Analysis

To observe the relative impact of the parameters, we evaluate the search time for a few scenarios where the switching time of the Tier 3 FSO link is faster than, equivalent to, or slower than the switching time of the Tier 2 mmWave link. To provide a more general analysis, all results are depicted in time relative to the switching time of the mmWave link (i.e., t_2). Without loss of generality, we also assume $T_d = 0$ since the Tier 1 discovery time is a constant offset independent of the other parameters.

The first results, depicted in Figure 6, illustrate the search time when the FSO tier can switch between configurations much faster than the mmWave tier (i.e., $t_2 = 100t_3$). The plot on the left shows the search time versus α_2 and the plot on the right shows the corresponding results as a function of β_3. In each plot, results are shown for combinations of $\beta_2 \in \{9, 25\}$ and $\alpha_3 \in \{625, 2500, 10,000\}$. These Tier 3 resolution values represent FSO links with instantaneous coverage of 16, 4, and 1 cm^2, respectively. When the Tier 2 resolution is low, we can see that the search time is heavily influenced by the Tier 3 resolution (i.e., the third component of Equation (2)). This implies that the Tier 2 scan completes quickly and most of the time is spent searching within Tier 3. Conversely, as the resolution increases there are more Tier 2 configurations in a given space and fewer Tier 3 configurations within the instantaneous coverage of a single Tier 2 configuration; therefore, the scan spends most of its time in Tier 2. Accordingly, the search time is heavily influenced by the Tier 2 range (i.e., the second component of Equation (2)). This is similarly seen in the right plot where search time is heavily influenced by Tier 2 range when Tier 3 range is small and heavily influenced by Tier 3 resolution when Tier 3 range is large.

In Figure 7, we demonstrate the impact of relative switching times—specifically, we show results for scenarios where: (a) switching between FSO configurations is 10× faster than switching between mmWave configurations; (b) FSO and mmWave switching times are equivalent; and (c) switching between FSO configurations is 10× slower than switching between mmWave configurations. The key observation here is that the trends are similar; however, increasing the time to switch between FSO configurations or reducing the time to switch between mmWave configurations shifts the optimal Tier 2 resolution to the right and, accordingly, reduces the optimal Tier 3 range. Intuitively, this makes sense since reducing the relative performance of Tier 3 implies that the system should increase the utilization of Tier 2.

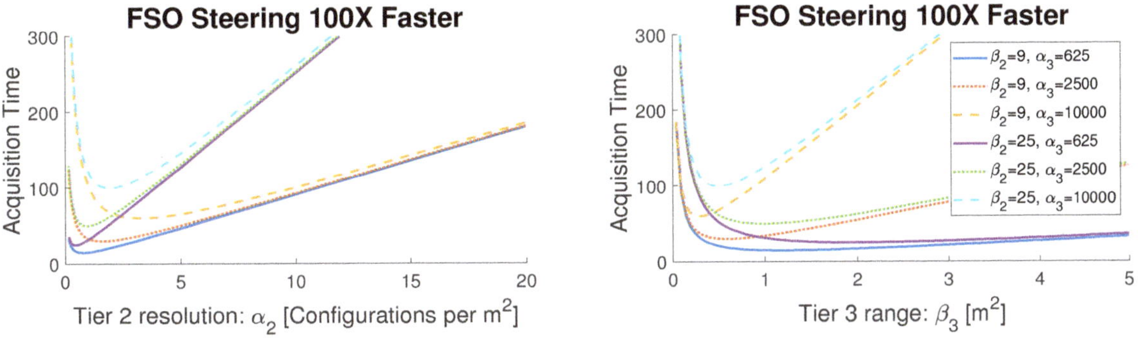

Figure 6. Maximum search time for MTH protocol as a function of Tier 2 resolution (i.e., α_2) and Tier 3 range (i.e., β_3) when $t_2/t_3 = 100$.

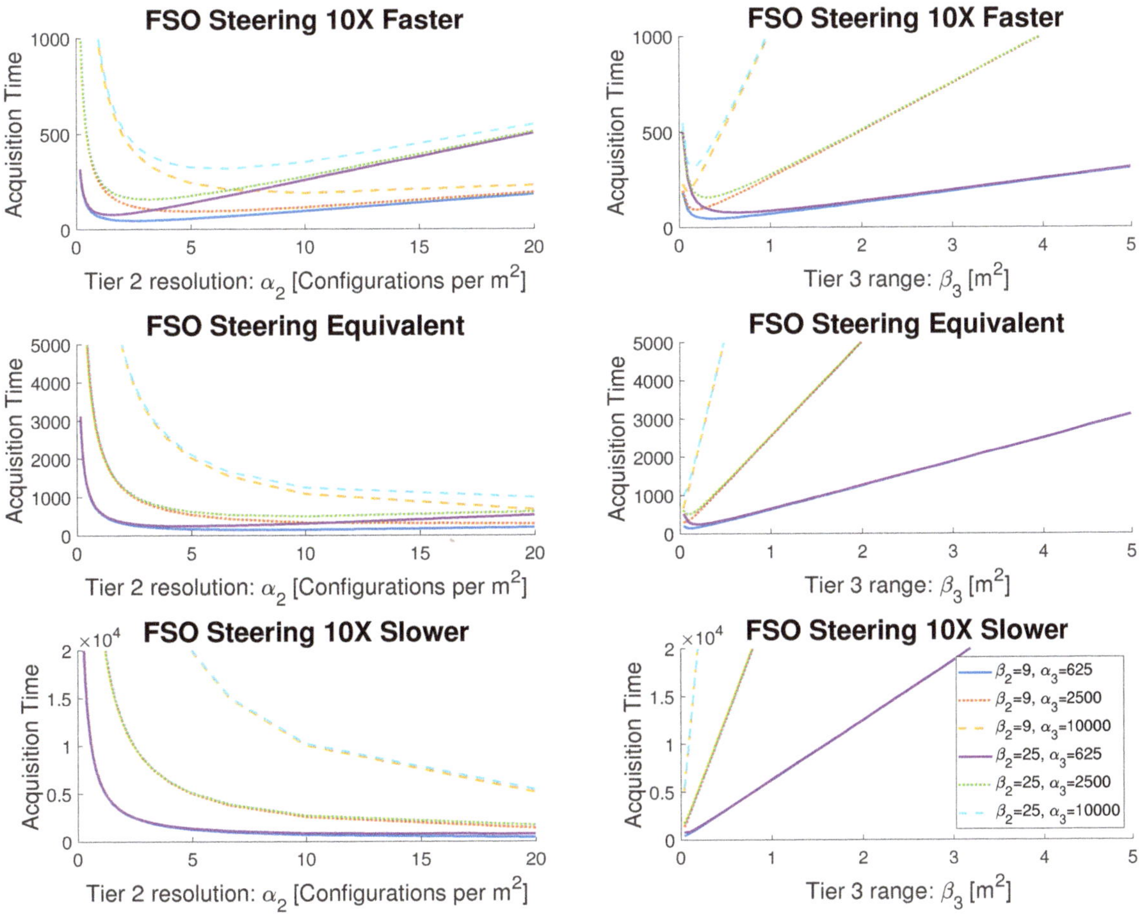

Figure 7. Maximum search time for MTH protocol as a function of Tier 2 resolution (i.e., α_2) and Tier 3 range (i.e., β_3). Results are shown for: $t_2/t_3 = 10$ (**top**); $t_2 = t_3$ (**middle**); and $t_2/t_3 = 0.1$ (**bottom**). The legend in the bottom right plot is common to all plots.

Lastly, Figure 8 highlights the impact that Tier 2 range, Tier 3 resolution, and relative switching times have on the optimal Tier 2 resolution (or, similarly, Tier 3 coverage). The obvious observation is that the three sets of figures appear similar; however, the color bar range shows that relative switching time has an impact on magnitude of the outcomes. As FSO switching becomes faster, Tier 3 can explore a larger space in a similar time, therefore the optimal Tier 2 resolution (i.e., α_2) decreases and the optimal Tier 3 range (i.e., β_3) increases. Observing the impact of β_2 and α_3, we can see that larger spaces (i.e., larger β_2 values) increase the optimal Tier 3 range in order to distribute the additional required search time across both tiers. Increased Tier 3 resolution drives the FSO link towards pencil-beam emission, requiring more Tier 3 configurations over a given area; thus, the optimal Tier 2 resolution increases to again better utilize Tier 2 in the search.

To reiterate the point from above, this analysis presents a qualitative example of the tradeoffs; however, quantitative analysis depends on the uniformity of coverage, hardware characteristics (e.g., steering velocity), and the feedback latency of the control messaging protocol between AP and UD, which may be pipelined. Accordingly, this analysis demonstrates tradeoffs in design parameters, but the optimal search time ultimately depends on the protocol characteristics.

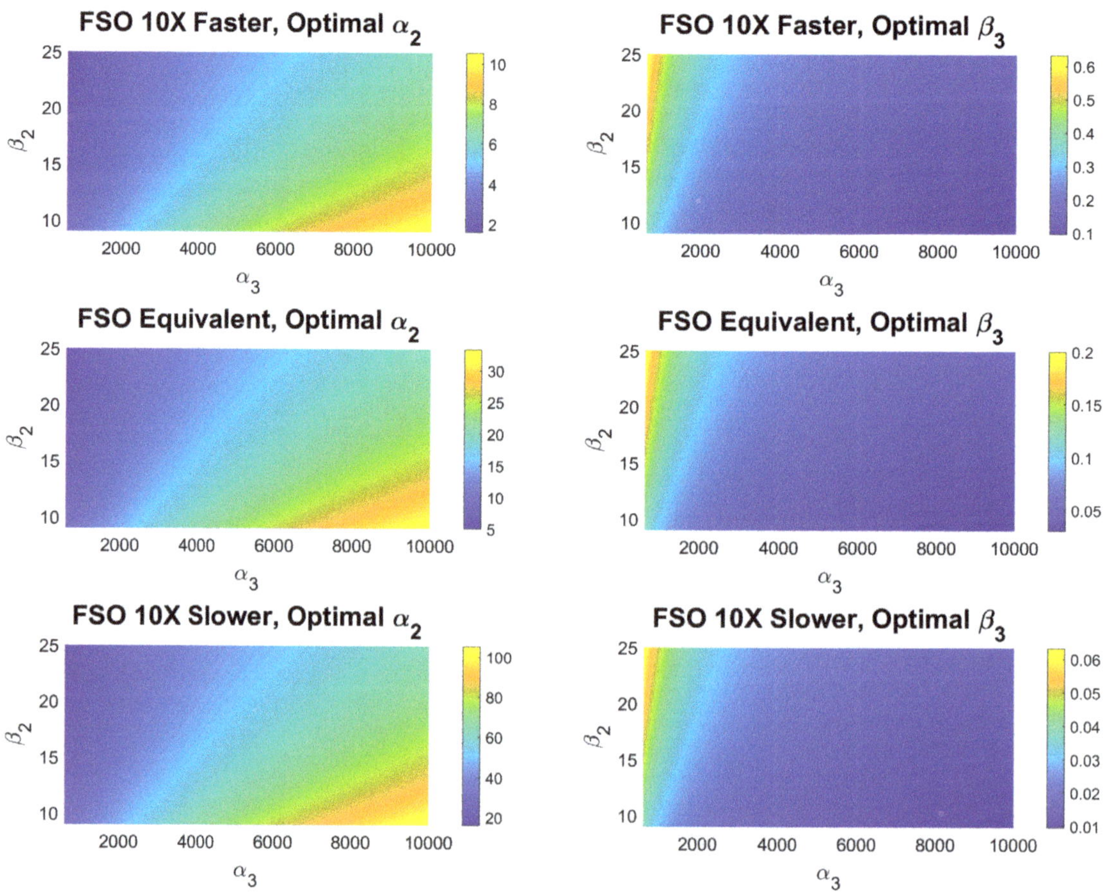

Figure 8. Optimal Tier 2 resolution (i.e., α_2) and Tier 3 range (i.e., β_3) for systems where relative switching times are related by: $t_2/t_3 = 10$ (**top**); $t_2 = t_3$ (**middle**); and $t_2/t_3 = 0.1$ (**bottom**).

6. Conclusions

In this paper, we provide a thorough overview of the requirements and considerations for a novel multi-tiered heterogeneous beam management strategy that promises to: (a) improve traffic distribution across technologies; and (b) improve initial access time of a steerable indoor FSO network. In particular, we introduce a high level beam refinement protocol that accounts for device characteristics and link reliability before moving to a tier with higher directionality. The protocol also aims to improve indoor FSO acquisition time by taking advantage of the unique characteristics of directional technologies operating at different tiers. To evaluate the relative impact of various system parameters, we developed a qualitative analysis framework for the proposed MTH beam management strategy. This framework was used to demonstrate tradeoffs in the design decisions and derive an optimization analysis that minimizes the worst-case search time in a three tier implementation. While the presented evaluation is intended primarily for relative comparison, the analysis motivates a tighter coupling and integration of various directional communication technologies within the future wireless communications ecosystem.

Author Contributions: Conceptualization, M.B.R., T.D.C.L. and M.H.; Formal analysis, M.B.R.; Writing—original draft, M.B.R.; Writing—review & editing, T.D.C.L. and M.H. All authors have read and agreed to the published version of the manuscript.

Funding: This research received no external funding.

Data Availability Statement: The data presented in this study are available on request from the corresponding author.

Conflicts of Interest: The authors declare no conflict of interest.

References

1. Petrov, V.; Moltchanov, D.; Koucheryavy, Y. On the Efficiency of Spatial Channel Reuse in Ultra-Dense THz Networks. In Proceedings of the 2015 IEEE Global Communications Conference (GLOBECOM), San Diego, CA, USA, 6–10 December 2015; pp. 1–7. doi:10.1109/GLOCOM.2015.7417199.
2. Perez, J.; Chicharro, F.I.; Ortega, B.; Mora, J. On the evaluation of an optical OFDM radio over FSO system with IM-DD for high-speed indoor communications. In Proceedings of the 2017 19th International Conference on Transparent Optical Networks (ICTON), Girona, Spain, 2–6 July 2017; pp. 1–4. doi:10.1109/ICTON.2017.8025112.
3. Boulogeorgos, A.A.; Goudos, S.K.; Alexiou, A. Users Association in Ultra Dense THz Networks. In Proceedings of the 2018 IEEE 19th International Workshop on Signal Processing Advances in Wireless Communications (SPAWC), Kalamata, Greece, 25–28 June 2018; pp. 1–5. doi:10.1109/SPAWC.2018.8445950.
4. Calvanese Strinati, E.; Barbarossa, S.; Gonzalez-Jimenez, J.L.; Ktenas, D.; Cassiau, N.; Maret, L.; Dehos, C. 6G: The Next Frontier: From Holographic Messaging to Artificial Intelligence Using Subterahertz and Visible Light Communication. *IEEE Veh. Technol. Mag.* **2019**, *14*, 42–50. doi:10.1109/MVT.2019.2921162.
5. Sarkar, T.S.; Sinha, B.; Mukherjee, S.; Joardar, I.; Mazumdar, S. Development of an FPGA based Indoor Free Space Optical (FSO) Communication System using 808 nm Infrared (IR) LASER Source. In Proceedings of the 2020 IEEE Calcutta Conference (CALCON), Kolkata, India, 28–29 February 2020; pp. 313–317. doi:10.1109/CALCON49167.2020.9106422.
6. Gomez, A.; Shi, K.; Quintana, C.; Maher, R.; Faulkner, G.; Bayvel, P.; Thomsen, B.C.; Obrien, D. Design and Demonstration of a 400 Gb/s Indoor Optical Wireless Communications Link. *J. Light. Technol.* **2016**, *34*, 5332–5339. doi:10.1109/JLT.2016.2616844.
7. O'Hara, J.F.; Ekin, S.; Choi, W.; Song, I. A Perspective on Terahertz Next-Generation Wireless Communications. *Technologies* **2019**, *7*. doi:10.3390/technologies7020043.
8. Aghasi, H.; Naghavi, S.; Tavakoli Taba, M.; Aseeri, M.; Cathelin, A.; Afshari, E. Terahertz electronics: Application of wave propagation and nonlinear processes. *Appl. Phys. Rev.* **2020**, *7*, 021302. doi:10.1063/1.5129403.
9. Tonkikh, E.V.; Burobina, K.D.; Shurakhov, A.A. Possible Applications of Sixth Generation Communication Networks. In Proceedings of the 2020 Systems of Signals Generating and Processing in the Field of on Board Communications, Moscow, Russia, 19–20 March 2020; pp. 1–6. doi:10.1109/IEEECONF48371.2020.9078581.
10. Firyaguna, F.; Kibilda, J.; Galiotto, C.; Marchetti, N. Coverage and Spectral Efficiency of Indoor mmWave Networks with Ceiling-Mounted Access Points. In Proceedings of the GLOBECOM 2017-2017 IEEE Global Communications Conference, Singapore, 4–8 December 2017; pp. 1–7. doi:10.1109/GLOCOM.2017.8254623.
11. Chowdhury, M.Z.; Hossan, M.T.; Islam, A.; Jang, Y.M. A Comparative Survey of Optical Wireless Technologies: Architectures and Applications. *IEEE Access* **2018**, *6*, 9819–9840. doi:10.1109/ACCESS.2018.2792419.
12. Chowdhury, M.Z.; Hasan, M.K.; Shahjalal, M.; Hossan, M.T.; Jang, Y.M. Optical Wireless Hybrid Networks: Trends, Opportunities, Challenges, and Research Directions. *IEEE Commun. Surv. Tutor.* **2020**, *22*, 930–966. doi:10.1109/COMST.2020.2966855.

13. Li, J.; Li, R.; Rahaim, M.B. Design and Implementation of a Heterogeneous Multi-Hop OWC/RF Relay Architecture for Opportunistic Ultra-Dense Wireless Networks. In Proceedings of the 2020 11th IEEE Annual Ubiquitous Computing, Electronics Mobile Communication Conference (UEMCON), New York, NY, USA, 28–31 October 2020; pp. 0684–0689. doi:10.1109/UEMCON51285.2020.9298126.
14. Gomez, A.; Shi, K.; Quintana, C.; Faulkner, G.; Thomsen, B.C.; O'Brien, D. A 50 Gb/s Transparent Indoor Optical Wireless Communications Link With an Integrated Localization and Tracking System. *J. Light. Technol.* **2016**, *34*, 2510–2517. doi:10.1109/JLT.2016.2542158.
15. Fahs, B.; Chowdhury, A.J.; Hella, M.M. A 12-m 2.5-Gb/s Lighting Compatible Integrated Receiver for OOK Visible Light Communication Links. *J. Light. Technol.* **2016**, *34*, 3768–3775. doi:10.1109/JLT.2016.2587598.
16. Fahs, B.; Romanowicz, M.; Hella, M.M. A Gbps Building-to-Building VLC Link Using Standard CMOS Avalanche Photodiodes. *IEEE Photonics J.* **2017**, *9*, 1–9. doi:10.1109/JPHOT.2017.2765499.
17. Fahs, B.; Hella, M.M. 3 Gb/s OOK VLC link using bandwidth-enhanced CMOS Avalanche photodiode. In *Optical Fiber Communications Conference and Exhibition (OFC)*; Optical Society of America: Location: Washington, DC, USA, 2017; pp. 1–3.
18. Oh, C.W.; Cao, Z.; Tangdiongga, E.; Koonen, T. Free-space transmission with passive two-dimensional beam steering for indoor optical wireless networks. *Opt. Express* **2017**, *24*, 19211–19227. doi:10.1364/OE.24.019211.
19. Rahaim, M.; Morrison, J.; Little, T. Beam Control for Indoor FSO and Dynamic Dual-Use VLC Lighting Systems. *J. Commun. Inf. Netw.* **2017**, *2*. doi:10.1007/s41650-017-0041-7.
20. Koonen, T. Indoor Optical Wireless Systems: Technology, Trends, and Applications. *J. Light. Technol.* **2018**, *36*, 1459–1467. doi:10.1109/JLT.2017.2787614.
21. Abdalla, I.; Rahaim, M.B.; Little, T.D.C. Dynamic FOV Tracking Receiver for Dense Optical Wireless Networks. In Proceedings of the 2019 IEEE Global Communications Conference (GLOBECOM), Waikoloa, HI, USA, 9–13 December 2019; pp. 1–6. doi:10.1109/GLOBECOM38437.2019.9014277.
22. Abdalla, I.; Rahaim, M.B.; Little, T.D.C. Interference Mitigation Through User Association and Receiver Field of View Optimization in a Multi-User Indoor Hybrid RF/VLC Illuminance-Constrained Network. *IEEE Access* **2020**, *8*, 228779–228797. doi:10.1109/ACCESS.2020.3045929.
23. Attiah, M.L.; Isa, A.A.M.; Zakaria, Z.; Abdulhameed, M.; Mohsen, M.K.; Ali, I. A survey of mmWave user association mechanisms and spectrum sharing approaches: an overview, open issues and challenges, future research trends. *Wirel. Netw.* **2020**, *26*, 2487–2514. doi:10.1007/s11276-019-01976-x.
24. Ayyash, M.; Elgala, H.; Khreishah, A.; Jungnickel, V.; Little, T.; Shao, S.; Rahaim, M.; Schulz, D.; Hilt, J.; Freund, R. Coexistence of WiFi and LiFi toward 5G: concepts, opportunities, and challenges. *IEEE Commun. Mag.* **2016**, *54*, 64–71. doi:10.1109/MCOM.2016.7402263.
25. Rahaim, M.B.; Little, T.D.C. Toward practical integration of dual-use VLC within 5G networks. *IEEE Wirel. Commun.* **2015**, *22*, 97–103. doi:10.1109/MWC.2015.7224733.
26. Rahaim, M.; Abdalla, I.; Ayyash, M.; Elgala, H.; Khreishah, A.; Little, T.D.C. Welcome to the CROWD: Design Decisions for Coexisting Radio and Optical Wireless Deployments. *IEEE Netw.* **2019**, 1–9. doi:10.1109/MNET.2019.1800297.
27. Yuan, F.; Yang, C.; Wang, G.; Lei, M. Adaptive Channel Feedback for Coordinated Beamforming in Heterogeneous Networks. *IEEE Trans. Wirel. Commun.* **2013**, *12*, 3980–3994. doi:10.1109/TWC.2013.071913.121553.
28. Tataria, H.; Shafi, M.; Smith, P.J.; Dmochowski, P.A. Coordinated two-tier heterogeneous cellular networks with leakage based beamforming. *arXiv* **2015**, arXiv:1503.01566.
29. Peng, H.; Moriwaki, K.; Suegara, Y. Macro-Controlled Beam Database-Based Beamforming Protocol for LTE-WiGig Aggregation in Millimeter-Wave Heterogeneous Networks. In Proceedings of the 2016 IEEE 83rd Vehicular Technology Conference (VTC Spring), Nanjing, China, 15–18 May 2016; pp. 1–6. doi:10.1109/VTCSpring.2016.7504335.
30. Wang, F.; Chen, W.; Wu, Q.; Tang, H. Robust MISO Beamforming Under the Deterministic Model in Two-Tier Heterogeneous Networks. *IEEE Access* **2017**, *5*, 14616–14625. doi:10.1109/ACCESS.2017.2731423.
31. Hefnawi, M. Hybrid Beamforming for Millimeter-Wave Heterogeneous Networks. *Electronics* **2019**, *8*, 133.
32. Ha, D.T.; Boukhatem, L.; Kaneko, M.; Martin, S. An Advanced Mobility-Aware Algorithm for Joint Beamforming and Clustering in Heterogeneous Cloud Radio Access Network. In Proceedings of the 21st ACM International Conference on Modeling, Analysis and Simulation of Wireless and Mobile Systems, New York, NY, USA, 25 October 2018; pp. 199–206. doi:10.1145/3242102.3242120.
33. Tabassum, H.; Salehi, M.; Hossain, E. Fundamentals of Mobility-Aware Performance Characterization of Cellular Networks: A Tutorial. *IEEE Commun. Surv. Tutor.* **2019**, *21*, 2288–2308. doi:10.1109/COMST.2019.2907195.
34. Dastgheib, M.A.; Beyranvand, H.; Salehi, J.A.; Maier, M. Mobility-Aware Resource Allocation in VLC Networks Using T-Step Look-Ahead Policy. *J. Light. Technol.* **2018**, *36*, 5358–5370. doi:10.1364/JLT.36.005358.
35. Wu, X.; Haas, H. Mobility-aware load balancing for hybrid LiFi and WiFi networks. *J. Opt. Commun. Netw.* **2019**, *11*, 588–597. doi:10.1364/JOCN.11.000588.
36. Giordani, M.; Polese, M.; Roy, A.; Castor, D.; Zorzi, M. A Tutorial on Beam Management for 3GPP NR at mmWave Frequencies. *IEEE Commun. Surv. Tutor.* **2019**, *21*, 173–196. doi:10.1109/COMST.2018.2869411.
37. Nor, A.M.; Mohamed, E.M. Millimeter Wave Beamforming Training Based on Li-Fi Localization in Indoor Environment. In Proceedings of the GLOBECOM 2017-2017 IEEE Global Communications Conference, Singapore, 4–8 December 2017; pp. 1–6. doi:10.1109/GLOCOM.2017.8254474.

38. Maletic, N.; Sark, V.; Gutiérrez, J.; Grass, E. Device Localization Using mmWave Ranging with Sub-6-Assisted Angle of Arrival Estimation. In Proceedings of the 2018 IEEE International Symposium on Broadband Multimedia Systems and Broadcasting (BMSB), Valencia, Spain, 6–8 June 2018; pp. 1–6. doi:10.1109/BMSB.2018.8436861.
39. Haider, M.K.; Ghasempour, Y.; Koutsonikolas, D.; Knightly, E.W. LiSteer: MmWave Beam Acquisition and Steering by Tracking Indicator LEDs on Wireless APs. In Proceedings of the 24th Annual International Conference on Mobile Computing and Networking (MobiCom '18), New Delhi, India, 29 October–2 November 2018; pp. 273–288. doi:10.1145/3241539.3241542.
40. Fahs, B.; Romanowicz, M.; Kim, J.; Hella, M.M. A Self-Alignment System for LOS Optical Wireless Communication Links. *IEEE Photonics Technol. Lett.* **2017**, *29*, 2207–2210. doi:10.1109/LPT.2017.2771303.

applied sciences

MDPI

Article

Low Complexity DSP for High Speed Optical Access Networking †

Jinlong Wei [1,2,*], Cedric F. Lam [3], Ji Zhou [4], Ivan Aldaya [5], Elias Giacoumidis [6], Andre Richter [6], Qixiang Cheng [7], Richard Penty [7] and Ian White [7,8,*]

1 Currently with Optical and Quantum Communication laboratory, European Research Center, Huawei Technologies Duesseldorf GmbH, Riesstr. 25, 80992 Munich, Germany
2 Was with ADVA Optical Networking SE, Märzenquelle 1-3, 98617 Meiningen-Dreissigacker, Germany
3 Google Access, 1600 Amphitheatre Pkwy, Mountain View, CA 94043, USA; cedricfunglam@gmail.com
4 Department of Electronic Engineering, College of Information Science and Technology, Jinan University, Guangzhou 510632, China; zhouji_jnu@163.com
5 Center for Advanced and Sustainable Technologies, São João da Boa Vista, State University of São Paulo, São João da Boa Vista 13876-750, Brazil; ivan.aldaya@sjbv.unesp.br
6 VPIphotonics GmbH, Carnotstraße 6, 10587 Berlin, Germany; ilias.giakoumidis@vpiphotonics.com (E.G.); andre.richter@vpiphotonics.com (A.R.)
7 Department of Engineering, University of Cambridge, 9 JJ Thomson Avenue, Cambridge CB3 0FA, UK; qc223@cam.ac.uk (Q.C.); rvp11@cam.ac.uk (R.P.)
8 University of Bath, Claverton Down, Bath BA2 7AY, UK
* Correspondence: jinlongwei2@huawei.com (J.W.); I.H.White@bath.ac.uk (I.W.)

† This paper is a significantly extended version of paper published in The Optical Networking and Communication Conference & Exhibition, held in Los Angeles, CA, USA, 19–23 March 2017.

Citation: Wei, J.; Lam, C.F.; Zhou, J.; Aldaya, I.; Giacoumidis, E.; Richter, A.; Cheng, Q.; Penty, R.; White, I. Low Complexity DSP for High Speed Optical Access Networking . *Appl. Sci.* **2021**, *11*, 3406. https://doi.org/10.3390/app11083406

Academic Editors: Fabio Cavaliere and Luca Potì

Received: 5 March 2021
Accepted: 7 April 2021
Published: 10 April 2021

Abstract: A novel low-cost and energy-efficient approach for reaching 40 Gb/s signals is proposed for cost-sensitive optical access networks. Our proposed design is constituted of an innovative low-complex high-performance digital signal processing (DSP) architecture for pulse amplitude modulation (PAM-4), reuses existing commercial cost-effective 10-G components and eliminates the need of a power-hungry radio frequency (RF) component in the transmitter. Using a multi-functional 17-tap reconfigurable adaptive Volterra-based nonlinear equalizer with noise suppression, significant improvement in receiver optical power sensitivity is achieved. Results show that over 30 km of single-mode fiber (SMF) a link power budget of 33 dB is feasible at a bit-error-rate (BER) threshold of 10^{-3}.

Keywords: passive optical network; digital signal processing; feedforward equalization; decision feedback equalization; noise suppression

1. Introduction

Today's fastest standardized passive optical network (PON) is the 25 Gb/s per lane Ethernet PON [1]. Driven by bandwidth hungry applications such as high definition video services and virtual reality, a higher lane rate PON is under study by Telecommunication Standardization Sector of the International Telecommunications Union (ITU-T) [2]. An alternative approach is to increase the lane count by ultra-dense wavelength division multiplexing (UWDM) with central office consolidation and extended geographic coverage. On this topic, the Institute of Electrical and Electronics Engineers (IEEE) 802.3cs super PON standard under development supports a 50 km reach and a larger than 1:64 split ratio but requires a wavelength router in the field. As the attractiveness of PONs is very sensitive to cost and power consumption, it is highly desirable that high speed PONs leverage the optical component ecosystem of today's mature 10 Gb/s PONs, where intensive efforts have been made by the research community [3–14]. Higher spectrally efficient modulation schemes and advanced digital signal processing (DSP) must be employed to achieve such targets. Examples include chirp managed non-return-to-zero (NRZ) [3], electrical/optical Duobinary [4–7], four-level pulse amplitude modulation (PAM-4) [3,7–10,13] and PAMs with even more amplitude levels [15], as well as more complex multi-carrier [5,9,11] or multi-band modulation schemes [12].

Notably, PAM-4 has been chosen as the format for high speed Ethernet standards [16] for data center (DC) applications due to its implementation simplicity. High-volume PAM-4 chips are commercially available. However, at >25 Gb/s bit rates, implementing PAM-4 with simple linear equalizers [7,8,17,18] to support a transmission distance of 20 km over a single-mode fiber (SMF)—as required in access networks—using only 10-G optoelectronics [7], is extremely challenging, unless higher bandwidth transmitters or receivers (e.g., 25 G photodiodes [8]), or a dedicated chromatic dispersion compensation module [4] are adopted. It has been shown that reach extension is possible by mitigating system and component nonlinearities using DSP without changing the optical infrastructure [18,19].

It is worth mentioning that, alternatively, there are research studies proposing to employ coherent-lite technologies [20–23]. Coherent reception enables a much higher achievable transmission capacity as well as longer transmission distances but requires a higher number of devices. However, the work presented in this paper is limited to simple direct detection (DD) applications using PAM-4 at the C-band. Although previous DD PAM-4 systems with improved capacity [24,25] have been demonstrated, they required the use of sophisticated neural network (NN) nonlinear equalization [24], or an O-band where chromatic dispersion is much smaller [25]. Moreover, none of the previous DD PAM-4 demonstrations considered a simpler nonlinear equalization architecture with noise suppression.

This work proposes a novel 40 Gb/s lane rate PON architecture that includes a high-performance DSP unit, 10-G optoelectronics and at the same time eliminates the need for a power-hungry radio frequency (RF) driver from the transmitter. Thanks to the adoption of shared bi-directional optical amplifiers (OAs) in the optical line terminal (OLT) site, as standardized in the 10-G lane PON, the optical launch power can be maintained at a high level in order to guarantee a high optical power budget. The technical challenge is that the nonlinearity attributed to signal–signal beating upon square-law detection must be tackled and traded off with the carrier-to-signal power ratio (CSPR) of the modulation scheme. Our novel DSP approach uses a nonlinear adaptive DSP with zero overheads at the receiver side to effectively tackle the nonlinearity issue. The nonlinear DSP can be further simplified and flexibly reconfigured, guaranteeing a low-cost and low-power receiver. Such a PON has the capability of full C-band tunability with 100 GHz channel spacing, positioning it as an attractive solution for 40 Gb/s lane rate PONs, and hence, enabling PON links with >1.6 Tb/s link capacity (e.g., by using 48 WDM channels of 40 Gb/s transceivers over the whole C-band). Here, experimental demonstrations were performed using a single channel 40 Gb/s PAM-4 signal transmission over 20 km (30 km) SMF with a link power budget of 38 dB (30.7 dB), at a pre- forward error correction (FEC) bit-error-rate (BER) of 10^{-3}. In addition, when a simpler KP-4 FEC with a BER threshold of 2.2×10^{-4} was considered, a link power budget of 33 dB (27 dB) was achieved after a 20 km (30 km) SMF transmission.

2. Experimental Setup

Figure 1a illustrates the experimental setup and Figure 1b depicts the detailed architecture of the nonlinear equalizer. In the transmitter, an offline PAM-4 signal is generated after a 2-tap symbol-spaced digital pre-distortion and a Nyquist shaping with a roll-off factor of 0.45. The generated signal is loaded into a 16 GHz, 80 GS/s digital-to-analog convertor (DAC) with an output swing of approximately 250 mV. Without using any driver amplifier, a 10-G tunable transmitter assembly (TTA) consisting of a tunable laser operating at 1543.3 nm and a differentially driven InP Mach–Zehnder intensity modulator (MZM) with a bandwidth of approximately 11 GHz [8] is driven by the DAC output directly. Following the MZM, a tunable optical band-pass filter (OBPF) mimicking an optical multiplexer with a 3 dB bandwidth of approximately 40 GHz is adopted and its output is amplified by a booster Erbium-doped fiber amplifier (EDFA). The OBPF has a 2nd order super Gaussian filtering profile. A variable optical attenuator (VOA) at the output of the

EDFA optimized the launched optical power into the fiber link, thus, always providing the lowest BER. After transmission over an SMF link of up to 30 km, a combined optical receiver consisting of a VOA, a pre-amplifier EDFA, a 40-GHz bandwidth OBPF mimicking an optical de-multiplexer with identical configuration of the transmitter OBPF, and a 10-G p–i–n photodiode- transimpedance amplifier (PIN-TIA) is utilized to detect the received optical signal. The PIN-TIA has a bandwidth of roughly 11 GHz and its input power is adjusted and optimized by a prior VOA. The detected PAM-4 signal is then converted into a digital signal by analog to digital convertor (ADC) sampling at 80 GS/s and the following DSP is conducted offline.

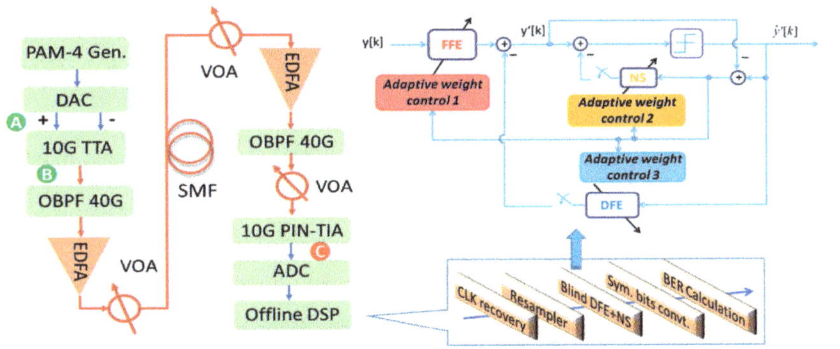

(a) (b)

Figure 1. (**a**) Experimental setup of a 40 Gb/s lane rate next generation passive optical network (NG-PON) downstream link using pulse amplitude modulation (PAM-4) and the diagram of the receiver digital signal processing (DSP). (**b**) Detailed architecture of the nonlinear equalizer.

The detailed process of offline DSPs for the receiver is also presented in Figure 1. The clock recovery obtains the optimum sampling phase and then a re-sampler down-samples the signal into 2 samples per symbol. A following blind nonlinear feedforward and decision feedback equalizer (FFE/DFE), together with a noise suppression (NS) filter, is then adopted to combat both linear and nonlinear effects from the low-bandwidth components and fiber link. After symbol-to-bit mapping, the BER prior to forward error correction (FEC) is the result of a one-on-one comparison between the transmitted and the recovered bits. The adaptive nonlinear equalizer (NLE) is the core of the receiver DSP, which is indicated in Figure 1b. The proposed reconfigurable NLE has the capability to combat both linear and nonlinear distortions as well as to suppress the signal noise. The DSP is operated in a completely blind manner with zero overhead. As indicated in Figure 1b, the nonlinear equalizer consists of both a Volterra FFE, a DFE and an NS filter to recover the symbols.

The cost function adopted for the optimization of the coefficients is the least mean square (LMS), which is expressed as:

$$\begin{aligned}\hat{y}'[k] = \ & \textstyle\sum_{i=0,j=1}^{M-1,N}\left(a_i y[k-i] - b_j \hat{y}'[k-j]\right) \\ & +\textstyle\sum_{i=0,j=i}^{M-1} a_{ij} y[k-i] y[k-j] - \textstyle\sum_{i=1,j=i+1}^{N} b_{ij} \hat{y}'[k-i] \hat{y}'[k-j] \\ & -\textstyle\sum_{h=1}^{H} p_h\left(y'[k-h] - \hat{y}'[k-h]\right)\end{aligned} \quad (1)$$

where M, N, and H are the FFE, DFE and NS filter orders, and y and y' represent the FFE input and output, respectively. $\hat{y}'$ is obtained from a hard decision of y'. Only linear and the 2nd-order Volterra kernels are considered to maintain a low level of complexity. $a_i(b_j)$ and $a_{ij}(b_{ij})$ are the linear and nonlinear FFE (DFE) tap coefficients, respectively. p_h is the NS filter tap coefficient. The idea of the NS filter is to predict the noise of the current FFE output sample based on the past H noise samples and the predicted noise is then subtracted from the signal sample. The past noise samples can be estimated via $y'[k\text{-}h]$ -$\hat{y}'[k\text{-}h]$ thanks to the decision feedback. The nonlinear DSP has excellent reconfigurability by switching ON or OFF the DFE and NS filter and part of the taps with lower weights can be disabled, depending on the channel conditions and transceiver power requirement.

A low complexity 10G-Ethernet PON (EPON) FEC Reed-Solomon [RS(255,223)] with 14.35% overhead, 6.56 dB net coding gain and a threshold BER of 10^{-3} is assumed here [26]. Considering the superior performance of the proposed NLE, a 5.84% overhead KP-4 RS(544, 514) FEC with a threshold BER of 2.2×10^{-4}, widely adopted in high speed Ethernet links, can also be used [27].

3. Experimental Results at 20 km Transmission

The eye diagrams of the modulator output shown in Figure 2 exhibit strong eye closure and asymmetry with smaller lower eyes. The degradation of the signal quality caused by the MZM is obvious by comparing the eye diagram with the DAC output eye diagram. The latter is much more open and the apertures are more uniform, although a slight eye closure is observed due to the DAC bandwidth limitation. The eye diagrams of the detected signals are also shown in Figure 2, in which they are completely closed, regardless of the fiber transmission distance. This is a direct result of the joint effect of limited transceiver bandwidth and fiber dispersion.

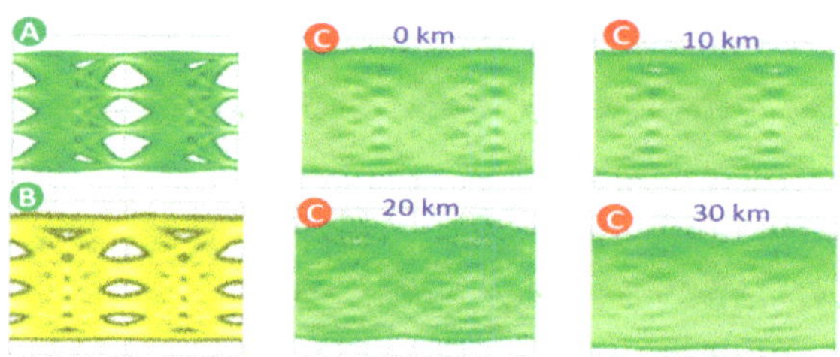

Figure 2. The eye diagrams (10 ps/div time resolution) obtained at the output of the DAC (marker **A**), the modulator (marker **B**), and the photodiode (marker **C**), respectively.

Figure 3a identifies an optimum operating point that is far below the linear regime, which is verified by the BER versus modulator bias voltage curve presented in Figure 3b. Considering a small differential driving signal swing of about 500 mV, which is about one ninth of the V_π of the MZM, a bias closer to the null point brings about a lower CSPR (namely, a higher signal to noise ratio (SNR) for a fixed optical power by suppressing optical carrier) but stronger signal–signal beating interference upon square-law detection at the same time. The optimum bias gives rise to the best trade-off between CSPR and beating interference. This leads to a low optical output power, but it can be amplified by a booster EDFA. This configuration is critical to reduce the transceiver cost and power due to the avoidance of an RF driver in each channel since the EDFA cost can be shared by all channels.

Figure 4a presents the BER versus the received optical power performance under various receiver DSP configurations for an SMF length of 20 km. With only a linear FFE with an identified optimum tap count of $M = 8$, the BER improves slowly with increasing optical power but an error floor appears at a BER level of 5×10^{-5}. The use of DFE with an optimum tap count $N = 4$, in addition to FFE, shows a performance enhancement, especially in lowering the error floor. Based on the optimum FFE and DFE tap counts, the 2nd-order nonlinear terms given in Equation (1) are additionally introduced, requiring a total number of 44/10 FFE/DFE taps according to Equation (1). It is obvious that nonlinear equalization brings a tremendous improvement in both the power sensitivity and error floor. Further improvement is achieved by using an NS filter, which is more significant at lower BERs. Overall, a total optical power sensitivity improvement of approximately 2 dB (4 dB) is obtained at the 10-G EPON FEC (KP-4 FEC) threshold BER of 1×10^{-3} (2×10^{-4}), respectively.

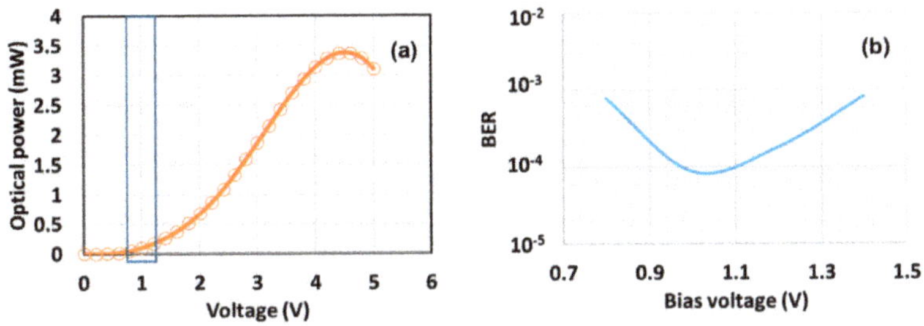

Figure 3. (**a**) The Mach–Zehnder intensity modulator (MZM) transfer function and the identified optimum modulation regime in the inset block and (**b**) bit-error-rate (BER) vs. modulator bias voltage for optical back to back case.

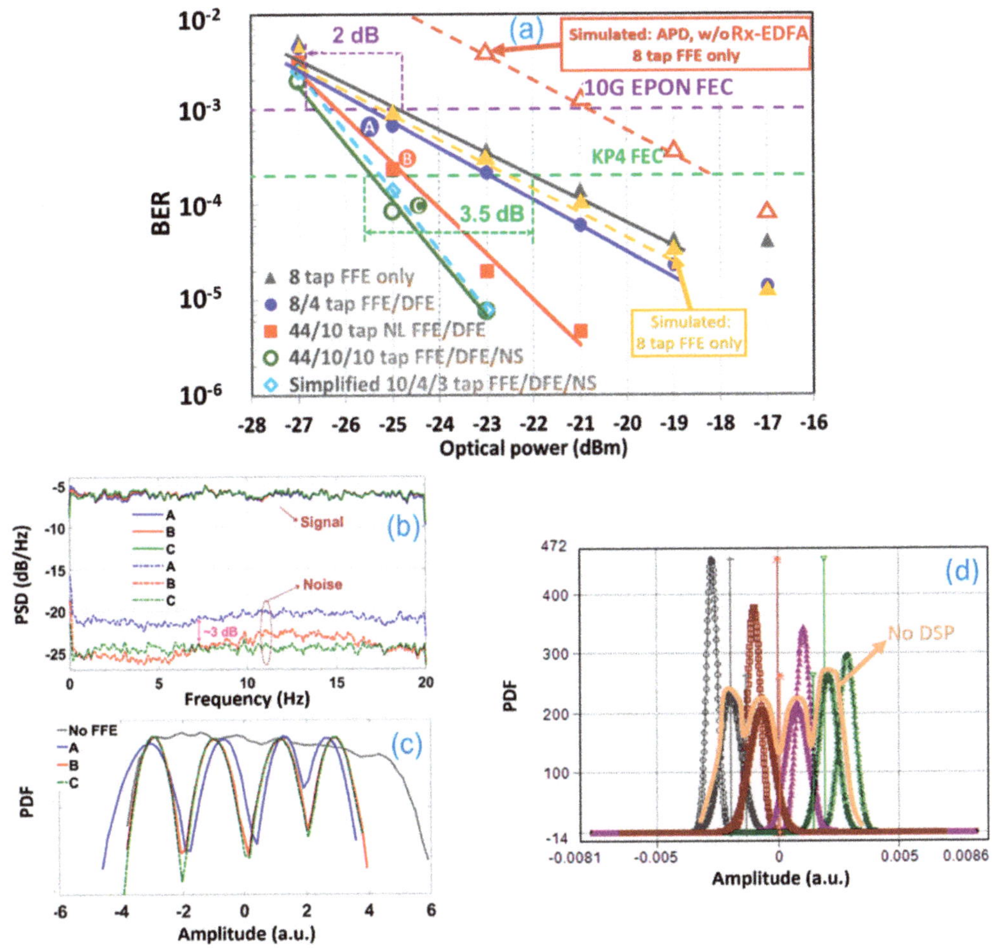

Figure 4. (**a**) BER vs. optical power performance using different equalization approaches for transmission of 20 km single-mode fiber (SMF) and (**b**) power spectral densities. (**c**) Histograms of signals corresponding to DSP configurations marked in BER curves with −25 dBm optical power. (**d**) Received simulated histograms for avalanche photodiode (APD) case (w/o receiver EDFA) with and without DSP at −25 dBm optical power.

The improvement enabled by the receiver adaptive DSP is firstly attributed to—as indicated in Figure 4b—the nonlinear equalization, that effectively mitigates the signal–signal beating interference due to fiber dispersion and square-law detection, leading to approximately a 3 dB reduction in noise power. Secondly, the nonlinear response of the MZM near to the null bias point can also be significantly compensated by nonlinear equalization. This is identified by comparing the histograms shown in Figure 4c of the equalized signals of linear and nonlinear equalization cases under the same optical power: nonlinear equalization reduces the frequency of amplitudes distributed in the threshold regimes, leading to reduced errors on the following hard decisions. Moreover, Figure 4c also indicates that the nonlinear equalized PAM-4 signal shows even more amplitude space between each level. The NS filter not only significantly suppresses the noise distributed in the frequency range beyond 10 GHz but also whitens the noise power spectral density.

It is also shown in Figure 4a that it is possible to significantly reduce the FFE/DFE and NS filter taps, and hence, the filter complexity, without sacrificing the performance of 20 km SMF transmission. This is achieved by disabling the taps whose values are below the average of the corresponding tap category.

The pre-amplified receiver could potentially be replaced by an avalanche photodiode (APD) in the downlink system for cost-efficiency reasons, but the corresponding sensitivity penalty will typically rise to ~5 dB in this case [7,8,12]. However, if dynamic biasing of the APD is used, it is expected that the sensitivity can be reasonably improved [28].

4. Simulation Results at 20 km Transmission

Due to the absence of a high-bandwidth APD in our laboratory, we performed realistic simulations using the VPIphotonics Design Suite platform (VPItransmissionMaker) to predict the impact of a realistic APD in the proposed setup. For our simulations, we used two cases, the first being identical to the experimental setup using the EDFA and a PIN with a transimpedance amplifier (TIA). While the second replaced the latter components with a single APD + TIA. For simplicity, we used an 8 tap FFE in the DSP unit for comparison with the corresponding experimental measurements with a simplified DSP. A total of 2^{20} symbols were used for effective BER Monte-Carlo simulation. The simulation parameters for the PAM-4 with the APD set-up were identical to [8], while the following additional parameters were considered in the simulation: a single-polarization system was used using an standard SMF (SSMF) with attenuation, chromatic dispersion, chromatic dispersion slope and a Kerr nonlinear index of 0.2 dB/km, 16 ps/(km $\times$ nm), 0.08 ps/km $\times$ nm^2, and 2.6×10^{-20}m^2/W, respectively. For both EDFAs, the noise figure was set 6 dB. For the PIN and APD, the thermal noise was set at 1 pA/$\sqrt{}$(Hz), the dark current at 10 nA and the responsivity at 0.9 A/W. The avalanche multiplication factor for the APD was 2.5. The bandwidth limitation of the modulator and PIN/APD was simulated using Bessel low-pass filters of the 3rd order, while the TIA was modelled with a 4th filter order. Gaussian OBPFs of the 2nd order were also simulated. Identical PAM-4 parameters to the experiment were used, with a current noise spectral density of 10 nA/$\sqrt{}$(Hz). The laser linewidth was set to 10 MHz and its relative intensity noise (RIN) was set to -130 dB/Hz at a measured power of 1 mW. The laser average power was set to 10 mW, while the side mode separation was set to 200 GHz and the side mode suppression ratio to 100 dB. The MZM was assumed to be ideal with an insertion loss of 6 dB and extinction ratio of 30 dB.

In Figure 4a, together with the experimental traces from Section 3, the simulated results are depicted for the case of an 8 tap FFE at 20 km of transmission. Figure 4d shows the received histograms for the afore-mentioned simulated DSP case without using an equalizer. Our simulated results designate the following: (1) An implementation penalty of ~0.4 dB is observed at the KP4 FEC-limit when compared to the experimental curve, mainly induced from the electro-optical components; (2) A ~4.3 dB sensitivity penalty is observed when using an APD without receiver pre-amplification, confirming the prediction reported in [7,8,12].

5. DSP Analysis and Experimental Transmission-Reach Limitations

The simplification of the NLE filter taps is illustrated in Figure 5. Note that for the simplified DSP case, the equalization runs again after the taps below the threshold are disabled from the complete DSP. On convergence, the survived taps show different values compared to their complete DSP counterparts since the interaction between taps changed. Taking the nonlinear FFE term as an example, Figure 5 shows that the most significant nonlinear terms are the square terms and the product terms between most adjacent symbols, which corresponds to the nonlinear FFE coefficient $a_{i,j}$ for $|i-j| \leq 1$. The product terms between symbols further away from each other are negligible and can be eliminated to reduce computational complexity. It is observed that the penalty of the simplified nonlinear FFE/DFE, relative to the full nonlinear FFE/DFE, increases slightly with increasing the fiber length beyond 20 km. This is because of the increased fiber dispersion and stronger distortion (as shown in Figure 2), and the fact that a small part of the eliminated nonlinear product terms shown above are no longer negligible. It is noticeable that both the FFE and DFE terms show that more weighted taps survived via simplification. While for the NS filter, the tap significance was changed slightly via simplification. This is because the change of the FFE/DFE taps has an influence on the noise correlation. Overall, the total tap count is reduced by 75% (from 44/10/10 to 10/4/3 taps). The simplified DSP makes the receiver comparable to a commercial simply equalized clock and the data recovery (CDR) of PAM-4 as regards the complexity and power dissipation.

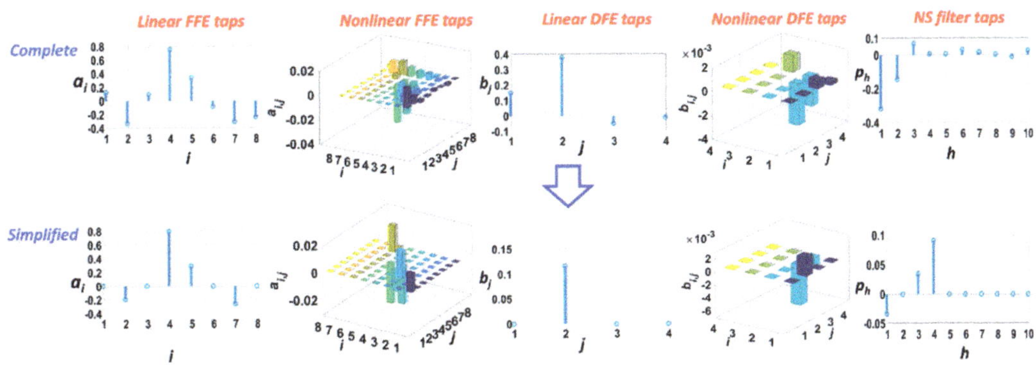

Figure 5. Feedforward and decision feedback equalizer (FFE/DFE) and noise suppression (NS) filter tap coefficients for full DSP and simplified DSP cases with 20 km SMF at an optical power of −21 dBm.

Figure 6a summarizes the optical power sensitivity at the threshold BERs of both of the proposed FECs for various fiber lengths under two receiver DSP configurations: 8/4 tap FFE/DFE without 2nd-order terms, and a simplified 10/4/3 tap FFE/DFE/NS filter with both linear and 2nd-order terms as well as the NS filter. It clearly shows that when compared with a linear FFE/DFE, 2nd-order Volterra nonlinear equalizations with an NS filter considerably enhance the receiver sensitivity and that the gain is more significant at longer fiber distances. The linear equalization fails to support 30 km SMF transmission for both FECs.

Considering a launch power of 12 dBm, Figure 6b presents the achievable link power budget, subject to the DSP configurations in Figure 6a. As references, three different high value power budget specifications from 10 G EPON standard are illustrated. It indicates that the proposed 40 Gb/s PAM-4 PON can achieve a link power budget of 39 dB (33 dB) and 37 dB (27 dB) for 20 km (30 km) SMF transmissions by using the 10 G EPON FEC and KP-4 FEC, respectively. The system can support PR30 and PR40 power budget specification with 20 km and 25 km distances, respectively, under all DSP configurations. Only the FFE/DFE/NS configuration together with 10 G EPON FEC can support PR50 by considering a 25 km reach. It is worth mentioning that the pre-amplified receiver may be replaced by an APD in the downlink system for cost reasons and the corresponding achievable power budgets will be ~7.5 dB less [7,8,12]. In this case, the PR30 power budget (20 km reach) can still be achieved. When dynamic biasing APD is used, it is expected that the power budget can be reasonably improved [28]. In the case of using an APD without a receiver EDFA, simulations in Figure 4 have revealed that 20 km reach can be achieved and so it can cover the PR30 power budget as depicted in Figure 6b.

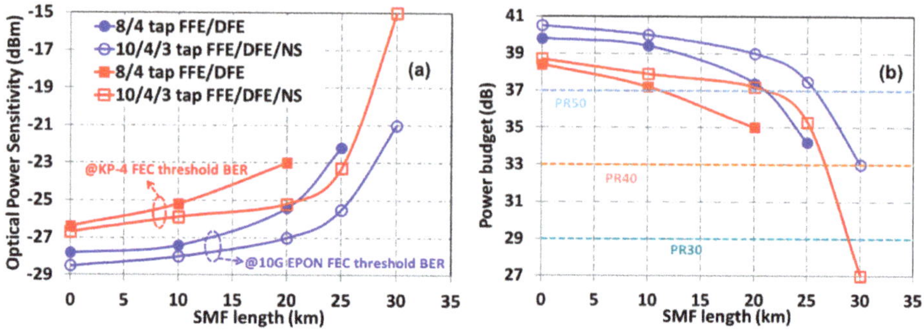

Figure 6. (**a**) Receiver sensitivity and (**b**) the corresponding power budget at different fiber lengths of the 40 Gb/s PAM-4 system.

6. Conclusions

We have demonstrated a unique low-complexity and energy-efficient 40 Gb/s PAM-4 PON system using a novel high-performance DSP unit and a 10-G transceiver without requiring a power-hungry RF-driver. The DSP consisted of a low complexity reconfigurable blind adaptive Volterra-based nonlinear FFE/DFE, integrated with an NS filter as a receiver equalizer to alleviate various linear and nonlinear effects from the components and the fiber. The results revealed that a link power budget of 39 dB and 37 dB for 20 km (30 km) SMF transmission was feasible at a BER threshold of 10^{-3}, respectively. Up to a 75% reduction in the DSP complexity is feasible without performance degradation. The 17-tap NLE has a complexity that is comparable with a commercial equalized CDR. Simulation results revealed that an APD can potentially be used to avoid pre-amplification, thus, enabling greater cost and energy efficiency in modern PONs, while simultaneously covering the PR30 power budget at 20 km.

Author Contributions: Conceptualization, experiment, writing—original draft preparation, J.W.; formal analysis, investigation, simulation, writing—review and editing, J.W., C.F.L., J.Z., I.A., E.G., A.R., Q.C., R.P. and I.W. All authors have read and agreed to the published version of the manuscript.

Funding: This research was funded by the European Union under the CEEOALAN project with grant No. 623,515, and by the German Ministry of Education and Research under the KIGLIS project with grant No. 16KIS1230.

Institutional Review Board Statement: Not applicable.

Informed Consent Statement: Not applicable.

Data Availability Statement: Data sharing is not applicable to this article.

Acknowledgments: The authors thank Nikolas Eiselt, Christoph Wagner, Klaus Grobe, and Helmut Griesser from ADVA Optical Networking SE, Germany for useful discussions.

Conflicts of Interest: The authors declare no conflict of interest.

References

1. IEEE. P802.3ca 25 Gb/s and 50 Gb/s Ethernet Passive Optical Networks. Available online: http://www.ieee802.org/3/ca (accessed on 8 April 2021).
2. ITU-T Higher Speed Passive Optical Networks: 50G PMD. Available online: https://www.itu.int/itu-t/workprog/ (accessed on 8 April 2021).
3. Ji, H.; Hu, W.; Yi, L.; Li, Z.; Xue, L.; Li, X.; Yang, Q.; Wang, S.; Yang, Y.; Yu, S. Field Demonstration of a Real-Time 100-Gb/s PON Based on 10G-Class Optical Devices. *J. Light. Technol.* **2016**, *35*, 1914–1921. [CrossRef]
4. Houtsma, V.; Van Veen, D.; Gnauck, A.; Iannone, P. APD-Based DuoBinary Direct Detection Receivers for 40 Gbps TDM-PON. In Proceedings of the Optical Fiber Communication Conference and Exposition (OFC) 2015, Los Angeles, CA, USA, 24–26 March 2015. Paper Th4H. 1.
5. Houtsma, V.; Veen, D.V. Investigation of Modulation Schemes for Flexible Line-Rate High-Speed TDM-PON. *J. Lightw. Technol.* **2020**, *38*, 3261–3267. [CrossRef]
6. Ye, Z.; Li, S.; Cheng, N.; Liu, X. Demonstration of High-Performance Cost-Effective 100-Gb/s TWDM-PON Using 4x 25-Gb/s. In Proceedings of the 2015 European Conference on Optical Communication (ECOC), Valencia, Spain, 27 September–1 October 2015. Paper Mo.3.4.4.
7. Yin, S.; Houtsma, V.; Van Veen, D.; Vetter, P. Optical Amplified 40-Gbps Symmetrical TDM-PON Using 10-Gbps Optics and DSP. *J. Light. Technol.* **2016**, *35*, 1067–1074. [CrossRef]
8. Wei, J.; Eiselt, N.; Griesser, H.; Grobe, K.; Eiselt, M.H.; Olmos, J.J.V.; Monroy, I.T.; Elbers, J.-P. Demonstration of the First Real-Time End-to-End 40-Gb/s PAM-4 for Next-Generation Access Applications Using 10-Gb/s Transmitter. *J. Light. Technol.* **2016**, *34*, 1628–1635. [CrossRef]
9. Minghui, T.; Zhou, L.; Zeng, H.; Li, S.; Liu, X. 50-Gb/s/λ TDM-PON Based on 10G DML and 10G APD Supporting PR10 Link Loss Budget after 20-km Downstream Transmission in the O-band. In Proceedings of the Optical Fiber Communication Conference and Exposition (OFC), Los Angeles, CA, USA, 19–23 March 2017. Paper Tu3G.2.
10. Wei, J.; Giacoumidis, E. 40 Gb/s/λ optical amplified PAM-4 PON with transmission over 30 km SMF using 10-G Optics and simple DSP. In Proceedings of the 2017 Optical Fiber Communications Conference and Exhibition (OFC), Los Angeles, CA, USA, 19–23 March 2017. Paper Tu3G. 3.
11. Zhou, J.; Zhang, L.; Zuo, T.; Zhang, Q.; Zhang, S.; Zhou, E.; Liu, G.N. Transmission of 100-Gb/s DSB-DMT over 80-km SMF Using 10-G class TTA and Direct-Detection. In Proceedings of the 42nd European Conference on Optical Communication, Düsseldorf, Germany, 18–22 September 2016. Paper P421.
12. Wei, J.; Giacoumidis, E. Multi-band CAP for Next-Generation Optical Access Networks Using 10-G Optics. *J. Light. Technol.* **2017**, *36*, 551–559. [CrossRef]
13. Zhang, J.; Wey, J.S.; Yu, J. Symmetrical 50-gb/s/λ PAM-4 TDM PON in O-band with DSP and Semiconductor Optical Ampli-fier Supporting PR-30 Link Loss Budget. In Proceedings of the Optical Fiber Communication Conference and Exposition (OFC), San Diego, CA, USA, 11–15 March 2018; Paper M1B.4.
14. Bonk, R. SOA for Future PONs. In Proceedings of the Optical Fiber Communication Conference and Exposition (OFC), San Diego, CA, USA, 11–15 March 2018; Paper Tu2B. 4.
15. Yi, L.; Li, P.; Liao, T.; Hu, W. 100Gb/s/ λ IM-DD PON Using 20G-Class Optical Devices by Machine Learning Based Equalization [Invited]. In Proceedings of the European Conference on Optical Communication (ECOC), Roma, Italy, 23–27 September 2018; Paper Mo4B.5.
16. IEEE. P802.3bs 200 Gb/s and 400 Gb/s Ethernet Task Force. Available online: http://www.ieee802.org/3/bs/ (accessed on 8 April 2021).
17. Wei, J.L.; Grobe, K.; Sanchez, C.; Giacoumidis, E.; Griesser, H. Comparison of cost- and energy-efficient signal modulations for next generation passive optical networks. *Opt. Express* **2015**, *23*, 28271–28281. [CrossRef] [PubMed]
18. Wei, J.L.; Grobe, K.; Wagner, C.; Giacoumidis, E.; Griesser, H. 40 Gb/s Lane Rate NG-PON using Electrical/Optical Duo-binary, PAM-4 and Low Complex Equalizations. In Proceedings of the Optical Fiber Communication Conference and Exposition (OFC), Anaheim, CA, USA, 20–22 March 2016; Paper Tu3C.5.
19. Wei, J.; Zhou, J.; Giacoumidis, E.; Haigh, P.A.; Tang, J. DSP-based 40 Gb/s lane rate next-generation access networks. *Future Internet* **2018**, *10*, 118. [CrossRef]
20. Zhang, J.; Yu, J.; Li, X.; Wang, K.; Zhou, W.; Xiao, J.; Zhao, L.; Pan, X.; Liu, B.; Xin, X. 200 Gbit/s/λ PDM-PAM-4 PON system based on intensity modulation and coherent detection. *J. Opt. Commun. Netw.* **2019**, *12*, A1–A8. [CrossRef]
21. Zhang, J.; Wey, J.S.; Shi, J.; Yu, J. Single-Wavelength 100-Gb/s PAM-4 TDM-PON Achieving Over 32-dB Power Budget Using Simplified and Phase Insensitive Coherent Detection. In Proceedings of the European Conference on Optical Communication (ECOC), Roma, Italy, 23–27 September 2018.
22. Rashidinejad, A.; Nguyen, A.; Olson, M.; Hand, S.; Welch, D. Real-Time Demonstration of 2.4Tbps (200Gbps/λ) Bidirec-tional Coherent DWDM-PON Enabled by Coherent Nyquist Subcarriers. In Proceedings of the Optical Fiber Communication Conference and Exposition (OFC) 2020, San Diego, CA, USA, 8–12 March 2020; Paper W2A.30.
23. Erkılınç, M.S.; Emmerich, R.; Habel, K.; Jungnickel, V.; Schmidt-Langhorst, C.; Schubert, C.; Freund, R. PON transceiver technologies for ≥50 Gbits/s per λ: Alamouti coding and heterodyne detection [Invited]. *J. Opt. Commun. Netw.* **2019**, *12*, A162–A170. [CrossRef]

24. Ye, C.; Zhang, D.; Huang, X.; Feng, H.; Zhang, K. Demonstration of 50Gbps IM/DD PAM4 PON over 10GHz Class Optics Using Neural Network Based Nonlinear Equalization. In Proceedings of the European Conference on Optical Communication (ECOC), Gothenburg, Sweden, 17–21 September 2017.
25. Choi, M.R.; Yu, Y.; Bae, S.H.; Kim, D.H.; Kim, H. Transmission of 107-Gb/s PAM-4 Signal over 25 km of SSMF using O-band Commercial Off-the-Shelf DML. In Proceedings of the 2019 24th OptoElectronics and Communications Conference (OECC) and 2019 International Conference on Photonics in Switching and Computing (PSC), Fukuoka, Japan, 7–11 July 2019.
26. *802.3av-2009. IEEE Standard for Information Technology, Local and Metropolitan Area Networks—Specific Requirements*, Part 3: CSMA/CD Access Method and Physical Layer Specifications; Amendment 1: Physical Layer Specifications and Management Parameters for 10 Gb/s Passive Optical Networks; IEEE: New York, NY, USA, 2009.
27. *IEEE Std 802.3bs-2017. IEEE Standard for Ethernet Amendment 10: Media Access Control Parameters, Physical Layers and Management Parameters for 200 Gb/s and 400 Gb/s Operation*; IEEE: New York, NY, USA, 2017.
28. Zarkesh-Ha, P.; Efroymson, R.; Fuller, E.; Campbell, J.C.; Hayat, M.M. 5.2dB Sensitivity Enhancement in 25Gbps APD-Based Optical Receiver Using Dynamic Biasing. In Proceedings of the Optical Fiber Communication Conference and Exposition (OFC) 2020, San Diego, CA, USA, 8–12 March 2020; Paper Th3J.2.

applied sciences

MDPI

Article

Emulating Software-Defined Disaggregated Optical Networks in a Containerized Framework

Francisco-Javier Moreno-Muro [1,*], Miquel Garrich [1], Ignacio Iglesias-Castreño [1], Safaa Zahir [1] and Pablo Pavón-Mariño [1,2]

1 Department of Information and Communication Technologies, GIRTEL Group, Universidad Politécnica de Cartagena, Plaza del Hospital S/N, 30202 Cartagena, Spain; miquel.garrich@gmail.com (M.G.); iglesias.castreno@ignacio.xyz (I.I.-C.); zahie.safaa@gmail.com (S.Z.); pablo.pavon@upct.es (P.P.-M.)

2 E-Lighthouse Network Solutions, 30202 Cartagena, Spain

* Correspondence: javier.moreno@upct.es

Abstract: Telecom operators' infrastructure is undergoing high pressure to keep the pace with the traffic demand generated by the societal need of remote communications, bandwidth-hungry applications, and the fulfilment of 5G requirements. Software-defined networking (SDN) entered in scene decoupling the data-plane forwarding actions from the control-plane decisions, hence boosting network programmability and innovation. Optical networks are also capitalizing on SDN benefits jointly with a disaggregation trend that holds the promise of overcoming traditional vendor-locked island limitations. In this work, we present our framework for disaggregated optical networks that leverages on SDN and container-based management for a realistic emulation of deployment scenarios. Our proposal relies on Kubernetes for the containers' control and management, while employing the NETCONF protocol for the interaction with the light-weight software entities, i.e., *agents*, which govern the emulated optical devices. Remarkably, our agents' structure relies on components that offer high versatility for accommodating the wide variety of components and systems in the optical domain. We showcase our proposal with the emulation of an 18-node European topology employing Cassini-compliant optical models, i.e., a state-of-the-art optical transponder proposed in the Telecom Infrastructure Project. The combination of our versatile framework based on containerized entities, the automatic creation of agents and the optical-layer characteristics represents a novel approach suitable for operationally complex carrier-grade transport infrastructure with SDN-based disaggregated optical systems.

Keywords: disaggregated optical networks; software defined networking; optical network emulation

Citation: Moreno-Muro, F.-J.; Garrich, M.; Iglesias-Castreño, I.; Zahir, S.; Pavón-Mariño, P. Emulating Software-Defined Disaggregated Optical Networks in a Containerized Framework. *Appl. Sci.* **2021**, *11*, 2081. https://doi.org/10.3390/app11052081

Academic Editor: Fabio Cavaliere

Received: 1 February 2021
Accepted: 22 February 2021
Published: 26 February 2021

1. Introduction

The rapid growth of Internet traffic that we are currently experiencing [1,2] is based on the expansion of cloud services and the huge amount of traffic supported by the content delivery networks (CDNs) [1]. This clearly increases congestion issues in communication networks, with particular emphasis in the core and backbone segments [3]. Furthermore, the global crisis of COVID-19 has revealed the need for remote communications in both the social and business spheres [2,4]. This crisis has become an incipient increase in video-call-related traffic through the use of applications such as Zoom [5], Microsoft Teams/Skype [6] or Cisco Webex [7]. Finally, these challenges are compounded by the need to meet the performance specifications for the fifth generation Internet (5G), where it is planned to obtain (a) a scalable management environment that enables the dynamic deployment of cloud-based applications, (b) a reduction of approximately 20% in operational costs and (c) total latency to the user of less than 1 ms [8].

The combination of all these challenges requires drastic changes in the way transport network resources are controlled and managed, moving away from traditional network management techniques. Under this umbrella, software defined networking (SDN) and network function virtualization (NFV) emerge as key technologies to address the above challenges in an efficient way. In this work we will put the light on SDN.

SDN is becoming a consolidated technology for network management that encompasses a set of techniques aimed at the management of networks that focus mainly on one basic principle: decoupling the decisions made at the control plane with respect to the actions taken at the data plane [9]. This principle allows unprecedented degrees of flexibility in the system, because the network switches and routers become simple forwarding devices that send and receive traffic, while the logical control of the network can easily be centralized on an external controller [10]. Conversely to traditional vendor-locked communication networks, SDN technology brings to network operators, among other benefits, the capability to: respond to challenges (such as big traffic fluctuations due to major sport or public events), operate their network in a cost-effective manner, evolve the network infrastructure, foster innovative solutions [10].

Relevant for optical networks, SDN is a powerful enabler to control and manage the wide variety of network elements with particular photonic transmission and switching characteristics inherent of the optical domain [11]. In this context, software-defined optical networks (SDONs) aim to exploit the flexibility of SDN control to support network applications with an underlying optical network infrastructure [12]. Within SDON, software models dedicated to the characterisation of optical elements, devices and systems are of utmost significance to enable the creation of *software agents*. In this regard, an agent is assumed to be a light-weight software that translates the necessary commands to permit communication between the SDN controller and the optical device placed in the optical data plane. Consequently, agents play a relevant role in disaggregated optical networks.

Horizontal disaggregation in optical network is aimed at decomposing the optical data in its single components [13]. Disaggregated optical networks (DONs) leverage this idea to be presented as a new technology in optical networks and brings synergies with the SDN paradigm. Under this umbrella, several cooperative projects, such as Open Config [14], Open ROADM [15] or Telecom Infra Project [16], have been arising for the last years, with the target of providing unified models of optical devices. Thus, allowing the deployment of complex optical network architectures through the use of vendor-independent devices. This fact is remarkable because it avoids one of the most challenging traditional issues in optical network implementations, the vendor islands. The inherent independency of vendor equipment obtained in the aforementioned projects is mainly based on using a specific modelling language, namely, the Yet Another Next Generation (YANG) one [17]. YANG offers the possibility to consolidate in a common language and a catalogue of rules the definition of the main characteristics of device models that can be used in software agents. In SDN-based DONs the communication between the agents generated based on YANG models and the SDN controller (via South Bound Interface, SBI, in the canonical SDN architecture) usually relies on the NETCONF protocol [18]. The YANG/NETCONF symbiosis provides the ideal breeding ground not only for the development of optical devices that facilitate vendor interoperability but also for the development of software tools that emulate the real behaviour of such devices, which is basic in the early stages of the development of new devices and network planning. Further details on the YANG/NETCONF usage for optical networks and additional alternatives for SBIs can be found in [19].

1.1. State of The Art

The literature has a growing number of works focused on SDN DON and optical network emulation. However, studies that consider both topics from the application layer to the data plane are scarce. One interesting starting point to link SDN and DON is to put the light on the SDN controller. Some works like [13,20,21], embrace this approach, first by proposing a model-driven design of a SDN controller and, second, upgrading the functionalities of the SDN controller ONOS to support disaggregated optical networks. Such new capabilities include YANG models compatible with NETCONF-based SBIs that communicate with the optical devices. In [22], this idea is expanded to not only be applied to the controller but also to a high level of abstraction that comprises the control plane in multi-domain disaggregated optical networks. Furthermore, a monitoring and data analytics architecture is proposed in [23] to monitor optical disaggregation.

On the other hand, several studies focus on the data plane, precisely on modelling devices in the context of DON. For instance, YANG models and their implementation, via agents, of (SDN-enabled) sliceable bandwidth/bitrate variable transceivers (S-BVTs) have been presented for partially [24] and fully disaggregated optical networks [25]. One of the most important actors in optical networks is the reconfigurable optical add/drop multiplexer (ROADM) that implements the add and drop capabilities in optical nodes. They are addressed within the DON scope in [26,27]. Additionally, focusing on the data plane, [28] reports a gap analysis on physical-layer parameters of YANG models available in the OpenConfig, OpenROADM and OpenDevice projects for estimating the quality of transmission (QoT). In addition, this work is complemented by a proof of concept focusing on the reception of the relevant parameters from an OpenConfig muxponder in the application layer for planning purposes.

Relevant in the context of data-plane initiatives, it is worthwhile mentioning GNPY [29]: the open-source planning tool for the data plane in optical networks. GNPY is based on the Gaussian Noise (GN) model [30], and reported validation results in large-scale optical transmission testbeds [31]. In particular, [31] reports the GNPY validation in a mixed-fibre test-bed at Microsoft labs in which transponders of eight different manufacturers operated in the C-band reaching a propagation distance of 1945 km. Relevantly, GNPY has also been considered as an assistance tool for the computation of impairments in optical-layer paths on top of SDN controllers [32]. Nonetheless, these efforts are scarce in the literature because optical-layer awareness is commonly considered as stand-alone planning tools separated from operational aspects addressed by SDN controllers.

The next stage towards emulating a DON environment is device virtualisation that exploits the benefits provided by YANG models of devices, which is also explored in the community. A network virtualisation architecture for open (partially) disaggregated network uses the concept of device hypervisor in virtual networks by using OpenROADM data models [33]. Furthermore, another common approach for adding virtualization to DON-based devices is to implement their related agent in a virtual machine (VM)/container that can emulate a realistic behaviour of the optical device. This eases the instantiation and the management of the virtual entities with a computer or a server. Those VMs/containers must be configured to enable the SBI communication implementing the interaction with the SDN controller [13].

Finally, it is also worthwhile mentioning comparable initiatives to our proposal such as the software-defined packet-optical network emulator [34], which is based on an optical extension of the popular packed-oriented network emulator Mininet [35]. Relevantly, ref. [34] proposes a software suite capable of emulating optical-layer QoT performances such as OSNR, gOSNR and BER, and report its impact to the SDN controller. This emulation is based on augmenting packet in traditional Mininet instances with wavelength/channel information, which is used in modified instances of Open vSwitch. In fact, the entire Mininet-Optical emulation relies on a software structure that invokes the Open vSwitch instances as Python processes. Consequently, this approach diverges from realistic deployment/production scenarios in which optical systems are managed/controlled by its own light-weight software agent in dedicated physical or virtual resources.

1.2. Motivation and Previous Work

The state of the art in SDN DON and optical devices' emulation reveals that the community is pushing considerably in this direction. YANG-based models in accordance with the NETCONF protocol are efficient approaches towards minimizing or even removing the drawbacks of the so-called vendor islands. Thus, there are ongoing projects to provide unified models for optical devices. However, to the best of our knowledge, optical device emulation is in its early stages. Only some of the analysed works assume a scenario where SDN DON emulation tackles from the highest application layer down to the data plane. Handing over control and management of emulation to the application layer will permit a more precise optical network design and an automatic deployment of software agents related from device catalogue at the request of the user. Moreover, as pointed out in the data-plane initiatives in the previous subsection, interactions between realistic agent-based SDN implementations and data-plane models are scarce. In fact, the challenge of these initiatives relies on the combination of physical-layer modelling commonly devoted to off-line planning tools with operation-based systems such as SDN controllers. The complementary nature of those systems makes the development of a DON emulator a difficult and challenging task. Such context is the starting point that motivates this paper with the aim at contributing to achieve a such challenging goal.

This work is a significant follow-up of an open line started in [36], in which we reported our preliminary progress towards an optical emulation framework built on automatic agent creation. In particular, ref. [36] focused on a specific use case of an 18-node European topology with nodes based on open packet transponder (Cassini [37]) in which we were able to modify parameters such as optical launch power, wavelength, and modulation format. Here, we extend that work and present a substantial evolution of our Containerized Framework for emulating SDN-DONs with special attention to the coexistence of the Kubernetes-based [38] master nodes server–client architecture and a new automatic agent creation framework using NETCONF in the SDN controller–optical agent interaction irrespective of the YANG model used. Additionally, to test the practical feasibility of our proposal, a proof-of-concept is also presented.

1.3. Structure of the Paper

The rest of the paper is structure as follows. In Section 2, all the software tools needed, and the methodology followed are exposed. Section 3 is aimed at showing the proof-of-concept proposed to evaluate the proposal. Section 4 reports the results obtained with our framework accompanied with their discussion. Concluding remarks and future directions of our work are highlighted in Section 5.

2. Proposed Architecture for SDN-DON Emulation

In this section, we present all the details required to define and configure our proposal for the SDN DON emulated framework. First, we expose a summary of the previous work that feeds this proposal. Then, we present the general architecture that permits the automatic deployment of emulated SDN DON, the core of this work. This section concludes with configuration considerations that are required to set up the framework.

2.1. Legacy Work

In the previous work presented in [36], we reported our preliminary advancements towards an optical emulation framework for automatic creation of agents in SDONs.

As seen in Figure 1a, this architecture comprises three main elements: the ONOS SDN controller [39]; a pool of software agents that emulate the interaction with optical devices, in this case, Cassini transponders; and the API gateway.

ONOS is chosen to play the role of the SDN controller, and is configured with the required modules and YANG drivers to interact with the optical devices or agents in the data plane via NETCONF protocol. A pool of agents emulates a realistic environment of a DON data plane; those agents are implemented in docker containers and they are created automatically by the API gateway REST-based interface according to the topology specification given by ONOS.

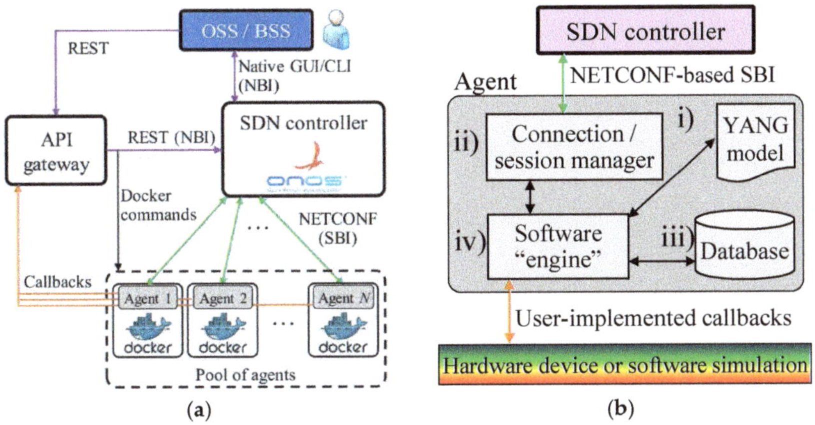

Figure 1. Architectural details of the previous work [36] (**a**) General overview of the architecture for automatic creation of agents in a software-defined optical networks (SDONs) environment; (**b**) Detailed view of the components in an optical network agent.

The other key point of this work is the structure of the optical network agent, depicted in Figure 1b. The software agents in the pool are virtualised entities within their respective Docker container. In addition, each agent includes four components: (i) the OpenConfig YANG model of Cassini transponders, (ii) the connection manager (i.e., the Netopeer2 NETCONF server) that enables connectivity with the controller, (iii) a database to store the configuration files and (iv) a software engine to detect and interact with changes in the configuration files. It is relevant to mention that the last three elements belong to the same subsysem, a Sysrepo framework aimed at automatic creation of YANG-based agent configuration. The Netopeer2 server is built in the Sysrepo framework and it provides the adequate NETCONF-based server for interacting with the SDN client. The Netopeer2-Sysrepo symbiosis provides benefits in the automatic creation of YANG agents but they have a closed and short catalogue of YANG-based optical device models that limits the scalabilty of a SDN DON emulation framework.

2.2. *Advanced Automatic Deployment of Emulated SDN DON*

The main target of this work is to enhance the dynamicity and the scalability of the automatic deployment of an emulated SDN DON environment. To achieve this challenging objective, we leverage the previous work focusing on two major and significative changes: (i) automatic container management by Kubernetes and (ii) a novel scalable agent definition.

2.2.1. General Architecture

Figure 2a showcases the general architecture of our proposal for an advanced automatic deployment of an emulated SDN DON framework. As it can be seen in the figure, the main difference with the legacy work is to leverage Kubernetes functionalities to manage and automatize the creation of agents, thus, minimizing the dependency of the API gateway. The other key contribution is the novel structure of the optical agent, shown in Figure 2b, facilitating the scalability and the emulation of a wide catalogue of YANG-based optical models. Further details of the structure of the agent are elaborated in next subsections.

This architecture is targeted to maximize the automation and the scalability of the emulation framework trying to exploit the benefits of a dynamic management and orchestration of the containerized optical agents. Moreover, two architectural decisions in our implementation deserve further description. First, it is worthwhile mentioning that we choose to employ docker containers because of they are more lightweight in terms of resources than VMs. We leverage on the fact that containers share the operating systems whereas VMs aim at emulating virtual hardware. Given that docker containers share the operating systems, docker applications consume a portion of the resources compared to a VM. Second, our architecture considers one docker container per agent for a realistic emulation of a carrier-grade deployed scenario. In particular, telecom operators commonly deploy control architectures that handle distinct geo-located optical nodes with dedicated computing resources for service deployment (e.g., in the edge computing paradigm) while also devoting part of those resources for the control of the network infrastructure. In this context, optical resources are managed and operated with a per-node computing element that receives and sends commands from/to the SDN controller while interfacing the hardware at the data plane. Hence, our proposal aims at emulating such conditions with an independent and isolated entity, i.e., docker container, per software agent as in the case of distinct geo-located optical nodes.

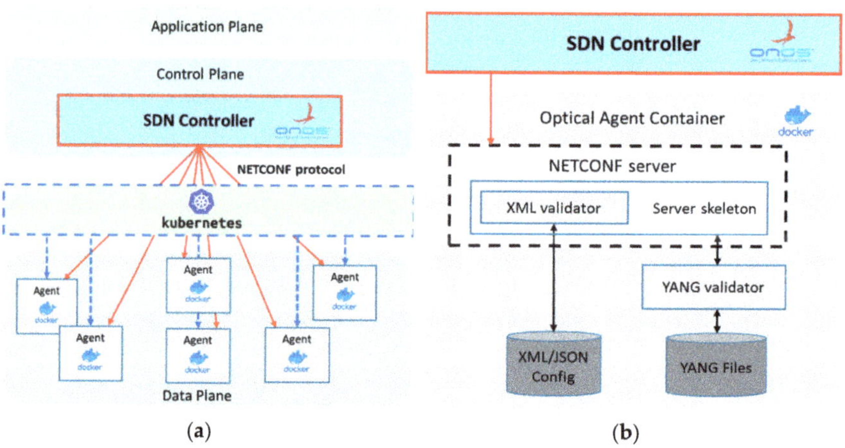

Figure 2. Architectural details of the proposal: (**a**) General overview of the architecture for the Advanced Automatic Deployment of an emulated software-defined networking (SDN) disaggregated optical network (DON) framework; (**b**) Schema of the elements in a generic optical network agent.

2.2.2. Kubernetes-Based Agent Management

In this work we propose to manage the agents' docker containers through Kubernetes. The choice of Kubernetes is based on the necessity of providing scalability and intelligent computational management in order to support large scale SDN DON emulated scenarios. Kubernetes has a wide range of benefits for virtualised systems that could help to accomplish the expected targets, e.g., service discovery and load balancing, storage orchestration, automated rollouts and rollbacks or self-healing just to mention a few. Moreover, Kubernetes is currently one of the most popular DevOps tools for container management, hence with a large support in the community [38]. Additionally, Kubernetes and Docker are complementary technologies, Kubernetes orchestrates one or more hosts that run containers, and Docker is the technology that starts, stops, and also manages those containers. In our model, Docker is a low-level technology orchestrated and managed by Kubernetes.

To complement the justification about why Kubernetes is key in our proposal, we present a present a representative use case. Consider a need to emulate a full SDN-based DON scenario assuming realistic functionalities, with multiple optical nodes and at each node multiple optical devices such as, transponders, wavelength selective switches, EDFA amplifiers or optical channel monitor. Just considering a 25-node optical network and for these four types of devices, $25 \times 4 = 100$ agents are needed, in the most favourable case assuming only 1 device per type and node (in reality this number is much higher). For this case alone, managing 100 agents instantiated in virtual machines or virtual containers without intelligent management, the scenario is completely unaffordable for one single personal computer, computationally speaking. This is where Kubernetes comes in; this tool not only provides intelligent management of containers (agents), but its automatic load balancing and the minimization of the usage of computational resources, both features related to scalability, among others, are basic to allow the emulation of realistic and large-scale SDN DON scenarios. Thus, one of the major benefits that Kubernetes brings to our work is that it allows the emulation of complex SDN DON scenarios running on a single laptop, this way exploiting the potential usability and exploitation of our framework (e.g., teaching, academia, SDN DON pre-deployment phases, etc.).

From an implementation perspective, we follow a traditional master nodes approach. A Kubernetes cluster is made up of one master and one or more nodes, as can be seen in Figure 3. Both master and nodes are Linux hosts that run on anything from VMs, bare metal servers, to private and public cloud instances.

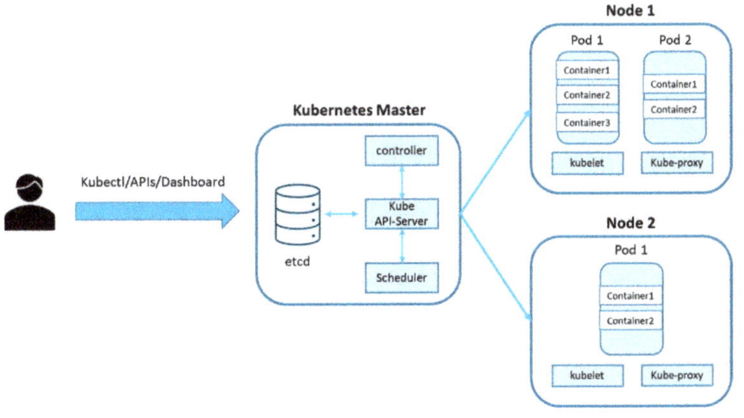

Figure 3. The Kubernetes master nodes-based architecture.

The Kubernetes master is composed of a set of small services that conform the control plane of the cluster and it can be externally accessed by the *kubectl* service, a RESTful API or a graphical dashboard, all managed by the *kube* API-server. The configuration and status of the cluster gets persistently stored in the *etcd*, while the controller implements the functionalities for controlling the Kubernetes cluster, such as, the node controller or the endpoints controller. Finally, the Kubernetes master offers the possibility to monitor new workloads and assign them to the nodes.

Still regarding Figure 3, the nodes can be seen as clients in a server–client architecture, with the master as server. The nodes are where the containers are hosted in correspondence to the agents in our framework. Within the nodes, there are independent spaces to run the containers, called pods. One node can have several pods and in each pod several containers. This approach is totally in line with network slicing, essential for 5G-aware deployments [40], being this fact another solid reason to choose Kubernetes as agent manager in a SDN DON emulation framework. Note that the server–client architecture, in correspondence with the Kubernetes master nodes (also illustrated with blue dashed arrows in Figure 2a), coexists with the client–server structure used in NETCONF protocol in correspondence with SDN controller–optical agent container (illustrated in solid red arrows in Figure 2).

Concerning Kubernetes integration as part of the Advanced Automatic Deployment of Emulated SDN DON oversees agents' management. First, the configuration for creating containers (e.g., agents) can be performed by three different ways: via script by using the *kubectl* service, manually from dashboard or remotely through the Kubernetes API.

Once the containers are up and running in one or multiple pods, the Kubernetes Master is in charge of managing the computational resources according to the workload and the specifications stated in the configuration phase.

2.2.3. Versatile Agent Definition

The second major contribution of this work is a novel definition of versatile optical agents that can be automatically instantiated. We present a generic definition of agents suitable not only for specific YANG models belonging to specific projects (e.g., Telecom Infra Project [37]) but also for all optical models that can be defined using the YANG language. This novel structure is decisive in the sense that a user can evaluate scenarios with a potentially infinite number of optical devices under the same emulation framework irrespective of their vendor, technology, or type. Therefore, this new functionality is critical not only to emulate DON scenarios with realistic current data models, but also to be compatible with all future data models using the YANG language.

Figure 2b depicts the structure proposed for a generic agent that emulates the role of a software entity in charge of governing optical devices. It is composed of four major components, a NETCONF server, a YANG validator, an XML (eXtensible Markup Language) /JSON (JavaScript Object Notation) config database and the files of the YANG model necessary to emulate an optical device, all coming from open-source initiatives and they are implemented in one single Docker container. Further details follow:

- NETCONF server: it provides NETCONF connectivity with the SDN controller with the aim at avoiding dependency of existing agent automation framework, similarly to the Sysrepo in the legacy work. In this work, the NETCONF skeleton is given by the Choppsv1 library [41] that supports both the creation of NETCONF clients and servers, however it does not provide the full implementation of the server. The *pyangbind* module complements the server, because it permits the generation of automatic code from YANG files, the validation of XML calls, essential in NETCONF-based communications and XML/JSON conversion amplifying the potential for connectivity [42].
- YANG validator: *pyang* is Python-based library playing the role of a validator, transformer, and generator of YANG code. It can be used for validating YANG modules, YANG modules transformation into different formats, and code generation from modules. This module is key to provide potential scalability and generalization [43].
- XML/JSON config database: it stores the configuration files related to the configuration of the agent for the different functional states, such as boot or running.
- YANG model files: finally, a set of YANG files that define the behaviour required to emulate an optical device. They are validated by the *pyang* module.

From a global point of view, it is worthwhile to indicate the interaction among the components described above. First, YANG models are provided, e.g., such as Cassini transponder models. Then, agents leverage on the YANG validator based on pyang for the creation of the NETCONF skeleton within the NETCONF server. Notably, it is worth recalling that SDN controllers commonly include a series of drivers that implement NETCONF client functionalities for interacting with the agents via NETCONF protocol. This interaction, on the agent side, will leverage on the pyangbind module for XML/JSON validation. Then, also in the NETCONF server side, custom instructions are implemented for interacting with the data plane. Finally, we report the interaction between control and data signals flow among the components in the next subsection. For instance, Figure 8 will showcase the interaction between agents (data plane) when connectivity is requested from the SDN controller (control plane).

3. Proof-of-Concept Configuration

In order to test and evaluate the emulation framework of this work, in this section we propose a proof-of-concept that integrates the key elements of the entire architecture. Here we detail the configuration and settings for performing such proof-of-concept.

The proof-of-concept follows the design exposed in Figure 2a, with three main actors: (i) ONOS playing the role of SDN controller that interacts and configure the emulated optical data plane (set of agents in this case), (ii) Kubernetes, in charge of the management of the Docker containers where the optical agents are implemented, and finally (iii) the data plane, represented by the set of agents emulating the software that governs the optical devices. For the sake of simplicity, we only use an agent of Cassini transponder according to the OpenConfig project. All the evaluation system has been deployed and executed in a high-performance laptop (Intel 8th generation i7 with 8 logical cores, 32 GB of 2400 MHz DDR4 RAM and 1 TB of SSD). Further details follow:

- Agent definition and integration: a specific docker container is built to integrate the architecture proposed and described in Section 2.2.3 for an automatic agent deployment. The set of YANG files that replicate the behaviour of a Cassini transponder are extracted from the GitHub repository of the OpenConfig project [44]. Precisely, the files *openconfig-terminal-device.yang*, *openconfig-if-ethernet.yang* and *openconfig-types.yang* are stored in the YANG file catalogue of the agent container.
- Kubernetes configuration: following the structure presented in Figure 3 Kubernetes Cluster is installed in Linux, an Ubuntu distribution, and it is composed by one master and one worker node and one pod in this worker node. Agents' creation and configuration are driven by the *deploy.yaml* file, see Figure 4, where it commands the deployment of the Docker containers. The service *kubectl* is the one to execute the deployment of the agents where results of such deployment are presented in the next section.

```
apiVersion: apps/v1
kind: Deployment
metadata:
  name: cassini-deployment
  labels:
    app: software-agent-cassini
spec:
  replicas: 18
  selector:
    matchLabels:
      app: software-agent-cassini
  template:
    metadata:
      labels:
        app: software-agent-cassini
    spec:
      containers:
      - name: software-agent-cassini
        image: software-agent-cassini
```

Figure 4. Containers deployment configuration script for Kubernetes.

- ONOS: we chose a virtualized distribution of ONOS (version 2.3.0) in a Docker container. The installation using Docker also permits to open the necessary ports for this proof-of-concept in one single command as presented below. Port 8181 is used for external communication with ONOS via GUI, port 5005 is a Java debugger port, port 8101 permits command line interface (CLI) communication via CLI and the port 830 permits NETCONF-based connectivity. Finally, once the container is up and running, pertinent "apps" must be activated to enable the expected functionalities within the SDN controller framework: "odtn-service", "roadm" and "optical-rest".

docker run -t -d -p 8181:8181 -p 8101:8101 -p 5005:5005 -p 830:830 –name onos onosproject/onos:2.3.0

- Topology definition: the topology is defined in a JSON file so that it can be interpreted by ONOS. Seven metrics can be set "Devices", "links", "hosts", "Apps", "ports", "regions" and "layouts". In this work we only use the objects "device" and "link". Each device object is linked to each container previously deployed by indicating some specific fields, such as IP of the container and port to enable NETCONF connectivity, ONOS driver or map location, as can be seen in the exemplary definition if Figure 5a. Moreover, the links define the interconnection between two agents by setting their IP and ports, type of connection or bidirectionality, among others as depicted in Figure 5b.

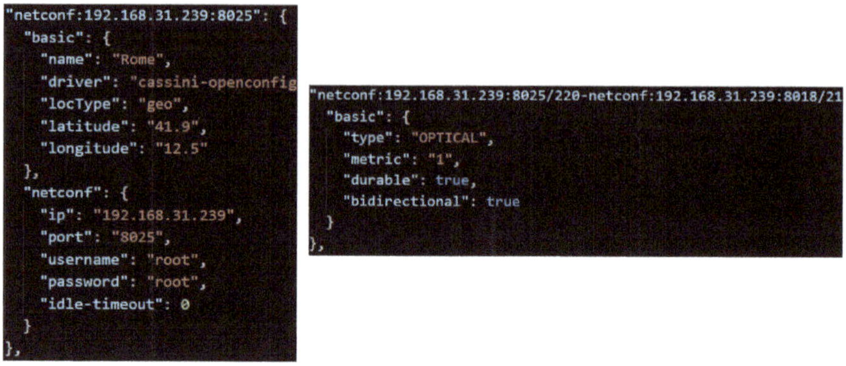

(a) (b)

Figure 5. Topology details in the JSON configuration file: (**a**) Device definition; (**b**) Link definition.

Once the file is ready, it is sent via ONOS NBI API to inform the details of the topology. Topology definition and JSON file creation are performed manually in this proof-of-concept emulating the 18-nodes European Optical Network. Future works will target the automatization of this procedure.

4. Results

This section collects and analyses the results obtained in the deployment of the proof-of-concept focusing on the main points of our proposal for an Advanced Automatic Deployment of Emulated SDN DON. The execution of this experiment is performed following the statements proposed in Section 2 and configured according to Section 3. Given the lack of complex SDN DON emulators in the literature (see Section 1.1) that can not only emulate such scenarios but also interact with it in a realistic way, the objective of this proof of concept is not aimed at evaluating the performance of our framework, but to highlight the feasibility and functionality of our proposal. All the results exposed in this section are obtained after the configuration stage is properly completed.

The first stage of the running workflow in this proof-of-concept is to show the performance of Kubernetes in the agent/container management. Figure 6 illustrates the 18 containers of the Cassini agents properly running according to the configuration file (*deploy.yaml*) defined under the Kubernetes umbrella.

One of the main benefits of managing Docker containers with Kubernetes is that the computational resources of the host are efficiently used to balance the workload of the containers according to their computational needs and performance. Once all the containers are up and running, the next stage is to bind the containers to the proposed optical topology by sending the JSON configuration file via ONOS API. At this point, ONOS recognizes the query and the topology is tied to the containers by assigning the existing agents to the nodes defined in the topology file and links are created to interconnect the nodes. As can be seen in Figure 7, ONOS shows the emulated 18-node European optical network topology in which each node corresponds to a Cassini agent.

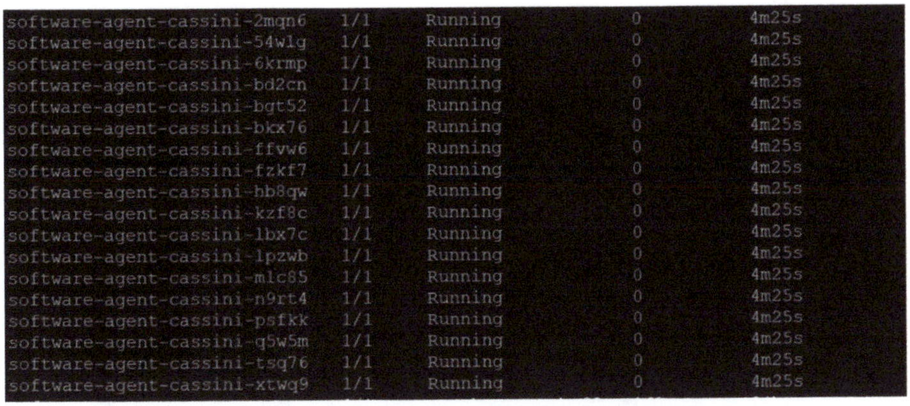

Figure 6. Screenshot of the CLI of Kubernetes showing 18 running agent containers.

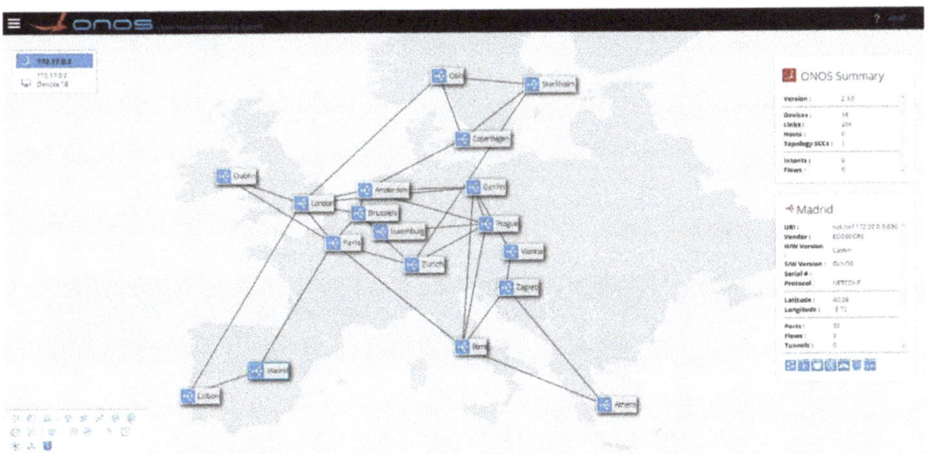

Figure 7. Emulated optical topology deployed on ONOS.

The functionality of the agents deployed in this emulated framework is not only related to the optical topology discovery from the SDN controller side, but also they can perform other common network tasks. Figure 8 illustrates how two containers establish real TCMP-based package traffic flows leveraging the deployment of the Emulated SDN DON framework.

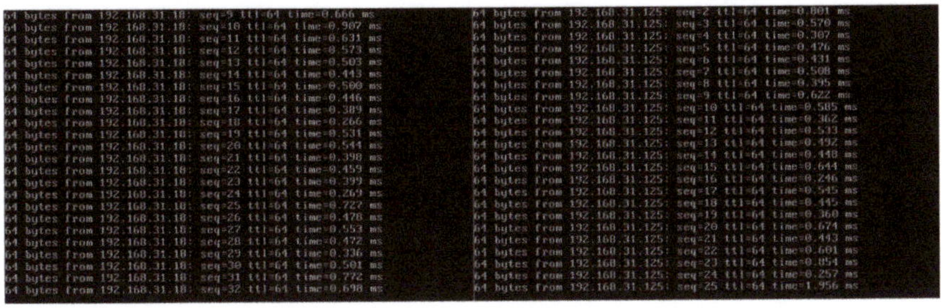

Figure 8. Real traffic flows between two ports of two different nodes.

In a similar way than a real optical device, our emulation framework permits to retrieve the status of the emulated devices in real time regarding their current configuration or other functional metrics. Figure 9 showcases an exemplary query requested by the SDN controller via NETCONF protocol about the status of the configuration of an optical agent (namely, emulated Cassini transponder) in a specific instant of time. In particular, the agent is queried about its current target-output, current-output, and current-input optical powers.

```
onos@root > power-config get netconf:192.168.31.239:8022/22

The target-output-power value in port 202 on device netconf:192.168.31.239:8022
is -5.100000.

The current-output-power value in port 202 on device netconf:192.168.31.239:8022
is 0.120000.

The current-input-power value in port 202 on device netconf:192.168.31.239:8022
is 0.120000.
```

Figure 9. Snapshot of NETCONF-based queries about the status of the configuration in an optical agent commanded from ONOS.

The functionality of the deployed agents is not only limited to configuration and status queries, the SDN controller can also change physical parameters of emulated devices in the same way as for a real optical device. Figure 10 is a proof of the performance of this procedure by using the CLI of ONOS to change the output power in an optical port for a specific Cassini transponder.

```
onos@root > power-config edit-config

netconf:192.168.31.239:8022/22 2 Set 2.00000 power on port
```

Figure 10. Snapshot of NETCONF-based queries to change the output power of the laser in an emulated optical agent commanded from ONOS.

5. Conclusions

In this work, we presented an emulation framework that leverages on a combination of SDN and container management with Kubernetes for the emulation of both control- and data-planes of disaggregated optical networks. Our legacy proposal was reviewed including the architectural details and the classical optical network agent structure that leveraged on the Netopeer2-Sysrepo framework. Then, we presented the Containerized Framework for emulating SDN-DONs paying special attention to the coexistence of the server–client architecture based on Kubernetes master nodes and the client–server structure that used NETCONF in the SDN controller–optical agent interaction. We further detailed the software agent structure which includes NETCONF server, YANG validator, XML/JSON config database and YANG models, whose combination offers versatility for accommodating the wide variety of optical components and systems. The implementation of the agents in this emulating system was deployed in Docker containers allowing advanced and efficient management by Kubernetes. We described all the configuration and settings for performing the proof-of-concept that consisted in an 18-node European topology with Cassini transponders. We reported the successful status request and configuration settings of optical-layer parameters such as transmitted and received optical power. Given the successful interaction between the control and data plane positions, our proposal demonstrates an adequate approach for managing an operationally complex carrier-grade transport infrastructure with SDN-based disaggregated optical systems.

Author Contributions: Conceptualization, F.-J.M.-M., M.G. and I.I.-C.; methodology, F.-J.M.-M. and M.G.; software, I.I.-C. and S.Z.; validation, F.-J.M.-M.; formal analysis, F.-J.M.-M. and P.P.-M.; investigation, F.-J.M.-M. and S.Z.; resources, P.P.-M.; writing—original draft preparation, F.-J.M.-M.; writing—review and editing, M.G., I.I.-C., S.Z., and P.P.-M.; supervision, P.P.-M.; funding acquisition, P.P.-M. All authors have read and agreed to the published version of the manuscript.

Funding: This research was funded Spanish Government: ONOFRE-2 project under Grant TEC2017-84423-C3-2-P (MINECO/AEI/FEDER, UE) and the Go2Edge project under Grant RED2018-102585-T; and by the European Commission: METRO-HAUL project (G.A. 761727).

Institutional Review Board Statement: Not applicable.

Informed Consent Statement: Informed consent was obtained from all subjects involved in the study.

Conflicts of Interest: The authors declare no conflict of interest.

References

1. Cisco. Cisco Annual Internet Report (2018–2023) White Paper. 2019. Available online: https://www.cisco.com/c/en/us/solutions/collateral/executive-perspectives/annual-internet-report/white-paper-c11-741490.pdf (accessed on 30 December 2020).
2. Sandvine. The Global Internet Phenomena Report COVID-19 Spotlight. 2020. Available online: https://www.sandvine.com/phenomena (accessed on 27 January 2021).
3. Napoli, A.; Bohn, M.; Rafique, D.; Stavdas, A.; Sambo, N.; Poti, L.; Noelle, M.; Fischer, J.K.; Riccardi, E.; Pagano, A.; et al. Next generation elastic optical networks: The vision of the European research project IDEALIST. *IEEE Comm. Mag.* **2015**, *53*, 152–162. [CrossRef]
4. Feldmann, A.; Gasser, O.; Lichtblau, F.; Pujol, E.; Poese, I.; Dietzel, C.; Wagner, D.; Wichtlhuber, M.; Tapiador, J.; Vallina-Rodriguez, N. The Lockdown Effect: Implications of the COVID-19 Pandemic on Internet Traffic. In Proceedings of the ACM Internet Measurement Conference, New York, NY, USA, 27–29 October 2020.
5. Zoom. Available online: https://zoom.us (accessed on 27 January 2021).
6. Microsoft Teams. Available online: https://www.microsoft.com/en-us/microsoft-365/microsoft-teams (accessed on 27 January 2021).
7. Cisco Webex. Available online: https://www.webex.com (accessed on 27 January 2021).
8. 5G Infrastructure Public Private Partnership (5G PPP). Key Performance Indicators (KPIs). Available online: https://5g-ppp.eu/kpis (accessed on 27 January 2021).
9. Open Networking Foundation. Software-Defined Networking: The New Norm for Networks. White Paper. 2012. Available online: https://opennetworking.org/category/sdn-resources/whitepapers (accessed on 27 January 2021).
10. Kreutz, D.; Ramos, F.M.V.; Verissimo, P.E.; Rothenberg, C.E.; Azodolmolky, S.; Uhlig, S. Software-Defined Networking: A Comprehensive Survey. *Proc. IEEE* **2015**, *103*, 14–76. [CrossRef]
11. Channegowda, M.; Nejabati, R.; Simeonidou, D. Software-defined optical networks technology and infrastructure: Enabling software-defined optical network operations. *IEEE/OSA J. Opt. Commun. Netw.* **2013**, *5*, A274–A282. [CrossRef]
12. Thyagaturu, A.S.; Mercian, A.; McGarry, M.P.; Reisslein, M.; Kellerer, W. Software Defined Optical Networks (SDONs): A Comprehensive Survey. *IEEE Commun. Surv. and Tut.* **2016**, *18*, 2738–2786. [CrossRef]
13. Giorgetti, A.; Casellas, R.; Morro, R.; Campanella, A.; Castoldi, P. ONOS-controlled Disaggregated Optical Networks. In Proceedings of the Optical Fiber Communication Conference (OFC), San Diego, CA, USA, 3–7 March 2019.
14. OpenConfig. Available online: http://www.openconfig.net/ (accessed on 27 January 2021).
15. OpenROADM. Available online: http://www.openroadm.org/ (accessed on 27 January 2021).
16. The Telecom Infra Project. Available online: https://telecominfraproject.com/ (accessed on 27 January 2021).
17. Bjorklund, M. *YANG—A Data Modeling Language for the Network Configuration Protocol (NETCONF)*; IETF RFC 6020; ISSN: 2070-1721; IETF: Fremont, CA, USA, 2010.
18. Enns, R.; Bjorklund, M.; Schoenwaelder, J.; Bierman, A. *Network Configuration Protocol (NETCONF)*; IETF RFC 6241; ISSN: 2070-1721; IETF: Fremont, CA, USA, 2011.
19. Garrich, M.; Bravalheri, A. Overview of South-Bound Interfaces for Software-Defined Optical Networks. In Proceedings of the International Conference on Transparent Optical Networks (ICTON), Bucharest, Romaina, 1–5 July 2018.
20. Casellas, R.; Martinez, R.; Vilalta, R.; Munoz, R. Metro-Haul: SDN Control and Orchestration of Disaggregated Optical Networks with Model-Driven Development. In Proceedings of the Optical Fiber Communication Conference (OFC), San Diego, CA, USA, 1–5 March 2018.
21. Giorgetti, A.; Sgambelluri, A.; Casellas, R.; Morro, R.; Campanella, A.; Castoldi, P. Control of open and disaggregated transport networks using the Open Network Operating System (ONOS). *IEEE J. Lightw. Techn.* **2020**, *12*, A171–A181. [CrossRef]
22. Casellas, R.; Martinez, R.; Vilalta, R.; Munoz, R. Abstraction and control of multi-domain disaggregated OpenROADM optical networks. In Proceedings of the European Conference on Optical Communication (ECOC), Dublin, Ireland, 22–26 September 2019.

23. Gifre, L.; Izquierdo-Zaragoza, J.; Ruiz, M.; Velasco, L. Autonomic Disaggregated Multilayer Networking. *IEEE/OSA J. Opt. Commun. Netw.* **2018**, *10*, 482–492. [CrossRef]
24. Nadal, L.; Casellas, R.; Fabrega, J.M.; Munoz, R.; Moreolo, M.S.; Rodriguez, L.; Vilalta, R.; Vilchez, F.J.; Martinez, R. Multi-vendor sliceable transceivers in partial disaggregated metro networks. In Proceedings of the European Conference on Optical Communication (ECOC), Dublin, Ireland, 22–26 September 2019.
25. Nadal, L.; Fabrega, J.M.; Moreolo, M.S.; Casellas, R.; Munoz, R.; Rodriguez, L.; Vilalta, R.; Vilchez, F.J.; Martinez, R. SDN-Enabled Sliceable Transceivers in Disaggregated Optical Networks. *IEEE J. Lightw. Techn.* **2019**, *37*, 6054–6062. [CrossRef]
26. Kundrat, j.; Havlivs, O.; Jedlinsky, J.; Vojtvech, J. Opening up ROADMs: Let Us Build a Disaggregated Open Optical Line System. *IEEE/OSA J. Opt. Commun. Netw.* **2019**, *37*, 4041–4051. [CrossRef]
27. Xie, C.; Wang, L.; Dou, L.; Xia, M.; Chen, S.; Zhang, H.; Sun, Z.; Cheng, J. Open and disaggregated optical transport networks for data center interconnects. *IEEE/OSA J. Opt. Commun. Netw.* **2020**, *12*, C12–C22. [CrossRef]
28. Garrich, M.; San-Nicolas-Martinez, C.; Moreno-Muro, F.; Lopez-de-Lerma, A.M.; de Dios, O.G.; Lopez, V.; Giorgetti, A.; Sgambelluri, A.; Tancevski, L.; Verchere, D.; et al. Gap Analysis on Open Models for Partially-Disaggregated SDN Optical Transport Environments. In Proceedings of the Optical Fiber Communication Conference (OFC), San Diego, CA, USA, 3–7 March 2019.
29. GNPY. Available online: https://gnpy.readthedocs.io/en/master/ (accessed on 10 February 2021).
30. Poggiolini, P. The GN model of non-linear propagation in uncompensated coherent optical systems. *IEEE/OSA J. Lightwave Technol.* **2012**, *30*, 3857–3879. [CrossRef]
31. Filer, M.; Cantono, M.; Ferrari, A.; Grammel, G.; Galimberti, G.; Curri, V. Multi-Vendor Experimental Validation of an Open Source QoT Estimator for Optical Networks. *IEEE/OSA J. Lightwave Technol.* **2018**, *36*, 3073–3082. [CrossRef]
32. Kundrat, j.; Campanella, A.; le Rouzic, E.; Ferrari, A.; Havlivs, O.; Havzlinsky, M.; Grammel, G.; Galimberti, G.; Curri, V. Physical-Layer Awareness: GNPy and ONOS for End-to-End Circuits in Disaggregated Networks. In Proceedings of the Optical Fiber Communication Conference (OFC), San Diego, CA, USA, 8–12 March 2020.
33. Casellas, R.; Giorgetti, A.; Morro, R.; Martinez, R.; Vilalta, R.; Munoz, R. Virtualization of disaggregated optical networks with open data models in support of network slicing. *IEEE/OSA J. Opt. Commun. Netw.* **2020**, *12*, A144–A154. [CrossRef]
34. Lantz, B.; Diaz-Montiel, A.A.; Yu, J.; Rios, C.; Ruffini, M.; Kilper, D. Demonstration of Software-Defined Packet-Optical Network Emulation with Mininet-Optical and ONOS. In Proceedings of the Optical Fiber Communications Conference (OFC), San Diego, CA, USA, 8–12 March 2020.
35. Mininet. Available online: http://mininet.org (accessed on 27 January 2021).
36. Iglesias-Castreno, I.; Alabarce, M.G.; Hernandez-Bastida, M.; Marino, P.P. Towards an Open-Source Framework for Jointly Emulating Control and Data Planes of Disaggregated Optical Networks. In Proceedings of the International Conference on Transparent Optical Networks (ICTON), Bari, Italy, 19–23 July 2020.
37. Cassini by TIP. Available online: https://cdn.brandfolder.io/D8DI15S7/as/q3wkdg-476u4o-8wg0g7/Cassini_at_a_Glance_-_Telecom_Infra_Project.pdf (accessed on 10 February 2021).
38. Kubernetes. Available online: https://kubernetes.io (accessed on 27 January 2021).
39. ONOS—Open Network Operating System. Available online: https://opennetworking.org/onos/ (accessed on 10 February 2021).
40. Foukas, X.; Patounas, G.; Elmokashfi, A.; Marina, M.K. Network Slicing in 5G: Survey and Challenges. *IEEE Comm. Mag.* **2017**, *55*, 94–100. [CrossRef]
41. Chopps v1 Netcon Client/Server Library. Available online: https://github.com/choppsv1/netconf (accessed on 27 January 2021).
42. PyangBind. Available online: https://github.com/robshakir/pyangbind (accessed on 27 January 2021).
43. Pyang—A YANG Validator. Available online: https://github.com/mbj4668/pyang (accessed on 27 January 2021).
44. OpenConfig Project Repository. Available online: http://openconfig.net/yang/terminal-device (accessed on 27 January 2021).

applied sciences

MDPI

Article

Analyses of the Key Technologies and Development Trends of Optical Networks from the Perspective of Standard Essential Patents

Shu-Hao Chang

Science and Technology Policy Research and Information Center, National Applied Research Laboratories, Taipei 10636, Taiwan; shchang@narlabs.org.tw; Tel.: +866-2-27377779

Featured Application: In this study, the researchers have constructed a technology network model to explore the development of key technologies in standard essential patents (SEPs) optical networks.

Abstract: Because of the advancement of 5G, mobile devices, and broadband, the development of optical network technologies has received increased research attention. However, studies have mostly focused on the technical or market aspects rather than employing a macroscopic perspective to reveal the technology distribution of optical networks and the development trends in their key fields. Given that technologies disclosed by standard essential patents (SEPs) are the key technologies that determine the standards in relevant industries, we used SEPs optical networks as the basis of analyses and constructed a technology network. Therefore, the contribution of this study lies in bridging the research gap between two studies, using the perspective of SEPs to observe the key technologies in leading industry standard-setting. In addition to the aspects of technologies and markets addressed in existing studies, this study primarily discusses technology distribution and trends in optical networks. The results reveal that transmission of digital information, wireless communication networks and multiple communication are the key technical fields in developing optical networks, with wireless communication devices and digital information transmission being the main focus in recent years. Considering the gradual standardization of future optical network technologies and the fierce competition between manufacturers in SEP applications, relevant industries and universities should cooperate in key technical fields for research and development as well as in talent cultivation to facilitate the development of key technologies and industrial standards in optical networks. The current results may serve as a reference for both industry and academia with regard to research and development resource allocation.

Keywords: optical networks; standard essential patents; patent analysis; network analysis; technical analysis

Citation: Chang, S.-H. Analyses of the Key Technologies and Development Trends of Optical Networks from the Perspective of Standard Essential Patents. *Appl. Sci.* **2021**, *11*, 1583. https://doi.org/10.3390/app11041583

Academic Editors: Fabio Cavaliere and Luca Potì
Received: 11 December 2020
Accepted: 5 February 2021
Published: 10 February 2021

1. Introduction

Optical networks are a means of communication that applies signals encoded onto light to transmit information in various telecommunication networks. According to a report by MarketsandMarkets, the optical interconnect market is projected to grow from USD 9.0 billion in 2020 to USD 17.1 billion by 2025; it is expected to grow at a compound average growth rate (CAGR) of 13.7% from 2020 to 2025 [1]. Optical networks have had revolutionary effects on the telecommunication industry and play an essential role in the digital age. They have large transmission capacity and provide excellent confidentiality; optical networks have become the main communication method in cloud computing and 'Internet of things' applications [2–4].

Optical networks have been extensively applied and incorporated in multiple industries, and this technology has progressed considerably. For example, it has been applied in quantum and optical communication [5,6], including quantum optics, a research field that discusses the quantum behavior of light. Photons are used to transmit wave functions to reduce the problem of information distortion caused by environmental interference. Another example is the application of SiO_2 to the field of optical materials [7,8]. The advancement of 5G, mobile devices, and broadband have also promoted the development of optical network technologies. The use of optical networks increases data bandwidth and reduces energy consumption per unit, which limits network latency and the number of disconnections, enables the connection of more networking devices, and improves the interaction and connection between them.

Accordingly, optical networks are widely applied in multiple technical fields, such as optics, communication, semiconductors, materials, artificial intelligence, and computing. However, relevant studies have mostly focused on technical discussions [5–8] or market research concerning optical networks [1,9] rather than on their technology distribution and development trends from a macroscopic perspective. Moreover, few studies have explored the technical trends in optical networks from the aspect of standard essential patents (SEPs). Given that technologies disclosed by SEPs are the foundation and also the key technologies that determine standards in relevant industries, we observed the development trends of these technologies from the perspective of SEPs.

In this study, we focused on the SEPs of optical networks and constructed a technology network of SEPs. Many studies have predicted technology development by using patent data [10,11] and applied them to conduct patent landscape analyses [12,13] because patent data are one of the most direct indicators for observing technology trends. Specifically, SEPs are where standards and patents overlap. The technologies disclosed are the indispensable standards in the industry; they are cited by subsequent technologies and have influence on the industry's technological development. For example, the telecommunications technology standards in optical communication are mostly established by the third generation partnership project (3GPP) members. The standard technologies approved by 3GPP members are submitted to the standard setting organizations (SSOs), which publish them as telecommunications technology standards. As such, we attempted to address two research gaps. First, we examined the key technologies that determined the standards in these industries. Second, unlike the literature that focuses on discussions from a technical or market perspective, we explored technology distribution and trends in optical networks and constructed a technology network of SEPs to more effectively elucidate the key technologies protected by SEPs.

In brief, we analyzed the technology trends and key technologies of optical network patents according to a technology network of SEPs and discussed the key technologies in optical networks as standards in the relevant industries. The current findings may serve as a reference for governments, academia, and relevant professional industries.

2. Literature Review

2.1. Current Development of Optical Networks

Because of the rapid development of mobile devices and the Internet of things, users have growing demands for network transmission. Hence, the provision of efficient transmission has become a major concern in network development, where optical networks play an essential role. Optical transmission technology has developed rapidly over the past thirty years through three main technological innovations: time division multiplexing (TDM) technology based on electrical multiplexing, optical amplification technology combined with wavelength division multiplexing (WDM) technology, and digital coherent technology [14]. Literature has used photonic technologies [15,16] and semiconductor optical amplifiers [17,18] to discuss 5G optical communication. However, at present, the technology development of 6G is gradually attracting attention. The development of optical communication technology should not be restricted to photonics or semiconductor optical amplifiers. Currently, the main focus in the development of optical networks is on optical fiber cables, optical devices (including chips), optical communication equipment, and optical modules. Optical fiber cables are the transmission channels for optical communication. Optical devices are the functional devices developed based on photoelectric effects, with functions such as sending and receiving information, WDM, amplifying, switching, and system management. They form the core of optical transmission systems and can be categorized into active optical devices and passive optical devices. Optical communication equipment is composed of optical devices, including an optical terminal transceiver and a switch. Optical modules are crucial devices that perform the conversion between optical signals and electrical signals in optical communication systems and play a major role in data centers and mobile broadband. The production process is as follows: first, optical chips and other components are combined into an optical device, and then the optical device is encapsuled as an optical module (e.g., a light source, an optical detector, or an amplifier).

Optical networks have vast future business potential in various application fields (e.g., 5G, artificial intelligence, data centers, and photonic integrated circuits.) How to combine optical network technologies with the development and application of 5G and artificial intelligence has attracted increasing attention in academia and the professional industry [19,20]. Currently, the focus is on the high transmission rate, low latency, and high bandwidth of 5G wireless communication, which can be applied to popular industries including artificial intelligence and unmanned self-driving automobiles. Furthermore, because of the immense growth of data, data centers are no longer limited to a few computer rooms; instead, they increasingly comprise groups of data center clusters. Such data centers function collaboratively and exchange tremendous amounts of data instantly, which creates a demand for data center interconnect. Data center interconnect equipment has unique requirements such as small size, low power consumption, large capacity, and high speed; optical interconnects have been introduced to provide said characteristics because of their high bandwidth, low power consumption, and reconfigurability [21]. Photonic integrated circuits have the potential to transmit vast amounts of data in a low-cost manner and have had a revolutionary influence on medical technology, photoelectric sensing, solar battery components, far-infrared light source, and light display. Photonic integrated circuit technologies combine optics, complementary metal-oxide-semiconductor technology, and advanced encapsulation techniques. If they are compatible with the semiconductor manufacturing process, the cost of optical integrated circuit chips can be reduced even further. In addition to optical communication, photonic integrated circuits can be widely applied for processor interconnections or the core interconnections of multicore processors [22]. Accordingly, optical networks have been employed in cross-field applications and exhibited output values with high growth. Therefore, in this study, we analyzed the technology development of optical networks and identified the current key technologies from the perspective of SEPs by conducted network analyses. The analysis method is elaborated in the following section.

2.2. Network Analyses of Technologies Licensed with SEPs

SEPs can be regarded as the intersection of standards and patents. The value of SEPs in both technological development and business highlights SEP-protected technologies as essential standards and key methodologies in relevant industries. This study differed from past studies—which only adopted the perspective of patents [23]—in that it adopts SEPs, closer to industry standards and industry technology trends, to observe the development of optical networks. In addition, research indicated that companies with SEPs possess considerable market power because they determine the technology standards in the market [24]. Thus, we conducted technology network analyses from the aspect of SEPs. Past studies have used the co-classification approach to identify key participants in the network, thereby identifying whether a participant is a star or a gatekeeper [25,26]. Considering that a single patent may correlate with multiple technical fields (patent classification numbers), we employed a co-classification approach to define the relationship between these fields [27,28] and established a technology network accordingly. Participants at the center of a technology network provide a key node, or a node that receives much attention. From the perspective of social networks, the value of a network node depends on how centralized the node is in the network. On the basis of the technology network of SEPs, we further conducted centrality analyses to examine the key technologies in optical networks.

3. Research Design

3.1. Search Strategy and Data Source

In this study, we adopted the IPlytics platform as the data infrastructure for SEPs data. IPlytics connects data on over 90 worldwide million patents, 4 million standards documents and 280k declared standard essential patents and pooled patents [29]. We examined the key technologies according to the SEPs from 2015 to 2019 by using the following search conditions: ((TTL/optical network) or (ABST/optical network) or (ACLM/optical network)), which yielded 465 SEPs. Among them, TTL, ABST, and ACLM, respectively, denote the optical network keywords in title, abstract, and claim in patent documents. We classified the technologies in the technology network by employing the Cooperative Patent Classification (CPC) system implemented by the United States Patent and Trademark Office (USPTO) and the European Patent Office (EPO) in early 2013.

3.2. Centrality Analyses

We searched for the key technologies in the technology network of SEPs by employing centrality analyses. The measurement methods are specified in the following sections.

3.2.1. Closeness Centrality

Closeness centrality is the reciprocal of the distance between a node and all other nodes. A node closer to other ones has a higher degree of closeness.

$$C_c(P_i)^{-1} = \sum_{j=1}^{n} d(P_i, P_j) \ with \ i \neq j$$

where $d(P_i, P_j)$ is the distance from Node i to Node j.

3.2.2. Eigenvector Centrality

In eigenvector centrality, the centrality of a node is related to the number of its adjacent nodes and also the centrality of its adjacent nodes. A node has higher centrality when it connects with those with higher centrality; that is, the eigenvector centrality of adjacent nodes is inequivalent.

$$C_e(P_i) = \lambda^{-1} \sum_{j=1}^{n} a_{ij} C_e(P_j)$$

where $C_e(P_i)$ and $C_e(P_j)$ are respectively the eigenvector centrality of Node i and Node j; a_{ij} is the node that enters the adjacency matrix A; λ is the maximum eigenvalue (a constant) of the adjacency matrix A.

In this linear function, eigenvector centrality is a linear combination where the centrality of a single node is regarded as that of all other nodes [30].

3.2.3. Fragmentation Centrality

Fragmentation centrality refers to a fragmentated network with reduced cohesion when a node within is removed. After the node is removed, if the proportion of nodes that cannot be connected to each other is small, the network remains stable, which implies that the node has less significance. In this study, we referred to Borgatti [25] and applied distance-weighted fragmentation for measurement.

$$C_f(P_i) = 1 - \frac{2\sum_{i>j} \frac{1}{d(P_i, P_j)}}{n(n-1)}$$

where $d(P_i, P_j)$ is the distance from Node i to Node j; n is the total number of nodes.

4. Results

4.1. Patent Search Results

With the advancement of optical network technologies and the increased attention to SEPs in various countries, the number of SEPs optical networks has grown yearly. The development trend is depicted in Figure 1.

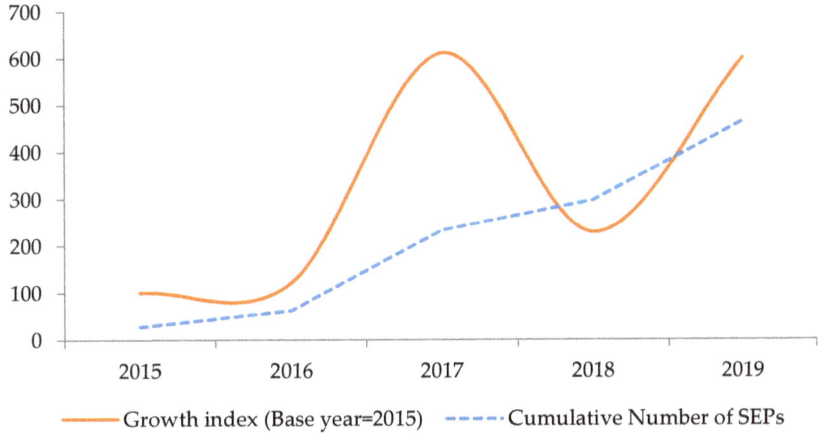

Figure 1. Growth trend of standard essential patents (SEPs) optical networks. Note: One hundred was applied as the number of SEPs in the base year (2015). In subsequent years, the growth index was (the number of SEPs of that year/the number of SEPs in the base year) ×100. Therefore, in 2019, the growth index was nearly 600 which means that it was approximately six times the growth index of 2015, which was 100. The cumulative number of SEPs denotes the accumulated number of SEPs each year.

Figure 1 illustrates that, compared with the number in 2015, SEPs optical networks have increased approximately six-fold by 2019. We further examined the distribution of the patent offices that granted SEP applications (Table 1).

Table 1. Patent offices that granted SEPs.

Ranking	Patent Office	Frequency	Percentage
1	WIPO	185	39.78%
2	CNIPA	96	20.65%
3	EPO	69	14.84%
4	USPTO	38	8.17%
5	KIPO	27	5.81%

Note: World Intellectual Property Organization (WIPO) patents refer to the patent cooperation treaty patents granted by the International Bureau of the WIPO; CNIPA: China National Intellectual Property Administration, Beijing; EPO: European Patent Office, Munich; USPTO: United States Patent and Trademark Office, Alexandria; KIPO: Korean Intellectual Property Office, Daejeon.

Table 1 demonstrates that approximately 40% of the SEPs were granted by the WIPO. This indicates that most applicants plan to file patent applications in multiple countries. To simplify the application procedures, they only apply to the WIPO rather than to the patent office of each country.

4.2. Technology Network Analysis

Figure 2 depicts the network model of the key technologies, and Table 2 lists the CPC code of each key technology.

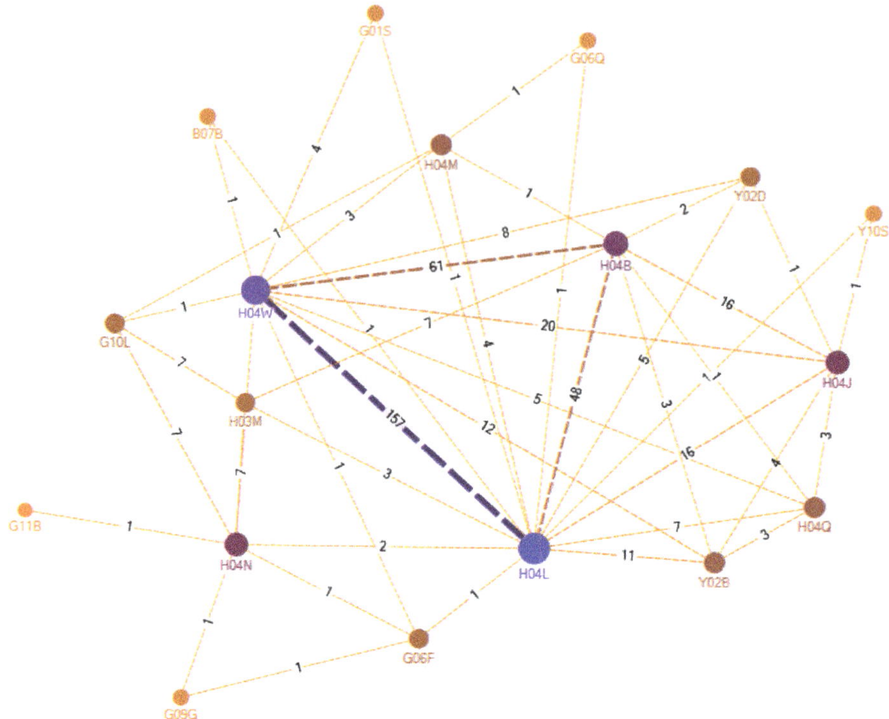

Figure 2. Cooperative Patent Classification (CPC) three-tier network diagram of laser and optical technologies. Note: The size of each node indicates the number of nodes connected to it; the thickness of the arc denotes the strength of the connection.

Table 2. Top 5 CPC codes for optical networks.

CPC	Closeness Centrality	CPC	Eigenvector Centrality	CPC	Fragmentation Centrality
H04L	15.500	H04L	0.448	H04L	0.552
H04W	14.500	H04W	0.419	H04N	0.551
H04B	12.167	H04B	0.340	H04W	0.532
H04N	12.000	H04J	0.304	H04B	0.513
H04J	11.667	H04Q	0.261	H04J	0.510

Table 2 reveals that H04L, H04W, H04B, and H04J are the top five technical fields in terms of closeness centrality, eigenvector centrality, and fragmentation centrality; Appendix A displays the definition of each CPC code. This suggests that transmission of digital information (H04L), wireless communication networks (H04W), transmission (H04B), and multiplex communication (H04J) were the key technologies in optical networks during the period under study.

4.3. Post Analysis: Yearly Changes of the Key Technologies in SEPs Optical Networks

We subsequently inspected the yearly changes in the key technologies in SEPs optical networks to understand the technology trends. The results of the analysis are depicted in Figure 3.

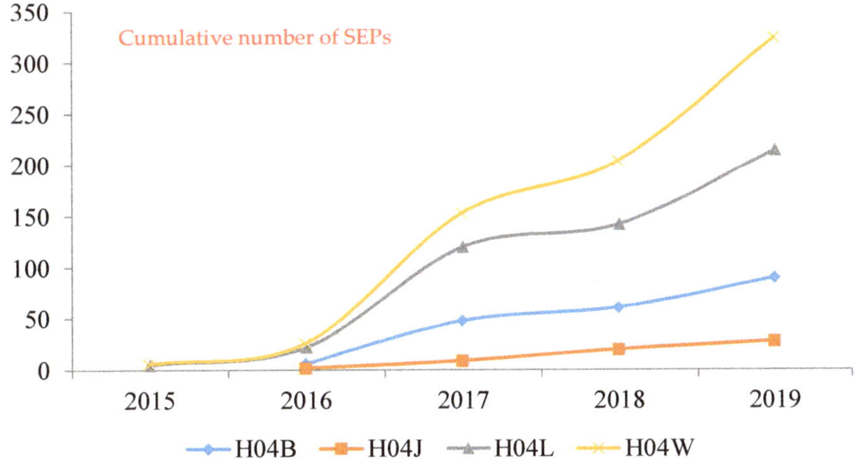

Figure 3. Yearly changes in the key technologies in SEPs optical networks. Note: The cumulative number of SEPs denotes the accumulated number of SEPs each year.

Figure 3 illustrates that H04W and H04L technologies have exhibited high growth in recent years, which signifies that wireless communication devices and digital information transmission have gained attention with the development of 5G and the Internet of things. Looking into the future, we can expect that, as 5G or even 6G develops, the application of communication technologies will become more common in people's lives, such as the Internet of Vehicle. Therefore, many companies co-established the Innovative Optical and Wireless Network, using photonics network and digital twin computing as their core technology. In sum, from Figure 3 and from the industry trend of forming alliances, we can predict that in the future how to integrate optical networks with wireless communication devices and digital information will be a major trend of the industry's technological development.

5. Conclusions

5.1. Discussion and Implications

In this study, we conducted network analyses to explore the key technologies in SEPs optical networks. The empirical results revealed that transmission of digital information, wireless communication networks, transmission, and multiplex communication are the key technologies in optical networks. Relevant professional industries have noted the increase in data volume and the demand for transmission speed and have thus emphasized the strengthening of digital information transmission. The number of hyperscale datacenters is increasing worldwide. These datacenters mainly provide services such as cloud computing, social media, software or platform services, and content transmission. Modules in optical networks that are capable of rapidly processing large amounts of information are thus crucial in such datacenters. At present, 100 G single channel has become the norm in the market, and the recent research and development of 400 G single channel is also promising [31,32].

In terms of transmission and multiplex communication, optical multiplexing is a perennial research focus [33,34]. Wavelengths used for optical multiplexing are specified in channel spacing grids issued by the International Organization for Standardization (ISO) and the International Telecommunication Union. This signifies that technologies in optical networks are highly valued by the ISO. In 2020, the European Telecommunications Standards Institute of the Industry Specification Group established the Fifth Generation Fixed Network (F5G) and proposed an industry prospect, Fiber to Everywhere, which defines the five work items, namely F5G use cases, landscape of F5G technology and standards, definition of fixed network generations, architecture of F5G, and F5G quality of experience [35]. Accordingly, optical network technologies are gradually being standardized, and the competition between manufacturers of SEPs is becoming fiercer.

In terms of theoretical contribution, research on optical networks have mostly focused on technical [5–8] or market [1,9] aspects rather than employing a macroscopic perspective to reveal technology distribution of optical networks and the development trends in their key fields. Additionally, few studies have explored the technology trends of optical networks from the perspective of SEPs. However, observing the technology distribution of optical networks from the aspect of SEPs has become increasingly relevant because such technologies are gradually being standardized. Therefore, we adopted SEPs as the research focus in expectation of addressing this research gap.

Furthermore, we introduced a technical map for optical networks to provide relevant industries and universities with valuable information. We analyzed the focus of technology development in a technology network of SEPs and disseminated information on research and development resource allocation and talent cultivation to industries and universities, respectively. With the development of optical networks, relevant industries should recruit technical talent in key technical fields and also cooperate with academia to increase relevant research and further cultivate technicians. The results of this study revealed that transmission of digital information, wireless communication networks, transmission, and multiplex communication are they key fields in optical networks. Relevant industries and universities can cooperate to enhance the development of optical networks and advance the key technologies that contribute to industrial standards.

5.2. Limitations and Future Research Directions

We examined the key technologies in optical networks from the perspective of SEPs. However, SEP applications proceed slowly; therefore, they only reveal the indispensable standard technologies but not the latest ones in the market, which is a considerable limitation of this study. We adopted the more mature CPC classification code to categorize the technical fields. The classification code only indicated the fields that require long-term funding and talent cultivation but neglected the detailed development process of the fields (e.g., the latest development trend of dense wavelength-division-multiplexed passive-optical-network, orthogonal frequency division multiplexing, and space division multiplexing). Hence, follow-up studies may conduct content analyses on individual patents to increase the depth of the research. Second, this study adopted the perspective of SEPs to discuss the focus of optical networks in the industry's standard stetting. Through the observations of this study, we can see that optical networks have had substantial growth in the industry standard setting. In recent years, there have been 465 SEPs. Regarding standard stetting, the number of SEPs on wireless communication devices and digital information transmission increased most quickly. However, future studies should investigate more detailed sub-topics and focus on expanding the key sub-topics reported by this study, such as wireless communication devices, digital information transmission, and multiplex communication, thereby integrating them into bigger categories. Consequently, the identification of the trend and need for talent development will attract the interest of a wider community. Moreover, we explored the global key technologies protected by SEPs from a macro-level perspective. Nevertheless, the motivation and necessity of filing SEP applications vary because technologies have different attributes and are developed under different industrial environments (e.g., electrical machinery, apparatus and energy have fewer SEPs [36]). We have only analyzed the technical fields disclosed by standard-setting organizations and SEPs. Future studies may employ data from other patent offices (e.g., the USPTO or EPO) to more precisely predict the development trends of novel technologies in optical networks.

Funding: This research was funded by the Ministry of Science and Technology of the Taiwan, grant number MOST 109-2410-H-492-001.

Informed Consent Statement: Not applicable.

Data Availability Statement: Data is contained within the article or supplementary material. The data presented in this study are available in [IPlytics platform].

Acknowledgments: The authors would like to thank the Ministry of Science and Technology of the Taiwan for financially supporting this research under Contract No. MOST 109-2410-H-492-001.

Conflicts of Interest: The authors declare no conflict of interest.

Appendix A

Table A1. Definition of CPC categories.

CPC Categories	Meaning
H04B	Transmission
H04J	Multiplex communication
H04L	Transmission of digital information, e.g., telegraphic communication
H04N	Pictorial communication, e.g., television
H04Q	Selecting
H04W	Wireless communication networks

References

1. *Optical Interconnect Market by Product Category, Interconnect Level, Fiber Mode, Data Rate, Distance, Application, Region-Global Forecast to 2025*; MarketsandMarkets: Northbrook, IL, USA, 2020.
2. Khan, I.; Ahmad, A.; Masood, M.U.; Malik, A.W.; Ahmed, N.; Curri, V. Impact of data center placement on the power consumption of flexible-grid optical networks. *Opt. Eng.* **2020**, *59*, 016115. [CrossRef]
3. Ma, J.; Israel, S. Virtualized networks and virtualized optical line terminal (vOLT). *Fiber Integr. Opt.* **2017**, *36*, 68–77. [CrossRef]
4. Yin, Y.; Liu, L.; Proietti, R.; Yoo, S.J.B. Software defined elastic optical networks for cloud computing. *IEEE Netw.* **2017**, *31*, 4–10. [CrossRef]
5. Losero, E.; Ruo-Berchera, I.; Meda, A.; Avella, A.; Genovese, M. Unbiased estimation of an optical loss at the ultimate quantum limit with twin-beams. *Sci. Rep.* **2018**, *8*, 7431. [CrossRef] [PubMed]
6. Tagantsev, A.K.; Fedorov, S.A. Quantum-limited measurements using an optical cavity with modulated intrinsic loss. *Phys. Rev. Lett.* **2019**, *123*, 043602. [CrossRef]
7. Chen, H.; Jia, H.; Yang, J.; Tian, Y.; Wang, T. Ultra-compact switchable mode converter based on silicon and optical phase change material hybrid metastructure. *Opt. Commun.* **2020**, *473*, 125889. [CrossRef]
8. Miller, K.J.; Hallman, K.A.; Haglund, R.F.; Weiss, S.M. Silicon waveguide optical switch with embedded phase change material. *Opt. Express* **2017**, *25*, 26527–26536. [CrossRef] [PubMed]
9. *Optical Networking Opportunities in the 5G Infrastructure Market: 2019 To 2028*; Communications Industry Researchers: Crozet, VA, USA, 2019.
10. Kim, J.; Lee, S. Forecasting and identifying multi-technology convergence based on patent data: The case of IT and BT industries in 2020. *Scientometrics* **2017**, *111*, 47–65. [CrossRef]
11. You, H.; Li, M.; Hipel, K.; Jiang, J.; Ge, B.; Duan, H. Development trend forecasting for coherent light generator technology based on patent citation network analysis. *Scientometrics* **2017**, *111*, 297–315. [CrossRef]
12. Li, Y.; Phelps, N.A.; Liu, Z.; Ma, H. Featured graphics: The landscape of Chinese invention patents: Quantity, density, and intensity. *Environ. Plan. A* **2019**, *51*, 823–826. [CrossRef]
13. Stoffels, M.A.; Klauck, F.J.R.; Hamadi, T.; Glorius, F.; Leker, J. Technology trends of catalysts in hydrogenation reactions: A patent landscape analysis. *Adv. Synth. Catal.* **2020**, *362*, 1258–1274. [CrossRef]
14. Morioka, T.; Jinno, M.; Takara, H.; Kubota, H. Innovative future optical transport network technologies. *NTT Tech. Rev.* **2011**, *9*, 1–8.
15. Chang, S.H. Revealing development trends and key 5G photonic technologies using patent analysis. *Appl. Sci.* **2019**, *9*, 2525. [CrossRef]
16. Katti, R.; Shanthi, P. A survey on role of photonic technologies in 5G communication systems. *Photonic Netw. Commun.* **2019**, *38*, 185–205. [CrossRef]
17. Chang, S.H. Patent analysis of the critical technology network of semiconductor optical amplifiers. *Appl. Sci.* **2020**, *10*, 1552. [CrossRef]
18. Saadaouil, F.; Fathallah, M.; Ragheb, A.M.; Memon, M.I.; Fathallah, H.; Alshebeili, S.A. Optimizing OSSB generation using semiconductor optical amplifier (SOA) for 5G millimeter wave switching. *IEEE Access* **2017**, *5*, 6715–6723.

19. Imtiaz, W.A.; Ahmed, H.Y.; Zeghid, M.; Sharief, Y. Two dimensional optimized enhanced multi diagonal code for OCDMA passive optical networks. *Opt. Quantum Electron.* **2020**, *52*, 1–17. [CrossRef]
20. Zhang, B.; Zhao, Y.; Rahman, S.; Li, Y.; Zhang, J. Alarm classification prediction based on cross-layer artificial intelligence interaction in self-optimized optical networks (SOON). *Opt. Fiber Technol.* **2020**, *57*, 102251. [CrossRef]
21. Lu, Y.; Gu, H. Flexible and scalable optical interconnects for data centers: Trends and challenges. *IEEE Commun. Mag.* **2019**, *57*, 27–33. [CrossRef]
22. *Silicon Photonics for Data Centers and Other Applications 2016*; Yole Développement: Lyon-Villeurbanne, France, 2016.
23. Chang, S.H. Key technologies and development trends of 5G optical networks. *Appl. Sci.* **2019**, *9*, 4835. [CrossRef]
24. Spulber, D.F. Standard setting organisations and standard essential patents: Voting and markets. *Econ. J.* **2019**, *129*, 1477–1509. [CrossRef]
25. Borgatti, S.P. Identifying sets of key players in a social network. *Comput. Math. Organ. Theory* **2006**, *12*, 21–34. [CrossRef]
26. Whelan, E.; Golden, W.; Donnellan, B. Digitising the R&D social network: Revisiting the technological gatekeeper. *Inf. Syst. J.* **2013**, *23*, 197–218.
27. Kwon, O.; An, Y.; Kim, M.; Lee, C. Anticipating technology-driven industry convergence: Evidence from large-scale patent analysis. *Technol. Anal. Strateg. Manag.* **2020**, *32*, 363–378. [CrossRef]
28. Choi, J.Y.; Jeong, S.; Jung, J.K. Evolution of technology convergence networks in Korea: Characteristics of temporal changes in R&D according to institution type. *PLoS ONE* **2018**, *13*, e0192195.
29. *SEP Analysis*; IPlytics: Berlin, Germany, 2019.
30. Borgatti, S.P.; Everett, M.G.A. Graph-theoretic perspective on centrality. *Soc. Netw.* **2006**, *28*, 466–484. [CrossRef]
31. Morsy-Osman, M.; Sowailem, M.; El-Fiky, E.; Goodwill, T.; Hoang, T.; Lessard, S.; Plant, D.V. DSP-free 'coherent-lite' transceiver for next generation single wavelength optical intra-datacenter interconnects. *Opt. Express* **2018**, *26*, 8890–8903. [CrossRef] [PubMed]
32. Sowailem, M.Y.S.; El-Fiky, E.; Morsy-Osman, M.; Zhuge, Q.; Hoang, T.M.; Paquet, S.; Paquet, C.; Woods, I.; Liboiron-Ladouceur, O.; Plant, D.V. Self-homodyne system for next generation intra-datacenter optical interconnects. *Opt. Express* **2017**, *25*, 27834–27844. [CrossRef]
33. Shimosawa, K.; Uenohara, H. Subchannel drop and add operation by using silicon photonic all-optical orthogonal frequency division multiplexing demultiplexers. *Opt. Lett.* **2020**, *45*, 3852–3855. [CrossRef] [PubMed]
34. Xu, S.; Wang, J.; Zou, W. Optical patching scheme for optical convolutional neural networks based on wavelength-division multiplexing and optical delay lines. *Opt. Lett.* **2020**, *45*, 3689–3692. [CrossRef] [PubMed]
35. ETST. ETSI Launches New Group on 5th Generation Fixed Network Shifting the Paradigm from Fibre to the Home to Fiber to Everything Everywhere. Available online: https://www.etsi.org/newsroom/press-releases/1723-2020-02-etsi-launches-new-group-on-5th-generation-fixed-network-shifting-the-paradigm-from-fibre-to-the-home-to-fiber-to-everything-everywhere (accessed on 28 July 2020).
36. *Landscaping Study on Standard Essential Patents (SEPs)*; IPlytics: Berlin, Germany, 2019.

applied sciences

MDPI

Review

Optical Technology for NFV Converged Networks

Paola Iovanna [1,*], Alessandra Bigongiari [1], Alberto Bianchi [1], Sylvie Menezo [2], Marco Romagnoli [3], Vito Sorianello [3] and Fabio Cavaliere [1]

1 Ericsson Research, Via Moruzzi 1, 56124 Pisa, Italy; alessandra.bigongiari@ericsson.com (A.B.); alberto.bianchi@ericsson.com (A.B.); fabio.cavaliere@ericsson.com (F.C.)
2 SCINTIL Photonics, Parvis Louis Néel 7, CEDEX 9, 38040 Grenoble, France; sylvie.menezo@scintil-photonics.com
3 CNIT—Photonic Networks and Technologies Nat'l Lab, Via Moruzzi 1, 56124 Pisa, Italy; marco.romagnoli@cnit.it (M.R.); vito.sorianello@cnit.it (V.S.)
* Correspondence: paola.iovanna@ericsson.com

Abstract: 5G and its evolution towards 6G is unlocking new use cases that will require the reconsideration of the existing network architectures and its operation. As the network will be required to support new service types and radio protocol splits, the traditional physical point to point connections will need to be replaced with a transport network up to the antenna site to guarantee low latency services and high bandwidth. Optical based transport is a key enabler to realize such a convergent network, where the traditional fixed infrastructure in use for mobile services and mobile infrastructure should also support enterprises services. The Software Defined Network (SDN) and Network Function Virtualization (NFV) technology plays a key role to evolve towards digitalization. It allows to simplify the creation of new services and to implement a real decoupling between the infrastructure and the network functions that run virtually, on generic processing units located everywhere in the network. Supporting automation is a key requirement that traditional optical technology is not able to meet. In this paper the reference scenarios for the access network are presented with the analysis of their requirements and the enabling optical solutions based on integrated silicon photonics.

Keywords: 5G; beyond 5G; SDN; NFV; convergent network; Xhaul; integrated silicon photonics; tunable filters; ROADM

Citation: Iovanna, P.; Bigongiari, A.; Bianchi, A.; Menezo, S.; Romagnoli, M.; Sorianello, V.; Cavaliere, F. Optical Technology for NFV Converged Networks. *Appl. Sci.* **2021**, *11*, 1522. https://doi.org/10.3390/app11041522

Academic Editor: Amalia Miliou
Received: 7 December 2020
Accepted: 29 January 2021
Published: 8 February 2021

1. Introduction

5G and its evolutions enable new use cases such as industrial automation, mission critical Internet of Things (IoT), and advanced consumer services like virtual reality, augmented reality, and gaming with outstanding user experience. These new use cases will create additional revenue streams for Communication Service Providers. However, these new services require an adaptation of the network and its operations. In order to provide a good level of connectivity everywhere and among every point, the traditional network segments are evolving towards "one network" where the mobile infrastructure will be able to support any service, including enterprises, and the fixed infrastructure will evolve to also support mobile traffic. Such a new network segment where 5G mobile transport, fixed access, and vertical services converge is often referred to as Xhaul [1,2].

To comply with low latency use cases, some mobile core network functions need to move to the access. Traditional big data centers, in regional network segments, will be replaced by smaller ones, located closer to the antenna sites. Moreover, the transport network is evolving, replacing point to point physical connections, common in the access area, with rings, tree o simple mesh transport networks.

The mobile core network is evolving as well, becoming cloud native to allow new services to be created faster and in a more agile way. However, while some network functions will run on generic processors, other functions will remain implemented on dedicated hardware and managed as regular physical network functions (PNF).

In addition to 5G and IoT, other technologies, such as the Software Defined Networking (SDN) and Network Function Virtualization (NFV), will contribute to deeply transform the business of new digitalized industries. SDN facilitates the management and configuration by applying the software approach in all network operations. NFV allows to decouple the network functions from the proprietary hardware appliance, by using functions as Virtual Machines (VMs). Moreover, NFV uses Information Technology (IT) to partition the networks into virtualized building blocks that link up to produce communication services. As soon as industries introduce virtualized and software-defined infrastructure to expand their business in the digital economy, network automation will become essential to re-scale and re-purpose their operations.

The concept of "one network" targets End-to-End (E2E) network management, encompassing the entire network: radio, transport, and cloud resources. The access network segment will require not only higher bandwidth and lower latency than today, but also a higher level of automation, comparable with that of the other segments. Today optical communication technology can meet in principle high bandwidth and low latency requirements, however an enhanced level of automation is needed to enable the realization of networks that are managed E2E automatically as "one network". Moreover, just scaling down in performance and cost the optical technology used in today metro and aggregation networks could be insufficient, since the capillary distributed terminations of the new access network will demand much smaller equipment size, energy consumption, and cost.

Silicon photonics are promising for this purpose, but they have been utilized, so far, mainly for datacom applications, e.g., to provide internal high bandwidth connectivity in data centers, that present different requirements with respect to telecom applications. Achieving a high level of automation at a suitable cost and size is one of these requirements. There are promising research results addressing this aspect [1] but not yet large-scale industrial investments.

In this paper, we will discuss the new transport requirements in the access network, focusing on an implementation based on integrated silicon photonics. In Section 2 the reference network scenario and the main requirements are reported; in Section 3 a survey of silicon photonics technology is presented with the main challenges and potentialities; Sections 4 and 5 provide two examples of novel enabling silicon photonics components and their design issues: tunable optical filters and Reconfigurable Optical Add Drop Multiplexers (ROADM), respectively. Conclusions are in Section 6.

2. Reference Network Scenario and Main Requirements

The transport network requirements depend on both the E2E services to support, such as IoT, and the type of radio architecture. In a radio split architecture, the different layers of the protocols stack are allocated in different nodes, instead of being all in the same base station. Figure 1 reports a reference scenario for an access network able to support mobile transport features with alternative radio splits, concurrently supported. As shown in Figure 1, common topologies in this access area are Passive Optical Networks (PON) and ring optical networks.

In this scenario, the antenna sites are connected to the Central Office (CO) using different radio interfaces, each one corresponding to a different protocol split: CPRI [3], which is the most critical in terms bandwidth and latency requirements; eCPRI [4], which requires less bandwidth but has similar latency requirements of CPRI; and S1 [5] and F1 [6] that have less demanding bandwidth and latency specifications. The CO hosts both VNFs and PNFs (Physical Network Functions). For the sake of simplicity, in Figure 1, the radio interfaces are shown in the ring topology case, but the same traffic applies to the PON.

An efficient transport infrastructure should be able to accommodate different services, as well as different radio split options. This concept is usually referred as Xhaul [7]. In [7], alternative Xhaul implementations are reported, including a case where PON physical infrastructure is used, but not its Time Division Multiplexing Access (TDMA) protocol. Wavelength Division Multiplexing (WDM) has several advantages for the implementation of an Xhaul infrastructure: it is transparent to the different radio interfaces, it maximizes the transported capacity over optical fiber and has low latency [7]. However, a key aspect for enabling the path towards NFV is to provide a suitable level of automation. Automation in PON and rings presents peculiar requirements that will be better detailed in the next sections.

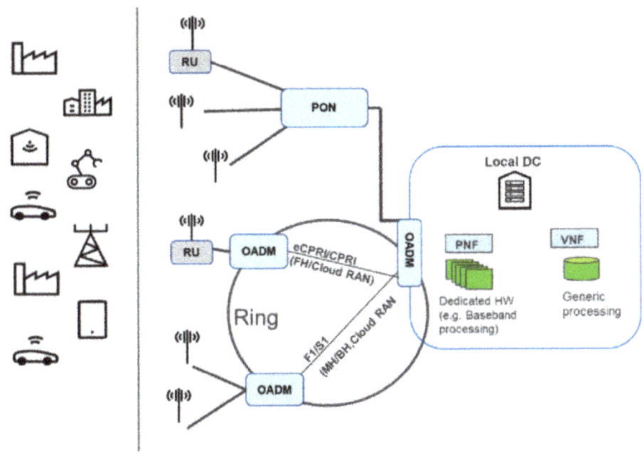

Figure 1. Reference scenario for access network segment.

2.1. WDM Overlay in PON

In most cases, network operators would like to reuse installed PON infrastructure to carry mobile traffic. This allows to exploit the existing infrastructure, saving the high cost of new installations. The topology of a PON is point-to-multipoint, i.e., a "tree" that connects the CO with end user sites through a power splitter. A WDM overlay on a PON [8] is a good solution to support several radio splits, especially in the case of low latency requirements and high bandwidth traffic, such as CPRI and eCPRI [9]. With the WDM overlay, the same fiber transmits the PON upstream and downstream channels and the WDM channels over a separate spectrum [8]. Figure 2 shows a PON with WDM overlay. The distance between CO and antenna site depends on the operator infrastructure: it is typically 5–10 km in densely populated areas.

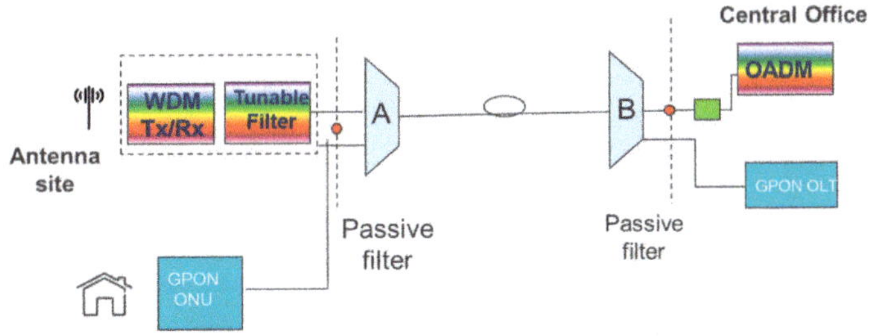

Figure 2. PON with WDM overlay [8].

Bidirectional operation (i.e., using a single fiber for both propagation directions) is another important feature of the network scenario in Figure 2. It simplifies the network installation and configuration, that can be done "plug and play", without taking care of the different fibers used in uplink and downlink.

The key device to provide a first level of automation in such a network is a full tunable Transceiver (TRX), at least at the PON termination site. As shown in Figure 2, the WDM comb is transmitted from the CO to the antenna site, where a tunable optical filter at the receiver selects the right wavelength. Today, most WDM networks make use of fixed optical filters, which impacts inventory and network configuration complexity. To reduce the TRX cost, direct detection is preferable to coherent transmission [10]. The bit rate of interest in an Xhaul network to carry CPRI and eCPRI channels is 25 Gb/s [11]. Tunable direct detection TRX at 25 Gb/s are commercially available but they support tunability only in the transmission function, by means of tunable lasers. All current technologies to realize tunable filters at the receiver, such as Micro Electro-Mechanical Systems (MEMS) [12] or liquid crystals [13], are not complaint with the TRX requirements in terms of size and cost. Tunable TRXs developed for the New Second Generation PON (NG-PON2) only support 4–8 WDM channels [14,15] while the number of channels should be much higher in a fully automated Xhaul network, for example 20 WDM channels with 100 GHz frequency spacing, according to the ITU-T standard frequency grid. As shown in Figure 3a, a conventional TRX, where only the transmitter is tunable, works only if it receives one wavelength that has been previously selected (e.g., from a MUX in the Figure 3a). This is due to the fact that it cannot select the wavelength from an ingress WDM channel comb. Hence a conventional TRX does not work in a PON infrastructure based on power splitters, where no wavelength selective optical devices are present (Figure 2). Figure 3 shows the compatibility of a fully tunable TRX with a power splitter. In this case a tunable filter is integrated in the TRX module in front of the receiver.

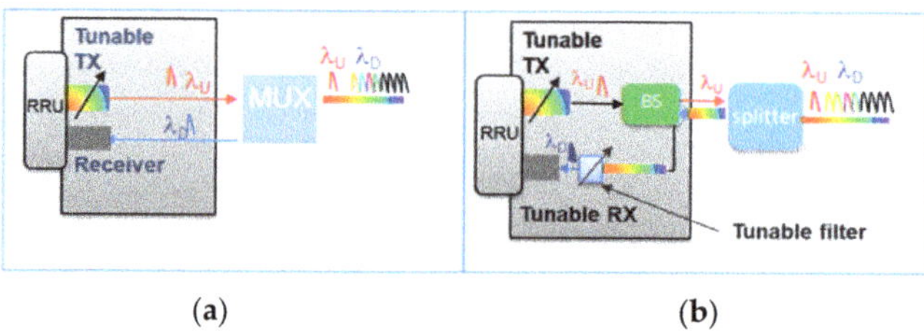

Figure 3. TRX functional architectures. (**a**) tunable TRX transmission only [15,16]; (**b**) Full tunable TRX.

Integrated silicon photonics is a promising technology for realizing cost effective tunable filters for fully tunable TRX. Details about this technology, design options and technological challenges, will be discussed in Sections 4 and 5.

2.2. Ring Networks

As shown in Figure 4, CO and antenna sites can be connected by a WDM ring. In this kind of network, to ensure the compatibility of well-established protection mechanisms and the optical components widely used in metro networks (transceivers, wavelength multiplexers, and demultiplexers, optical amplifiers, etc.), the use of two fibers, one for each propagation direction, is common [16].

Also, in this Xhaul case, different radio split options must be supported concurrently. Since this scenario is expected to be used for centralized radio access network deployments, where based-band processing is pooled among geographically distant Remote Units (RU), the distance between the CO and the antenna site is usually 10–20 km, compatibly with the latency requirements of the fronthaul interfaces.

In such centralized scenario, 25 Gbit/s channels from the antenna sites may be aggregated (using a packet or circuit switch) in 100 Gbit/s WDM channels. With Dual-Polarization Quadrature Phase Shift Keying (DP-QPSK) this requires a bandwidth (BW) of at least 37 GHz to the ROADMs in the ring. In Figure 4, only the ROADMs are shown;

the electrical switch is out of the scope of this analysis, more details about it can be found in [7].

ROADMs are unusual in traditional access networks but here they are essential to assure the level of automation in network operation that SDN and NFV technology require. De facto the ROADMs allow to dynamically and automatically configure functions of wavelength bypass for traffic not to be processed locally, wavelength add or drop for local traffic and wavelength routing toward adjacent rings, e.g., used as further stage of aggregation. Up now, ROADMs are used in aggregation and long-haul networks, to deal with high amounts of aggregated traffic and long distances.

Hence, the ROADMs used in current optical metro long-haul networks are based on high performance but expensive Wavelength Selective Switches (WSS) [17]. Alternative, less expensive and power efficient technologies are required to address the use case of the access network. Integrated silicon photonics is a promising technology to realize small size, low cost ROADMs that operate over a distance much shorter than long-haul. In Sections 3 and 5 details about the use of silicon photonics to realize ROADMs for radio-access segment are disclosed.

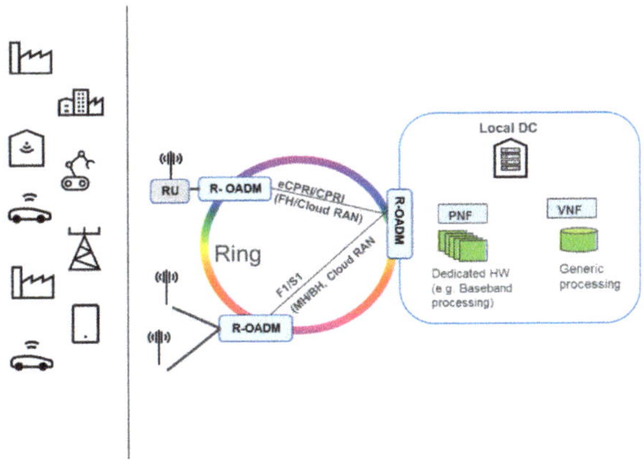

Figure 4. Ring topology [7].

3. Silicon Photonics Devices in Optical Transport: Challenges and Opportunities

Silicon photonics is a photonic integration technology that can exploit standardized Complementary Metal Oxide Semiconductor (CMOS) fabrication processes to produce optical components in high volumes with low manufacturing costs. Its transparency to wavelengths longer than 1.1 μm makes silicon an attractive material for optical devices that work in the 1.3 and 1.55 μm wavelength windows. Silicon photonic chips can include various functional blocks such as optical waveguides, variable optical attenuators, optical bandpass filters, optical switches, amplitude and phase optical modulators, and wavelength multiplexers and demultiplexers. However, there is a long road ahead before silicon can achieve, in photonics, the pervasiveness it has in digital integrated circuits. In particular, considerable research effort is spent on the following interwoven design issues: achieving low coupling loss between a silicon photonics chip and input and output optical fibers; coupling input light having an unknown polarization state to silicon photonics chips that can only sustain linearly polarized light, along the Transversal Electric (TE) propagation mode; including optical amplification functionalities on chip to mitigate coupling and propagation losses, and embed light sources; realizing cost effective packaging processes suitable of productions in high volumes.

3.1. Coupling and Polarization Control Issues in Silicon Photonics Devices

There are two techniques commonly used to connect a silicon photonic component to an optical fiber: edge coupling and surface grating coupling [18]. An edge coupler adapts the spot size of the light beam in optical fiber to the smaller spot size in the silicon circuit. Edge couplers have high coupling efficiency (i.e., low coupling loss), large bandwidth and are independent of the polarization state of the input light. However, since the polarization state of the input light is unknown in many applications, a dual polarization design for the photonic circuitry is often necessary, enabled by a polarization splitter and rotator [19]. In a Grating Coupler (GC), a diffraction grating is realized by periodically notching a silicon waveguide, so that light within a certain wavelength range is diffracted off in the direction of the connected optical fiber [20,21]. This mitigates the need for accurate alignment between silicon chip and fiber, facilitating the testing of the circuit and, consequently, its fabrication process. The combination of GCs and micro-lenses has been proposed [22] to further decrease the required precision to alignment and bonding, that adds significant cost to the packaging. Several GC designs exist, some of them able to deal with an arbitrary input polarization state. An example is the Polarization Splitting Grating Couplers [21], where input light in two orthogonal linear polarization states is directed into two different orthogonal waveguides in the silicon photonic circuit. These may commonly have insertion losses in a range of 3–5 dB but state of the art components may reach values close to 2 dB [23,24].

Active polarization alignment, based on silicon photonics polarization controllers, was also proposed as alternative to the dual polarization design, e.g., based Mach–Zehnder interferometers with phase shifters [25]. However, no "endless" operation, i.e., the capability to deal with any arbitrary polarization variation without causing service interruption, was demonstrated so far with practical system configurations. Moreover, fiber Polarization Mode Dispersion (PMD) makes impossible for a single controller to simultaneously recover the polarization of all channels in a WDM system.

3.2. Integration Issues of Optical Amplifiers and Active Sources in Silicon Photonics Devices

One issue with silicon is that it is a bad light emitter, due to its indirect bandgap, making it very difficult the realization of light sources and optical amplifiers. The integration of silicon with materials of the III–V groups, such as Indium Phosphide (InP), is a solution commonly adopted for the realization of Semiconductor Optical Amplifiers (SOA) and lasers on chip. Amplifiers provide additional power budget and allows to compensate for or mitigate the device insertion loss, which is primarily due to the coupling with input and output optical fibers. The high insertion loss is an issue that so far prevented silicon photonics to become widespread in optical networks, although it offers tremendous opportunities for cost reduction and enhanced flexibility. CMOS-compatible ROADMs are an example of silicon-based optical device enabling such a new network [26]. The integration of silicon with III-V materials for achieving optical gain is made it difficult by the mismatch of the crystal lattice in the two materials. Despite this, techniques have been developed for integrating InP SOAs on silicon [26]. In the so-called flip-chip integration, micro-packaged SOA assemblies are individually aligned by a flip-chip bonder with a silicon grating coupler, by means of active alignment techniques that use an auxiliary laser beam. A disadvantage of the flip-chip bonding is that a sub-micrometric alignment precision is required for coupling the laser beam, slowing down the production process and leading to significant impact on the device cost. Silicon III-V bonding at wafer level is an alternative integration technique. It consists of molecular bonding of III-V materials on top of a patterned Si-on-Insulator (SOI) layer. This process is more efficient than flip-chip bonding because it can be performed at die or wafer level, but it complicates the device structure and requires enhancements to the standard CMOS-compatible processes.

3.3. System Applications Enabled by SOAs Integrated in Silicon Photonics Devices

Since SOAs can be integrated in a silicon photonic circuits, they allow to replace bulky and expensive Erbium Doped Fiber Amplifiers (EDFA) in optical networks. In an ideal implementation, optical amplifiers will be no longer present as separate blades or boxes, but they will be fully integrated in multi-functional silicon photonics subsystems including optical switches, wavelength multiplexers, optical receivers, and so on. Examples of applications enabled by silicon photonics circuits containing SOAs are optical receivers with integrated SOAs, to operate at low input powers in a Passive Optical Network (PON) [27,28], and Dense WDM (DWDM) systems with SOAs used as line optical amplifiers in ROADMs [29]. However, this comes at a cost. It is known that SOAs typically have fast, sub-nanosecond, response times, which, combined with their non-linearity, may cause signal distortions [30]. For similar reasons, the use of SOAs is problematic in DWDM systems, where they introduce significant Cross-Gain Modulation (XGM) penalty [31]. Due to XGM, the gain experienced by a DWDM channel depends on amplitude and bit pattern of the other channels. Multi-channel non-linear effects, such as cross-phase modulation (XPM) and Four-Wave-Mixing (FWM), can further impair the crosstalk between channels [32]. For all these reasons, SOAs are typically used in single channel systems, at the transmitter output or the receiver input of a single channel transceiver, to increase the transmitted optical power or improve the receiver sensitivity, respectively. In DWDM systems, the XGM can be avoided operating in the small signal gain regime, i.e., with low input powers, where the SOA gain is insensitive to fluctuations of the input power. However, with low values of input power, the Amplified Spontaneous Emission (ASE) noise generated by the SOA is significant, reducing optical signal to noise ratio (OSNR) and system power budget. An example of SOA operating in small signal gain regime is its use at the receiver input in a PON, where the received optical power is limited by the presence of lossy passive splitters. This case will be analyzed in more detail in Section 4.

Designing an amplified silicon photonic circuit requires to carefully model how the SOA behavior is affected by the silicon cavity characteristics, i.e., cavity size, facet reflectivity, maximum pumping current that is possible to use, etc. This poses additional challenges with respect to stand-alone SOAs. Design guidelines for integrating SOAs in silicon photonic circuits are provided in [30], for a SOI chip with the flip chip bonded SOAs. The numerical analysis used a multi section cavity model and was then experimentally validated in a 32 channels DWDM system. In [28], it is shown that the usual travelling-wave assumption for the SOA is no longer valid, due to unavoidable non-perfect mode matching at the interface between the spot-size converter and the SOA, and even small reflections need to be taken into account. So, a Nearly Traveling Wave (NTW) model was developed. Polarization insensitiveness was achieved by means of dual a polarization design with two SOAs placed at the two output of a polarization splitting (GC). The impact of the ASE noise was estimated in the worst-case polarization condition, i.e., when the signal is all in one of the input polarizations and only noise was present in the orthogonal one. The results in [29] indicate that a maximum residual facet reflectivity of 4×10^{-4} for a chip-bonded SOA leads to a power penalty below 2 dB in a polarization-diversity twin SOAs receiver.

Working in small signal gain regime is not always possible when SOAs are used as line amplifiers and in this case special care must be paid to limit the XGM. Design guidelines are given in [27], using as an example a 12 channels DWDM system with five ROADMs node in a ring network. The ROADMs are silicon photonics chipsets that use silicon micro-ring resonators as wavelength switching elements, controlled Mach–Zehnder interferometers for protection switching and silicon variable optical attenuators for loss equalization. Input SOAs are flip chip bonded to the silicon photonic chipset to compensate for device and fiber loss. The total power at SOA input was −1.2 dBm, corresponding to −12 dBm of channel power. The results in [27] show that it possible to use SOAs as integrated line amplifiers, but with limitations in terms of number of SOAs in the network (up to 5, in the example in [27]) and their gain (about 10 dB in the same example). This reduces maximum number and length of the fiber spans, but these limitations remain compatible with the requirements discussed in Section 2. For example, a SOA gain of 10 dB would allow to allocate 5 dB for the ROADM insertion loss of 5 dB and further 5 dB of fiber span loss, corresponding to a fiber length 20 km, with an attenuation coefficient of 0.25 dB/km. With 5 ROADM nodes, the total length of the ring would be 100 km, well beyond the typical distance of a metro access network [30].

3.4. Silicon Photonics in Optical Transceivers and Laser Integration Issues

An area where silicon photonics is already successful in the market is that of optical transceivers. However, even with a silicon photonic implementation, significant advances are required to meet the needs of 5G deployments that will require a throughput in the order of Terabit/s and power consumption of few pJ/bit/s [1,31]. This will probably call for co-packaged solutions where photonic transceivers are integrated in a multi-chip module with digital Integrated Circuits (IC). Integration of silicon photonics with III-V lasers is crucial also here: in [33], a co-packaged 1.6 Tb/s (4 × 400 Gb/s) silicon photonic engine is described. In [34], a co-packaged engine based on silicon photonics integration with III-V quantum dot lasers on a large-scale wafer is proposed. The throughput ranges from 800 Gb/s to 3.2 Tb/s, using 100 Gb/s Non-Return to Zero (NRZ) or 200 Gb/s Order—4 Pulse Amplitude Modulation (PAM4). Silicon photonics can also be helpful to decrease the cost of DWDM high-speed TRX. For example, a cost-effective coherent receiver based on a silicon photonics 120° hybrid has been demonstrated in [35]. As a second example, in [36] 224 Gb/s were transmitted on a single wavelength over 10 km of fiber using a Stokes receiver and an integrated silicon nitride optical dispersion compensator.

To support computationally insensitive applications, such as baseband processing in massive multi-input multi-output (MIMO) systems and Terabit/s packet switching, the density of Integrated circuits (IC) per surface units is expected to remarkably increase in modern Printed Circuit Boards (PCB). This leads to an increase in the devices operating temperature, especially in applications where active cooling is problematic for cost, size, or power consumption constraints, such as outdoor radio units. One known advantage of silicon photonics is its capability of operating at high temperatures [37]. However, lasers are needed anyway in optical transceivers and they are notoriously very temperature sensitive. Quantum dot lasers can in principle operate at high temperatures and can be integrated with silicon photonics devices [38] but in field operation and high yield need to be demonstrated yet. Using Remote Laser Sources (RLS), where the laser is placed far from the PCB thermal hotspots and is externally coupled to an optical transceiver hosting a silicon photonics modulator, is an alternative to quantum dot lasers. Separating the laser from the rest of the transceiver leads to negligible additional cost over short link distances (in the order of 1 m). The main cost contribution is due to the Polarization Maintaining Fiber (PMF) needed to connect laser and modulator. Over longer distances (in the order of 10 m or higher), using PMF is instead unpractical, due to cost. In those cases, special RLS arrangements [39] can be used to face the changes of the light polarization state during the propagation over a Standard Single Mode Fiber (SSMF) link. They unavoidably lead to an increase in cost that anyway remains lower compared to alternatives (e.g., active cooling), at least until quantum dot lasers will become available in large scale.

3.5. Thermal Management in Silicon Photonics

Thermal management represents one of the main challenges of silicon photonics. This aspect has important impacts in the use of tunable optical components, since lasers and optical filters based on resonant optical elements are controlled via the thermo optic effect.

Currently, the approach adopted in transceivers with a tunable laser source is the use of solid state thermo-electric coolers (TEC), also indicated as macro-TEC, together with resistive heating elements for fine-tuning each laser. The TEC acts to maintain a reference temperature of the photonic chip: a temperature sensor monitors the temperature at a given location in the chip and the measure is used in a feedback control, generally based on a Proportional-Integral-Derivative (PID) algorithm, that operates the TEC. The heaters then are used to independently set the local temperature that corresponds to the desired frequency of operation of each laser in the chip. This method is not optimal from the point of view of power consumption and packaging, especially in the perspective of obtaining a higher integration level. A different approach that promises a more efficient temperature control and a higher potential for dense integration consists in the use of micro-TECs [40,41]. These are micron-size coolers that are placed in proximity of the hot-spots, minimizing the parasitic thermal resistance between the heat source and the thermoelectric cold junction. The coefficient of performance (COP) of micro-TECs is estimated to be about 0.8 COP, against the 0.3 value of traditional macro-TECs.

An alternative approach to the use of TECs is the use of a more complex feedback loop based on the monitoring of the local temperature and the adaptation of local heater's power. Temperature monitoring can be performed directly, with thermal photodiodes [42], or indirectly by inferring the thermal shift from the output power [43] or the data signal [44].

The tuning precision of the micro-heaters can be conveniently set to meet the maximum tolerated deviation from the laser nominal frequency. For example, an optical filter or switch based on a micro-ring resonator (MRR), having a typical radius size of 6 μm, has a 0.01 nm precision (1.25 GHz at 1550 nm) on the central frequency, for a temperature control precision of 0.1 K. This calculation refers to a Silicon on Insulator platform (SOI), the value of the thermo-optic coefficient of silicon is $CS = 1.86 \times 10^{-4}\ K^{-1}$. However, the actual frequency deviation depends on the speed of the feedback control mechanism, i.e., in the actuation of the heater power change following a monitoring measurement. The control speed must be able to follow the temperature fluctuations that, for various reasons, the chip can undergo. It is possible to achieve a stronger stability to temperature fluctuations with materials such silicon nitride, which has thermo optic coefficient $CN = 2.51 \times 10^{-5}\ K^{-1}$ [45]. Alternatively, larger radius MRRs can be designed. Both these options allow to achieve a higher precision of the tuning mechanism, but at the expenses of a reduction in the tuning range.

3.6. Is Silicon Photonics Promise of Cost Reduction and Ease of Integration, When All Issues Are Considered, Significant Enough to Make Silicon Photonics a Game Changer?

In the previous sections, we have introduced silicon photonics as the photonic integration technology with the highest potential for reducing the cost of optical components in communications systems, ideally making costs and volumes comparable to those of ICs in consumer devices. We also discussed the performance gaps that remain to be filled before this goal can be achieved, in terms of coupling issues with optical fiber, polarization control issues, integration issues with semiconductor optical amplifiers, and issues in interfacing silicon photonics devices with laser sources. Solving those issues will require significant technology advances and, indeed, if silicon photonics will really be able to meet its cost reduction promises is still an open question across the scientific community and, even more, across the industry. We think that, despite the outstanding difficulty of some of those challenges, they can and will be eventually won, based on the solutions we outlined in the previous sections. These will be further detailed in the next sections for two specific use cases (full tunable transceivers and reconfigurable optical add-drop multiplexers). The major obstacles to the success of silicon photonics are not technical, in our opinion. The lack of widely accepted standards and design libraries, contrary to electrical ICs, and the difficulties of setting up an economically sustainable business model, appear to play a major role. Silicon fabs are reluctant to divert resources from a profitable consumer market to products with much lower volumes and more uncertain business. On the other hand, telecommunication systems vendors seldom have all the competences in house. Even so, a vertical production model, where everything is made in house, from design to production, passing through prototyping and sampling, could hardly be sustainable in absence of certain selling volumes (that, in turn, cannot be guaranteed in absence of products, creating a detrimental vicious circle). Avoiding an impasse calls for alternatives, creative multi-player models, where: (i) industry can leverage on design competences provided by universities and research centers, (ii) sampling and prototyping are supported by publicly funded pilot lines, and (iii) standardization organizations facilitate the availability of multi-vendor interoperable solutions based on standard libraries and design practices.

4. Tunable Filters in Silicon Photonics for Full Tunable Transceivers

It has been discussed in Section 2 how tunable optical filters have a key role in PON networks with WDN overlay, when embedded in a WDM transceiver before the photodetector. Tunable optical filters introduce flexibility in the planning of the network and its upgrade, and enable software (SW) reconfiguration, reducing inventory costs.

Presently, commercial tunable filters are mainly based on Micro-Electro-Mechanical Systems (MEMS), miniaturized electro-mechanical elements that allow wavelength selection by moving a micro-mirror. However, the power consumption of MEMS based filters may be excessive for integration in pluggable modules [12]. Furthermore, the cost of MEMS based filter is high for the scale of the application in scenarios such as 5G access networks and data centers. The high cost is due to due to their complex mechanical structures based on free space optics and 3-dimensional movements of micro-mirrors. In addition, there are few solutions allowing the fabrication of MEMS based filters through CMOS compatible processes (which are available in a standard electronic production line). This may prevent a reduction in costs even for large volume fabrication.

A second solution available in commercial products is thin film filters. These are stacks of dielectric layers with thickness equal to a quarter of the central passband wavelength, where wavelength tuning is achieved by varying the incident angle of the incoming light beam. The properties of the filter are determined by the number of layers and the optical properties of the dielectrics. Commonly used materials are silica (SiO_2) as the low-index layer and tantalum pentoxide (Ta_2O_5) as the high-index layer. The typical size is 2 mm^2. The three main deposition techniques are used to achieve performances compatible with DWDM filtering applications, they are: ion beam assisted deposition (IBAD), plasma-assisted deposition (PAD), and ion beam sputtering (IBS). However, the power consumption of thin film filters may be excessive for the scale of the application scenario and the cost of thin film filters (due to the fabrication process and the cost of the controls associated to incidence angle variation) is relatively high for 5G access networks [46]. In addition, thin film filters with tunable functions cannot be integrated in a silicon photonic chip with standard CMOS compatible processes, and the footprint of the filter is large compared to the total area of a photonic chip.

For this reason there is the need to explore new solutions for the realization of tunable optics that are compliant with size and consumption requirements of, e.g., common pluggable modules, such as SFP28 [47] and are compatible with production in volumes at a cost comparable with the current module price.

Silicon photonics provides a viable solution to produce tunable filters with small footprint and reduced consumption that are potentially integrable in small form factor transceivers. Furthermore, the compatibility with CMOS production lines in use for integrated electronics allows the reduction in production costs with respect to commercial filter solutions based on MEMS and thin film filters. Potentially, SOAs can be integrated for increased power budget in an efficient way [30,31].

The requirements for the use of tunable optics in the scenarios outlined in Section 2 are to operate on a wide spectral range of deployed WDM network: to achieve full functionality the offer of channels should be increased from the actual number of 4–8 channels [15] to about 20 bidirectional channels. As discussed in Section 2, the spacing of the channels should be 100 GHz with a minimum filter BW of 37 GHZ.

For the deployment of WDM on the existing system apparatus, the filtering function should not impact the power budget of the line, therefore optical amplifiers such as SOA may be included in the chip design to compensate for the insertion loss introduced by the filter. This is a fundamental implementation point since the integration of SOA must be compatible with volume production. Issues related to SOA integration on silicon were widely discussed in Section 3.

Examples of optical parameters for a 25 Gbit/s tunable transceiver are given in Table 1, based on the reference scenarios set out in Section 2.

Table 1. Optical parameters for a 25 Gbit/s tunable transceiver.

System Parameters	Value	Notes
Data Rate	25.78125 Gbit/s	
Modulation format	NRZ OOK	
DWDM channel spacing	100 GHz	
DWDM channels	191.4–193.5 THz Downstream 194.0–196.1 THz Upstream	1550–1570 nm
TRX Form factor	SFP28	
TRX Power consumption	1.5 W	
BER	5 10–5	
Rx Parameters		
Min channel input power	−18 dBm	
Min Optical Channel BW (−0.5 dB)	37 GHz	
Interchannel cross-talk	−20 dB	Defined as in [48]; all DWDM channels on; all channels set at equal power.
Min OSNR at photodiode (0.1 nm)	25 dB	Dependent on SOA noise figure
Min PDL	1 dB	
Tx parameters		
Min Tx output power	0 dBm	Launch power in fiber
Min extinction ratio	6 dB	

Different approaches are possible for the design of the silicon photonic filter in the TRX. In the following we consider different TRX architectures with alternative options for the filter elements and the method employed for coupling with the input–output fiber. Figure 5 shows two possible architectures for a tunable transceiver including fiber coupling, a polarization diversity scheme, optical amplifiers and add and drop tunable resonators. The WDM input is split into two sub-bands that correspond to the portion of the band dedicated to the transmission channels from the transmitter (TX) and the reception channels at receiver (RX).

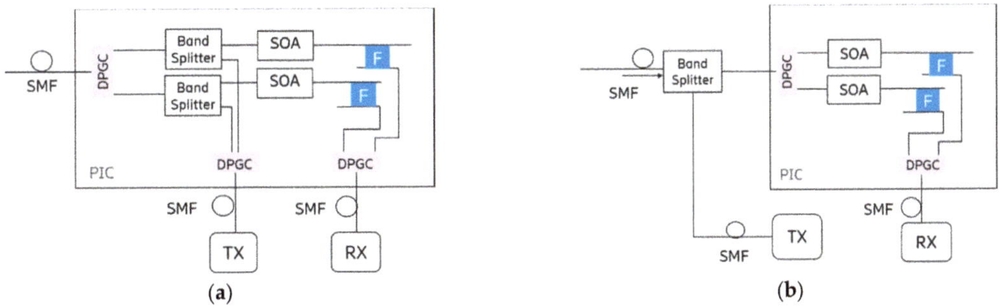

Figure 5. Tunable transceiver architectures with polarization diversity schemes and fiber coupling with integrated band splitter (**a**) and external band splitter (**b**). EC: edge coupler, PS: polarization splitter, PR: polarization rotator, F: add and drop resonator filter, DPGC: dual polarization grating coupler, PIC: photonic integrated circuit.

In this scheme the two sub-bands represent two halves of the operating BW of the transceiver. This operation can be performed by an element integrated in the photonic chip (PIC), as in Figure 5a, or an external band splitter, as in Figure 5b, e.g., a commercial band splitter based on thin films [49].

The advantage of the external splitter is in the reduction in the BW that must be coupled to the photonic chip, which can represent an advantage in the fabrication of the fiber coupler. Additionally, integrated band splitters are a type of photonic element that has not yet been implemented. However, the design of integrated solutions is desirable since the use of an external band splitter might increase the footprint and consumption of the tunable transceiver.

The resonators accomplish the filtering function, adding or dropping a resonant wavelength that can be tuned, e.g., via thermo-optic variation of the refraction index of the resonator [50]. These can be micro ring resonators (MRR), comprising multiple micro rings [51], reflector resonators based on Bragg Gratings, such as Distributed Feedback Bragg Reflectors (DFBR) [52] which comprise coupled Bragg gratings and a multi-mode interferometer (MMI), or ring assisted Mach–Zehnder interferometers [53]. The requirement for these resonators is to have a free spectral range (FSR) that is larger than the operating range of the filter so to select a single wavelength within the range. This requirement is challenging for the design since it may require tight fabrication tolerances to the solutions based on micro rings, with precise control of the coupling in the add and drop waveguides [54].

Figure 5 shows two options for the fiber-PIC coupling: the input fiber in Figure 5a is coupled to the chip via edge coupler, whereas the output fibers in Figure 5b are coupled via dual polarization grating couplers. The choice of the type of couplers in this architecture is determined by the bandwidth of the input and output spectrum; this is approximately 40 nm for the input coupler, therefore wide-band element, such as an edge coupler, may be preferred. On the other hand, the output couplers in this configuration operate just on one half of the BW (TX or RX) and have more relaxed constraints on the passband, that can be easily ensured by a DPGC. It must be noted that state of the art DPGC based on SiN-on-Si can presently reach a 1 dB bandwidth of at least 29 nm, with insertion losses of around 2.4 dB and 0.2 dB polarization dependent loss (PDL) [24]. The BW can be further extended over 40 nm by exploiting a design based on SiN-on-Si platform and back mirrors as shown in [25], where a value of 70 nm was achieved for a single polarization grating coupler. Therefore, the functional group comprising Edge Coupler (EC), Polarization Splitter (PS), and Polarization Rotator (PR) in Figure 5a can be substituted with a wide bandwidth DPGC (WB-DPGC) as shown in Figure 6.

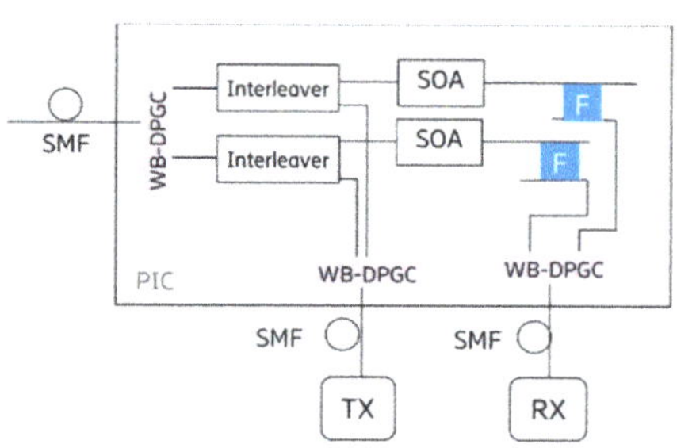

(a)

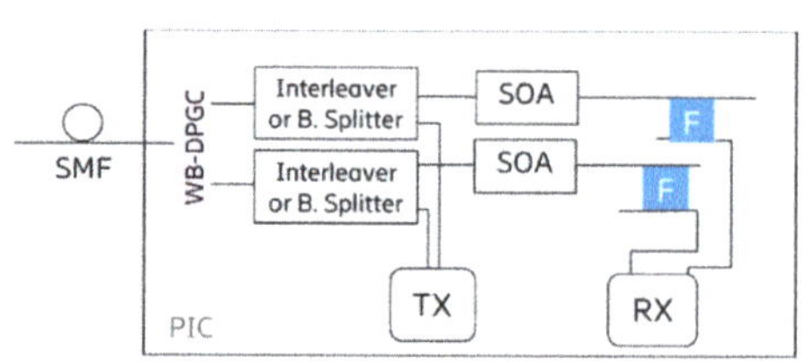

(b)

Figure 6. Tunable transceiver configuration with interleavers, wideband DPGC and (**a**) external TX and RX or (**b**) integrated TX and RX. WB-DPGC: wide band dual polarization grating coupler.

Another design option consists in replacing the band splitter elements with solutions that operate with two interleaved sub bands. This requires an appropriate spectrum management: it is possible, e.g., to associate all odd channels to the TX and the even channels to the RX, using a filter with the appropriate FSR (equal to twice the channel spacing). This architecture solution is shown in Figure 6a.

This kind of interleaved solutions requires both the input and the output couplers to be wideband. However, band splitters based on this scheme may consume very low power since they need minimal temperature stabilization to keep the interleaver comb filter in place.

Finally, Figure 6b shows an architectural solution where TX and RX are integrated in the transceiver. In this architecture, laser source, modulators, and photodiode are integrated in the photonic chip and electronic divers and amplifiers can be connected to these elements through wire bonding or integrated on top of the PIC e.g., in a 3D integration scheme [55].

In the following, it is described a design implementation of an architecture of the type shown in Figure 6b, which does not include the integrated SOA. The SOA can be easily integrated according to the integration process described in [56,57], enabling III–V integration at the backside of standard silicon photonics wafers. A detailed implementation is given in Figure 7. The filtering function at the receiver input is operated by a set of 4 coupled micro-ring resonators, whose parameters have been optimized to achieve a flat central BW of 37 GHz at −1 dB and a rejection BW of more than 25 dB. The resonator has FSR of 25 × 0.8 nm = 20 nm, corresponding to 25 WDM channels with 0.8 nm (100 GHz) spacing.

Coupling of the Photonic Integrated Circuit (PIC) with fibers is performed using surface grating couplers:

1. The external modulated tunable laser is fed to the chip via a Polarization Maintaining (PM) fiber, coupled with a Single Polarization Grating Coupler (SPGC) to the integrated TX ring.
2. The Single Mode (SM) fiber duplex, comprising the TX output and the RX input to the tunable filter is coupled to the chip via a DPGC.
3. The output of the RX tunable filter is coupled out from the chip to a Single Mode fiber (SMF) pigtailed with a DPGC, to reach an external avalanche photo detector (APD).
4. The dual polarization grating couplers separate the two orthogonal polarization components (named TE and TM in Figure 7) of an input light into two TE polarization components propagated by two separate waveguides.

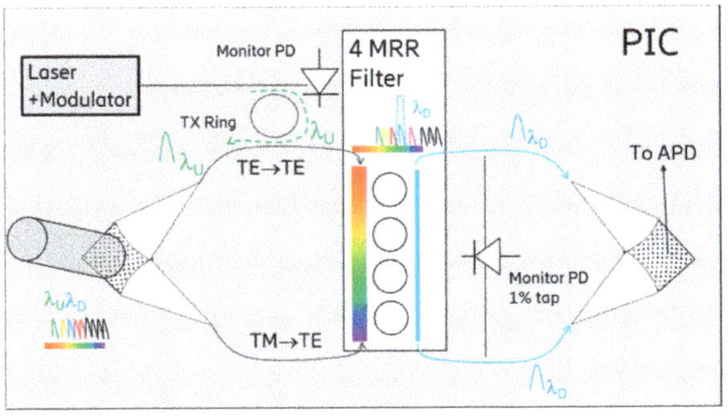

Figure 7. Scheme of a design based on a 4 rings resonator for the RX filter and a simple ring resonator for the routing of wavelengths from the TX. The DPGC on the left is a duplex fiber port: UL wavelengths are fed to the fiber and DL wavelengths are fed to the PIC. The DPGC on the right couples the output of the filter to the APD.

In this design, the input from the modulated external tunable laser is sent to the SM duplex fiber port via a ring filter, through the drop port ('TX Ring') that is tuned to the optical wavelength of the laser. The TX Ring has a FSR of 6 × 0.8 nm; this low FSR design is possible thanks to a wavelength management strategy that is implemented according to the wavelength plan described in Figure 8. The wavelength plan comprises two groups of channels for the Uplink (UL) and the Downlink (DL): a first group of 20 T channels, from λ_1 to λ_{20}, are spaced by 0.8 nm on the lower C band spectrum, and a second group of twenty Rx channels, from λ_{21} to λ_{40}, are spaced by 0.8 nm in the upper C band spectrum.

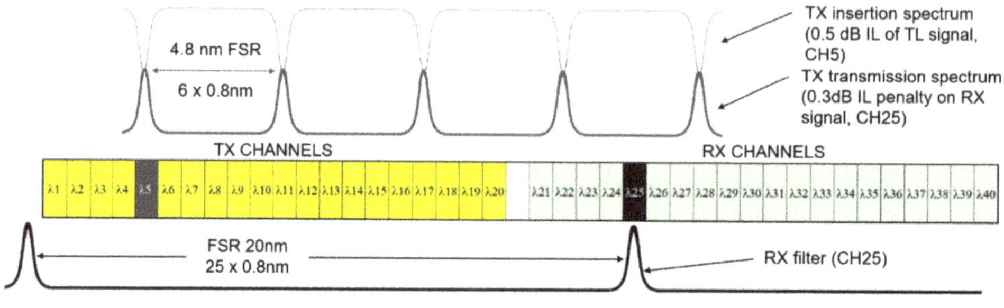

Figure 8. Scheme of the spectrum management for the design solution of Figure 7 with a TX Ring filter having FSR of 4.8 nm and RX filter having an FSR of 20 nm.

A 0.8 nm guard band, consisting in one unused channel, separates the highest TX channel λ_{20} from the lowest Rx channel λ_{21}. The wavelength plan assumes that Tx and Rx wavelengths from a given Optical Network Unit (ONU), e.g., ONUi, are paired as λ_i/λ_{I+20}, respectively, for the TX and RX channels (i.e., λ_5 TX paired with λ_{25} RX, λ_7 Tx paired with λ_{27} RX).

As shown on Figure 8 for the case of ONU$_5$, when the TX filter is tuned to λ_5 for inserting the external laser, there will not be any impact on the light received at the RX, having wavelength λ_{25}.

With such a small FSR, a large micro-ring radius of 20 μm can be used. The coupling coefficients of the rings are set to minimize the laser add loss to 0.5 dB, while the through loss seen by the RX signal on its way to the tunable receiver is kept below 0.3 dB (see Figure 5a). When using a ring filter for adding the laser light, one should minimize the nonlinearities of the ring: this is achieved by using a low Q ring (2000) which minimizes the power within the ring.

The RX filter is more complicated as it requires to have a 20 nm FSR. A single ring cannot be used due to the flatness requirements, BW and extinction specifications; the design is based on 4 coupled rings.

For the realization of both TX and RX filters, one needs to consider the dispersions that characterize the fabrication process. The Silicon on Insulator (SOI) thickness variation (±10% of the SOI thickness) accounts the most in the ring performances variations. This adds on to the width variation of the waveguides. Figures 9 and 10 show the range of performance variation calculated via Monte Carlo (MC) simulations for each filter, assuming a dispersion in the coupling coefficients of ±10%. These assumptions were verified in a first fabrication run with a set of rings with different coupling coefficients.

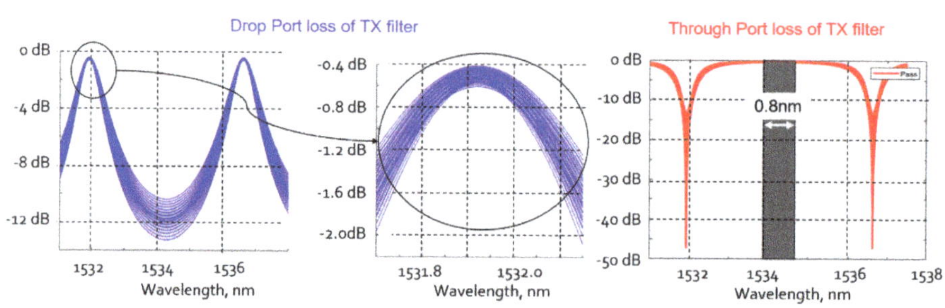

Figure 9. Spectral response of the TX drop and through filter ports according to MC simulations.

The estimated footprint for the photonic chip comprising filters, couplers and additional monitor photodiodes is 2.5 × 3.7 mm, where the RX (APD and amplifiers) and TX (modulators and tunable laser) are external.

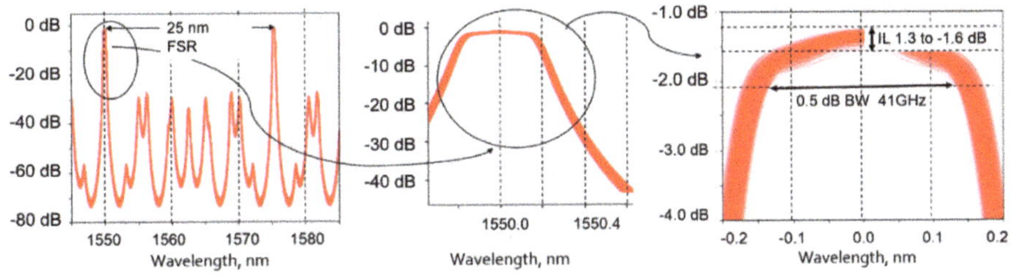

Figure 10. Spectral response of the RX filter according to MC simulations.

5. ROADM in Silicon Photonics

As reported in Section 2, low cost, reduced size ROADMs are a key technology for providing the right level of automation in the access network segment. Tunable filters can be used to operate the wavelength selection in a ROADM [58]. When used in ROADMs, tunable optical filters relieve operators from deploying and storing many variants of fixed optical add drop multiplexers (OADM) where each fixed OADM corresponds to a specific group of wavelengths by replacing the fixed OADMs with a single reconfigurable device. This leads to advantages in network planning simplification and saving of costs for the acquisition and maintenance of backup components, which are necessary to cope with possible failures, as failures can be addressed with a single tunable device.

Different WSS technologies exist, based on Microelectromechanical Mirrors (MEMS), Liquid Crystal (LC) on glass and Liquid Crystal on Silicon (LCoS) [59,60]. LCoS technology dominates current WSS shipments because it can support flexible channel plans. However, LCoS devices are polarization dependent, they need a polarization diversity configuration. Additionally, cost, size, port isolation, and crosstalk remain major design challenges. ROADMs architectures based on WSS, that reduce cost and fiber connections, with respect to traditional solutions, are reported in [61]. However, they are far to meet the requirements of the network scenarios presented in Section 2, for which a truly low-cost ROADM node is essential.

Silicon photonics are the best candidates for such networks due to its characteristics of easy integration with control electronics, miniaturization, mass producibility, and potential for high yield and low cost in the well-established CMOS production infrastructure. In [1,25,62] examples of system-on-chip ROADMs, referred as Mini-ROADM, are presented. An example of architecture for the Mini-ROADM is depicted in Figure 11.

This device has two-line ports, referred as east and west line port, for operation over two fibers rings. It presents two independent structures, one used for adding WDM channels to send the output line port, and the other used for dropping channels from the input line port. In a prototype realized at Ericsson Research, each structure comprises 8 ports to add and drop an equal number of 200 GHz spaced WDM channels in C band. A similar design can be generalized to a higher number of 100 GHz spaced channels. Such a two-way Mini-ROADM can be also enhanced to operate over a larger number of ways: for instance, a 4-way ROADM can be realized by connecting 4 of these structures [63]. In the Mini-ROADM, optical channels are added and dropped by silicon micro-ring resonators (MRR) as wavelength-selective switching elements that also acts as an optical filter.

Each wavelength switching element is implemented by two-coupled MRRs with a slight resonance offset, set to obtain 40 GHz optical BW at −1 dB. The measured adjacent channel isolation is larger than 20 dB.

Silicon photonics MRRoptical filters have multiple advantages respect to other types of filter technologies since they are low cost, have low power consumption, and have a very small footprint in the range of micrometers. However, they are resonant devices, quite sensitive to temperature variations, and for operating correctly in the system the wavelength resonance needs to be stabilized.

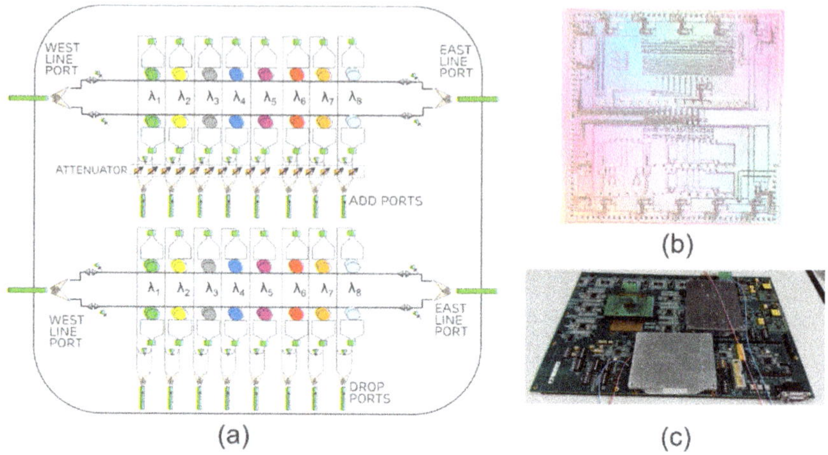

Figure 11. (**a**) Block diagram of the realized mini-ROADM, (**b**) photonic chip, (**c**) control board.

Two possible methods can be adopted for MRR resonance stabilization:

1. Use of a temperature control circuit to keep the temperature of the entire PIC at a pre-determined value by using an off-chip thermo-electric cooler (TEC). The TEC is used as the actuator that is driven by a feedback control (usually a PID), together with an integrated temperature sensor
2. Use of an active tuning circuit to lock the MRR resonance to the carrier wavelength by using a micro-heater as actuator on top of the MRR.

Method (1) has the advantage of requiring electronic circuits with limited complexity but it requires more power to stabilize the whole PIC temperature.

Method (2) requires control circuits with higher complexity, with respect to approach (1), but has the advantage of requiring very low power, of the order of few mWatt per MRR. Additionally, the package is simplified by the absence of the TEC.

In the design of the mini-ROADM the approach (1) was preferred for its implementation simplicity. The temperature stabilization circuit required 1–2 W power to provide an accuracy of 0.1 C and a response time of the order of 1 ms.

Each structure in the mini-ROADM has a polarization diversity architecture, with a double-polarization grating coupler at the input that adapts the random polarization of the input signal to the device's main propagation mode. At the output ports, the two polarization components are recombined before coupling to the optical fibers. A previous version of this device was realized with a single polarization, as it was made insensitive to the input polarization with an integrated polarization controller [62,63]. However, with a single polarization controller the operation requires a low PMD, since different WDM channels, though launched with equal polarization, arrive with random polarization alignment at the input like port. Therefore, a dual polarization architecture is preferred.

The PIC of the mini-ROADM prototype has a size of 5 × 5 mm (see Figure 11b). The PIC is mounted on a control board (see Figure 11c) to provide functionalities, such as the adjustment of MRR heater currents for performing channel add or drop, and setting the propagation direction of each individual wavelength at the 1 × 2 switch placed at the add port (Figure 11a). Each switch reverts the propagation direction, e.g., for traffic protection in case of fiber break. This allows to implement well-known ring protection mechanisms without duplication of the optical interfaces at each add or drop port.

The control board is equipped with two Erbium Doped Fiber Amplifiers (EDFA) with a gain of 23 dB and maximum output power of 20 dBm, one for each bus in function of the working direction of the wavelengths along the network. The high gain and output power compensate the high coupling loss of the prototype, due to the non-ideal grating coupler fabrication process. They can be considerably lower in an engineered device. Integrated SOAs can be also considered but they have the limitations discussed in Section 2.

The spectrum of one the MRR is shown in Figure 12a, for a channel passing through the MRR and a dropped channel. The measured drop loss is 1.8 dB, excluding the polarization grating coupler losses. The maximum drop loss difference from the first to the last channels is about 1.5 dB.

The resonance frequency is plotted in Figure 12b versus different voltage values applied on the MRR heater. The curve in Figure 12b shows that the efficiency of tunability is 35 μW/GHz.

In Figure 12c is shown the spectral output of a single channel in three different operating conditions: (i) all heaters are switched off ('no heaters through/drop' in the legend) (ii) only the channel heaters are enabled, all the other channels are switched off ('only double ring heaters through/drop' in the legend); and (iii) both the channels heaters and all the other channels heaters are enabled ('all heaters through/drop' in the legend).

With the current silicon photonics technology, it is possible to expand the number of WDM channels up to 24 but supporting denser frequency spacing (e.g., 50 GHz) would require technological advances to improve the wavelength stability of the micro-rings.

As final remark, optical attenuators can also be integrated at the input of the add channel ports to adjust the transmitted optical power so that the WDM channels are equalized at the line output port (attenuators shown in Figure 5 but not implemented in the prototype discussed here). This would allow to avoid big power mismatch between bypass signals and added signals, and related crosstalk issues.

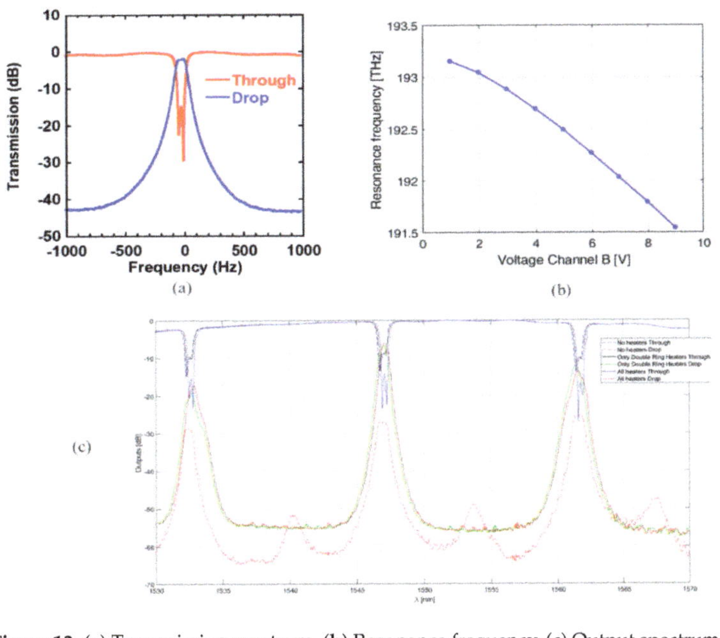

Figure 12. (**a**) Transmission spectrum, (**b**) Resonance frequency, (**c**) Output spectrum in through or drop filter conditions.

6. Conclusions

The evolution to 5G and beyond promotes novel technologies for the network such as SDN and NFV that allow a decoupling between the hardware resources and the services. This paradigm enables the digitalization of the network and facilitates the creation of services. The low latency requirements of the E2E services has enabled the replacement of big data centers located in regional area with smaller ones located near the access. Hence the access network is evolving in a converged network to support 5G mobile transport, fixed access, and vertical services. This network is typically named as Xhaul. A higher level of automation is required in all segments of the network, which is evolving to a converged network, able to support all service types. These will include the Xhaul segment and will be managed as "one network". Optics is the promising technology to meet the requirements of high bandwidth and low latency of 5G and beyond and offer a high level of automation suitable for SDN and NFV paradigm. However, due to the high capillarity of the equipment used in the Xhaul segment, the cost and the size of the optical nodes needs to be reduced, with respect to the case of the metro network segment. Integrated silicon photonics is a promising technology to achieve the cost and power consumption targets in the access segment. This technology was born for data center applications and has now reached a maturity level such to be suitable for telecom applications. De facto the SDN and NFV paradigm for Xhaul represents a concrete business case to push the use of integrated silicon photonics and accelerate the necessary investments to assess its use in the telecom segment. In this paper two main reference architectures for RAN access network are considered: PON and ring topology. Concrete technology solutions based on integrated silicon photonics are described for each architecture, that enable the realization of key optical components such as tunable transceivers and ROADMs with reduced cost and size. Overall, two network scenarios have been considered, such as WDM overlay PON and rings. For DWDM over PON, the use of silicon photonics is essential to realize pluggable transceivers, fully tunable both in transmission and receiver directions. For DWDM ring, the integrated silicon photonics is the key technology to realize a pluggable ROADM supporting 20 bidirectional channels. The main challenges are to realize such equipment have been discussed such as: size, power consumption, and cost to support 20 channels, and be integrable in a suitable form factor. Moreover, for the ROADM the open issue is the use of SOAs which can limit the number of channels.

The technical analysis carried out in the paper demonstrates the applicability of integrated silicon photonics in telecom applications and highlights the main challenges to be addressed.

Author Contributions: P.I., A.B. (Alessandra Bigongiari), A.B. (Alberto Bianchi), S.M., M.R., V.S. and F.C.: Writing—original draft, review & editing. All authors have read and agreed to the published version of the manuscript.

Funding: This work has been partially supported by European Commission H2020 5GPPP 5Growth project (Grant 856709).

Institutional Review Board Statement: Not applicable.

Informed Consent Statement: Not applicable.

Data Availability Statement: Not applicable.

Acknowledgments: We kindly acknowledge Francesco Testa for providing valuable information on the temperature management of the mini-ROADM prototype.

Conflicts of Interest: The authors declare no conflict of interest.

References

1. Iovanna, P.; Cavaliere, F.; Testa, F.; Stracca, S.; Bottari, G.; Ponzini, F.; Bianchi, A.; Sabella, R. Future Proof Optical Network Infrastructure for 5G Transport. *J. Opt. Commun. Netw.* **2016**, *8*, B80–B92. [CrossRef]

2. Sabella, R.; Iovanna, P.; Bottari, G.; Cavaliere, F. Optical transport for Industry 4.0. *J. Opt. Commun. Netw.* **2020**, *12*, 264–276. [CrossRef]
3. CPRI Specification V7.0 (9 October 2015). Available online: http://www.cpri.info/downloads/CPRI_v_7_0_2015-10-09.pdf (accessed on 10 November 2020).
4. eCPRI Specification V2.0 (10 May 2019). Available online: http://www.cpri.info/downloads/eCPRI_v_2.0_2019_05_10c.pdf (accessed on 10 November 2020).
5. 3GPP TS 36.413 Evolved Universal Terrestrial Radio Access Network (E-UTRAN); S1 Application Protocol, September 2014. Available online: https://www.etsi.org/deliver/etsi_ts/136400_136499/136413/12.03.00_60/ts_136413v120300p.pdf (accessed on 10 November 2020).
6. 3GPP TS 38.473, NG-RAN; F1 Application Protocol (F1AP), October 2018. Available online: https://www.etsi.org/deliver/etsi_ts/138400_138499/138473/15.03.00_60/ts_138473v150300p.pdf (accessed on 10 November 2020).
7. Iovanna, P.; Cavaliere, F.; Stracca, S.; Giorgi, L.; Ubaldi, F. 5G Xhaul and Service Convergence: Transmission, Switching and Automation Enabling Technologies. *J. Lightwave Technol.* **2020**, *38*, 2799–2806. [CrossRef]
8. ITU-TG.989 Series Recommendations, 3 September 2013. Available online: http://handle.itu.int/11.1002/1000/11810 (accessed on 10 November 2020).
9. Park, S.J.; Lee, C.-H.; Jeong, K.-T.; Park, H.-J.; Ahn, J.-G.; Song, K.-H. Fiber-to-the-home services based on wavelength-division-multiplexing passive optical network. *J. Lightw. Technol.* **2004**, *22*, 2582–2591. [CrossRef]
10. Smolorz, S.; Rohde, H.; Gottwald, E.; Smith, D.W.; Poustie, A. Demonstration of a coherent UDWDM-PON with real-time processing. In *2011 Optical Fiber Communication Conference and Exposition and the National Fiber Optic Engineers Conference*; Paper PDP-D4; IEEE: Los Angeles, CA, USA, 2011.
11. Houtsma, V.; Van Veen, D.; Porto, S.; Basavanhally, N.; Bolle, C.; Schmuck, H. Investigation of 100 G (4 × 25 G) NG-PON2 Upgrade using a Burst Mode Laser based on a Multi-Electrode Laser to enable 100 GHz Wavelength Grid. In *2018 Optical Fiber Communications Conference and Exposition (OFC)*; IEEE: San Diego, CA, USA, 2018; pp. 1–3.
12. Jeong, J.W.; Jung, I.W.; Baney, D.M.; Solgaard, O. *Tunable Optical Bandpass Filter with High-Quality Vertical Mirrors Microassembled on Movable MEMS Platforms, International Solid-State Sensors, Actuators and Microsystems Conference*; IEEE: Denver, CO, USA, 2009; pp. 2318–2321.
13. Sneh, A.; Johnson, K.M. High-speed continuously tunable liquid crystal filter for WDM networks. *J. Lightw. Technol.* **1996**, *14*, 1067–1080. [CrossRef]
14. Nesset, D. NG-PON2 technology and standards. *J. Lightw. Technol.* **2015**, *33*, 1136–1143. [CrossRef]
15. Asaka, K. What will be killer devices and components for NG-PON2? In *The European Conference on Optical Communication (ECOC)*; IEEE: Cannes, France, 2014; pp. 1–3.
16. ITU-T Recommendation G.8300 Characteristics of Transport Networks to Support IMT-2020/5G. May 2020. Available online: https://www.itu.int/rec/T-REC-G.8300 (accessed on 10 November 2020).
17. Kaspar, P.; De Valicourt, G.; Brenot, R.; Mestre, M.; Jenneve, P.; Accard, A.; Make, D.; Lelarge, F.; Duan, G.-H.; Pavarelli, N.; et al. Hybrid III-V/Silicon SOA in Optical Network Based on Advanced Modulation Formats. *Photon. Technol. Lett.* **2015**, *27*, 2383. [CrossRef]
18. Marchetti, R.; Lacava, C.; Carroll, L.; Gradkowski, K.; Minzioni, P. Coupling strategies for silicon photonics integrated chips. *Photonics Res.* **2019**, *7*, 231–2039. [CrossRef]
19. Socci, L.; Sorianello, V.; Romagnoli, M. 300 nm bandwidth adiabatic SOI polarization splitter-rotators exploiting continuous symmetry breaking. *Opt. Express* **2015**, *23*, 19261–19271. [CrossRef] [PubMed]
20. Marchetti, R.; Lacava, C.; Khokhar, A.; Chen, X.; Cristiani, I.; Richardson, D.J.; Reed, G.T.; Periklis, P.; Minzioni, P. High-efficiency grating-couplers: Demonstration of a new design strategy. *Sci. Rep.* **2017**, *7*, 16670. [CrossRef]
21. Cheng, L.; Mao, S.; Li, Z.; Han, Y.; Fu, H.Y. Grating couplers on silicon photonics: Design principles, emerging trends and practical issues. *Micromachines* **2020**, *11*, 666. [CrossRef] [PubMed]
22. Scarcella, C.; Gradkowski, K.; Carroll, L.; Lee, J.; Duperron, M.; Fowler, D.; O'Brien, P. Pluggable Single-Mode Fiber-Array-to-PIC Coupling Using Micro-Lenses. *Photon. Technol. Lett.* **2017**, *29*, 1943–1946. [CrossRef]
23. Chen, B.; Zhang, X.; Hu, J.; Zhu, Y.; Cai, X.; Chen, P.; Liu, L. Two-dimensional grating coupler on silicon with a high coupling efficiency and a low polarization-dependent loss. *Opt. Express* **2020**, *28*, 4001–4009. [CrossRef]
24. Mak, J.; Wilmart, Q.; Olivier, S.; Menezo, S.; Poon, J. Silicon nitride-on-silicon bi-layer grating couplers designed by a global optimization method. *Opt. Express* **2018**, *26*, 13656. [CrossRef]
25. Sorianello, V.; De Angelis, G.; Cassese, T.; Preite, M.V.; Velha, P.; Bianchi, A.; Romagnoli, M.; Testa, F. Polarization insensitive silicon photonic ROADM with selectable communication direction for radio access networks. *Opt. Lett.* **2016**, *41*, 5688–5691. [CrossRef]
26. Komljenovic, T.; Huang, D.; Pintus, P.; Tran, M.A.; Davenport, M.L.; Bowers, J.E. Photonic Integrated Circuits Using Heterogeneous Integration on Silicon. *Proc. IEEE* **2018**, *106*, 2246–2257. [CrossRef]
27. Nadimi Goki, P.; Imran, M.; Porzi, C.; Toccafondo, V.; Fresi, F.; Cavaliere, F.; Potì, L. Lossless WDM PON Photonic Integrated Receivers Including SOAs. *Appl. Sci.* **2019**, *9*, 2457. [CrossRef]
28. Nadimi Goki, P.; Imran, M.; Porzi, C.; Toccafondo, V.; Fresi, F.; Cavaliere, F.; Potì, L. WDM PON Receivers Enhanced by Integrated SOAs. In Proceedings of the Photonics in Switching and Computing (PSC), Limassol, Cyprus, 19–21 September 2018.

29. Nadimi Goki, P.; Imran, M.; Fresi, F.; Cavaliere, F.; Potì, L. Lossless ROADM by Exploiting low gain SOAs in fronthaul network. In Proceedings of the 2019 24th OptoElectronics and Communications Conference (OECC) and 2019 International Conference on Photonics in Switching and Computing (PSC), Fukuoka, Japan, 7–11 July 2019.
30. Saleh, A.A.M.; Habbab, I.M.I. Effects of semiconductor-optical-amplifier nonlinearity on the performance of high-speed intensity-modulation lightwave systems. *Trans. Commun.* **1990**, *38*, 839–846. [CrossRef]
31. Ramaswami, R. Amplifier Induced Crosstalk in Multichannel Optical Amplifier. *J. Lightw. Technol.* **1990**, *8*, 1882–1896. [CrossRef]
32. Singh, S.; Singh, A.; Kaler, R.S. Performance evaluation of EDFA, RAMAN and SOA optical amplifier for WDM systems. *Optik* **2013**, *124*, 95–101. [CrossRef]
33. Available online: https://newsroom.intel.com/news/intel-demonstrates-industry-first-co-packaged-optics-ethernet-switch/#gs.k5n3b4 (accessed on 10 November 2020).
34. Available online: https://www.businesswire.com/news/home/20200305005157/en/Ranovus-Launches-Single-Chip-ODIN%E2%84%A2-Silicon-Photonic (accessed on 8 February 2021).
35. Saber, M.G.; Xing, E.Z.; Morsy-Osman, M.; Patel, D.; Samani, A.; Alam, M.S.; Shahriar, K.A.; Xu, L.; Vall-Llosera, G.; Dortschy, B.; et al. Plant D. V. 25 and 50 Gb/s/λ PAM-4 transmission over 43 and 21 km using a simplified coherent receiver on SOI. *Photon. Technol. Lett.* **2019**, *31*, 799–802. [CrossRef]
36. Alam, M.S.; Morsy-Osman, M.; Shahriar, K.A.; El-Fiky, E.; Lessard, S.; De Angelis, G.; Sorianello, V.; Fresi, F.; Cavaliere, F.; Potì, L.; et al. 224 Gb/s transmission over 10 km of SMF at 1550 nm enabled by a SiN optical dispersion compensator and stokes vector direct detect receiver. In Proceedings of the Signal Processing in Photonic Communications (SPPCom) Conference, Washington, DC, USA, 13–16 July 2020.
37. Testa, F.; Giorgi, L.; Bigongiari, A.; Bianchi, A. Experimental evaluation of silicon photonics transceiver operating at 120 °C for 5G antenna array systems. *Electron. Lett.* **2018**, *54*, 1391–1393. [CrossRef]
38. Kwoen, J.; Jang, B.; Watanabe, K.; Arakawa, Y. High-temperature continuous-wave operation of directly grown InAs/GaAs quantum dot lasers on on-axis Si. *Opt. Express* **2019**, *27*, 2681–2688. [CrossRef] [PubMed]
39. Giorgi, L.; D'Errico, A.; Presi, M.; Ciaramella, E.; Testa, F. Remote light source for silicon photonic transceivers in mobile fronthaul applications. *Electron. Lett.* **2015**, *51*, 355–357. [CrossRef]
40. Enright, R.; Lei, S.; Nolan, K.; Mathews, I.; Shen, A.; Levaufre, G.; Frizzell, R.; Duan, G.-H.; Hernon, D. A Vision for Thermally Integrated Photonics Systems. *Bell Labs Tech. J.* **2014**, *19*, 31–45. [CrossRef]
41. Enright, R.; Lei, S.; Cunningham, G.; Mathews, I.; Frizzell, R.; Shen, A. Integrated, Thermoelectric Cooling for Silicon Photonics. *Ecs J. Solid State Sci. Technol.* **2017**, *6*, 3103–3112. [CrossRef]
42. DeRose, C.T.; Watts, M.R.; Trotter, D.C.; Luck, D.L.; Nielson, G.N.; Young, R.W. Silicon microring modulator with integrated heater and temperature sensor for thermal control. In *Proceedings Conference on Lasers and Electro-Optics*; paper CThJ3; Optical Society of America: San Jose, CA, USA, 2010.
43. Padmaraju, K.; Chan, J.; Chen, L.; Lipson, M.; Bergman, K. Thermal stabilization of a microring modulator using feedback control. *Opt. Express* **2012**, *20*, 27999–28008. [CrossRef] [PubMed]
44. Sun, C.; Wade, M.; Georgas, M.; Lin, S.; Alloatti, L.; Moss, B.; Kumar, R.; Atabaki, A.H.; Pavanello, F.; Shainline, J.M.; et al. A 45 nm CMOS-SOI Monolithic Photonics Platform with Bit-Statistics-Based Resonant Microring Thermal Tuning. *J. Solid-State Circuits* **2016**, *51*, 893–907. [CrossRef]
45. Elshaari, A.W.; Zadeh, I.E.; Jöns, K.D.; Zwiller, V. Thermo-Optic Characterization of Silicon Nitride Resonators for Cryogenic Photonic Circuits. *IEEE Photonics J.* **2016**, *8*, 1–9. [CrossRef]
46. Luo, Y.; Roberts, H.; Grobe, K.; Valvo, M.; Nesset, D.; Asaka, K.; Rohde, H.; Smith, J.; Wey, J.S.; Effenberger, F. Physical Layer Aspects of NG-PON2 Standards—Part 2: System Design and Technology Feasibility [Invited]. *J. Opt. Commun. Netw.* **2016**, *8*, 43–52. [CrossRef]
47. SFF Committee. INF-8074i Specification for SFP (Small Formfactor Pluggable) Transceiver. Available online: https://www.fluxlight.com/content/Tech-Docs/25GBASE-ER-S-SFP28%20Datasheet.pdf (accessed on 12 May 2001).
48. ITU-T, Optical System Design and Engineering Considerations, Recommendation G.Sup39, February 2016. Available online: https://www.itu.int/rec/T-REC-G.Sup39/en (accessed on 10 October 2020).
49. Available online: https://resource.lumentum.com/s3fs-public/technical-library-items/bandsplitter100_ds_cc_ae.pdf (accessed on 10 November 2020).
50. Gan, F.; Barwicz, T.; Popovic, M.A.; Dahlem, M.S.; Holzwarth, C.W.; Rakich, P.T.; Smith, H.I.; Ippen, E.P.; Kärtner, F. Maximizing the Thermo-Optic Tuning Range of Silicon Photonic Structures. In *Photonics in Switching*; IEEE: San Francisco, CA, USA, 2007; pp. 67–68.
51. Bogaerts, W.; De Heyn, P.; Van Vaerenbergh, T.; De Vos, K.; Kumar Selvaraja, S.; Claes, T.; Dumon, P.; Bienstman, P.; Van Thourhout, D.; Baets, R. Silicon microring resonators. *Laser Photonics Rev.* **2012**, *6*, 47–73. [CrossRef]
52. Porzi, C.; Sharp, G.J.; Sorel, M.; Bogoni, A. Silicon Photonics High-Order Distributed Feedback Resonators Filters. *IEEE J. Quantum Electron.* **2020**, *56*, 1–9. [CrossRef]
53. Liu, W.; Gong, Y.; Dong, X.; Niu, C. Research of Double Ring Resonant Assisted Mach-Zehnder Interferometer Filter. *Bandaoti Guangdian/Semicond. Optoelectron.* **2017**, *38*, 174–179.
54. Xu, Q.; Fattal, D.; Beausoleil, R.G. Silicon microring resonators with 1.5-μm radius. *Opt. Express* **2008**, *16*, 4309–4315. [CrossRef] [PubMed]

55. Saeedi, S.; Menezo, S.; Pares, G.; Emami, A. A 25 Gb/s 3D-Integrated CMOS/Silicon-Photonic Receiver for Low-Power High-Sensitivity Optical Communication. *J. Lightw. Technol.* **2015**, *4*, 2924–2933. [CrossRef]
56. Thiessen, T.; Mak, J.; Fonseca, J.; Ribaud, K.; Jany, C.; Poon, J.; Menezo, S. Back-Side-on-BOX Heterogeneously Integrated III-V-on-Silicon O-Band Distributed Feedback Lasers. *J. Lightw. Technol.* **2020**, *38*, 3000–3006. [CrossRef]
57. Menezo, S.; Thiessen, T.; Mak, J.; Fonseca, J.; Ribaud, K.; Yon, Z.; Jany, C.; Poon, J. Integration of III-V on Silicon Gain Devices at the Backside of Silicon-on-Insulator Wafers For Photonic Fully Integrated Circuits. In *Conference on Lasers and Electro-Optics OSA Technical Digest*; paper SM4J.7; Optical Society of America: Washington, DC, USA, 2020.
58. H2020 European Project TERABOARD. Available online: http://www.teraboard.eu/ (accessed on 10 October 2020).
59. Strasser, T.A.; Wagener, J.L. Wavelength-Selective Switches for ROADM Applications. *IEEE J. Sel. Top. Quant. Electron.* **2010**, *16*, 1150–1157. [CrossRef]
60. Yang, H.; Wilkinson, P.; Robertson, B.; Giltrap, S.; Snowdon, O.; Prudden, H.; Chu, D.P. *24 [1×12] Wavelength Selective Switches Integrated on a Single 4k LCoS Device;* OFC: San Diego, CA, USA, 2020.
61. Zong, L.; Zhao, H.; Feng, Z.; Yan, Y. *Low-Cost, Degree-Expandable and Contention-Free ROADM Architecture Based on M × N WSS;* OFC: Anaheim, CA, USA, 2016.
62. Velha, P.; Sorianello, V.; Preite, M.V.; De Angelis, G.; Cassese, T.; Bianchi, A.; Testa, F.; Romagnoli, M. Wide-band polarization controller for Si photonic integrated circuits. *Opt. Lett.* **2016**, *41*, 5656–5659. [CrossRef] [PubMed]
63. Available online: https://patents.google.com/patent/US20160204892 (accessed on 10 October 2020).

applied sciences

MDPI

Article

An Experimental OpenFlow Proposal over Legacy GPONs to Allow Real-Time Service Reconfiguration Policies

Noemí Merayo *, David de Pintos, Juan C. Aguado, Ignacio de Miguel, Ramón J. Durán, Patricia Fernández, Rubén M. Lorenzo and Evaristo J. Abril

Optical Communications Group of the Department of Signal Theory, Communications and Telematic Engineering. E.T.S.I. Telecomunicación, Campus Miguel Delibes, Universidad de Valladolid (Spain), Paseo de Belén 15, 47011 Valladolid, Spain; david9.krd@gmail.com (D.d.P.); jaguado@tel.uva.es (J.C.A.); ignacio.miguel@tel.uva.es (I.d.M.); rduran@tel.uva.es (R.J.D.); patfer@tel.uva.es (P.F.); rublor@tel.uva.es (R.M.L.); ejad@tel.uva.es (E.J.A.)
* Correspondence: noemer@tel.uva.es; Tel.: +34-983-423-000 (ext. 5549)

Abstract: The integration of Software Defined Networking (SDN) technologies in Passive Optical Networks (PONs) would provide great advantages to Internet Service Providers (ISPs) and Network Operators, since they can optimize the network operation and reduce its complexity. However, some tasks regarding online service and network configuration strategies are difficult to move to external SDN-controllers since they are time-critical operations. However, the control of some of these policies by SDN techniques could lead to better network and management configuration in a centralized and automatic way. As a consequence, we propose and experimentally test the integration of an OpenFlow approach over legacy Gigabit Passive Optical Networks (GPONs), which allows moving some global service configuration policies to an external SDN controller implementing an SDN management layer that adjust these strategies according to dynamic Quality of Service (QoS) requirements of services in residential users. The viability and efficiency of our approach are demonstrated using a GPON testbed and proposing a new business scenario for ISPs and Network Operators.

Keywords: software defined networking (SDN); gigabit passive optical networks (GPONs); online service configuration policies; experimental validation

Citation: Merayo, N.; de Pintos, D.; Aguado, J.C.; de Miguel, I.; Durán, R.J.; Fernández, P.; Lorenzo, R.M.; Abril, E.J. An Experimental OpenFlow Proposal over Legacy GPONs to Allow Real-Time Service Reconfiguration Policies. *Appl. Sci.* **2021**, *11*, 903. https://doi.org/10.3390/app11030903

Academic Editor: Fabio Cavaliere
Received: 22 December 2020
Accepted: 18 January 2021
Published: 20 January 2021

1. Introduction

Software Defined Networking (SDN) is an emerging and open communications technology that allows an abstraction and separation of the control and data planes. This separation permits an efficient control of the network and higher scalability and management automation [1,2]. Then, SDN allows the network to be intelligently and centrally controlled using software applications, so operators can manage the network consistently and comprehensively, regardless of the underlying network technology. Then, with SDN a part of the datapath resides on the switches and routers of the network, but there is a central controller that makes high-level routing decisions over them using protocols such as OpenFlow [3], NETCONF [4], or RESTCONF [5].

Due to these advantages, the integration of SDN in Passive Optical Networks (PONs) is gaining high importance, and its implementation to provide different functionalities is being proposed extensively by many authors. In addition, it is worth noting the importance of PON technologies in the deployment of current access networks worldwide. Indeed, the FTTH Council in the latest news of December 2020 states that 202 million homes passed with FTTH/B by 2026 in Europe compared to 88.1 million in 2019 [6]. Moreover, the FFTH/B penetration is expected 73.3% in 2026 (43.3% in 2019). In this network context, PON technologies will become predominant in the coming years, going from 48.4% in September 2018 to 73% in 2025 [7].

Due to these factors, much attention is being paid to investigate the integration of SDN technologies in PON infrastructures. In this way, many research papers try to emulate an SDN behavior on PON legacy equipment, so that PON devices can become SDN-controllable devices using SDN protocols [8–14]. In contrast, other proposals focus on moving certain features of current PON devices to external SDN controllers to achieve a centralized SDN control of the network, for example, bandwidth allocation policies or PON devices registration [15–23]. In this way, PONs are Point to MultiPoint (P2MP) networks based on a tree topology between the Optical Line Terminal (OLT) and the Optical Network Terminals (ONTs). In the upstream channel, from the ONTs to the OLT, PONs show a multipoint to point topology and every ONT shares the same transmission channel. Therefore, a MAC (Medium Access Control) protocol is required inside this channel in order to avoid collisions among data from different users (ONTs). Dynamic Bandwidth Allocation (DBA) based on Time Division Multiple Access (TDMA), is the most used strategy to share the channel capacity among ONTs (users) cycle by cycle, since the OLT dynamically distributes the available bandwidth depending on the current demand of users (ONTs) and its contracted services. The cycle time is the total time in which all ONTs connected to the PON transmit in a round robin discipline (TDMA).

As it can be noticed, the allocation process inside PON infrastructures is made between the OLT and the ONTs cycle by cycle, so its global outsourcing to an external SDN controller can cause an inefficient network performance in the bandwidth allocation process due to the latency between the SDN controller and the OLT, since SDN controllers are outside PONs at a certain distance. However, it could be worth that some decisions regarding DBA policies, such as the maximum bandwidth assigned to specific services (such as voice, video or data), can be managed by the SDN controller to provide an efficient and global SDN network management and configuration of the PON, although these policies do not imply the cycle-by-cycle control of the bandwidth allocation process, which would cause low efficiency due to the distance between the SDN controller and the OLT.

As a consequence, in this paper, we propose designing and implementing an SDN management layer over a GPON testbed able to deal with some service configuration strategies, such as the maximum allowed bandwidth to services, according to the real-time QoS requirements of network subscribers (ONTs). These global policies do not require a cycle-by-cycle communication between the SDN management layer and the OLT, thereby not degrading performance despite the distance between the SDN infrastructure and the PON. In contrast to other existing approaches, that most of them propose service provisioning or DBA policies by means of simulation models, we demonstrate the feasibility of moving specific strategies over a legacy GPON testbed which has implemented an end-to-end SDN solution using the OpenFlow protocol and SDN virtual switches (Open Virtual Switch, OVS) [24] connected to OLTs and ONTs, all of them controlled by an external SDN controller, in our case OpenDayLight (ODL) [25]. Therefore, we propose both a SDN solution over a GPON testbed and the implementation of an SDN management layer that deals with service configuration policies inside the GPON testbed. In fact, we propose that the SDN management layer can be able to provide and configure real-time Internet services and their associated maximum bandwidth to users that contract Internet plans according to their real-time demand and contracted bandwidth requirements using the SDN controller and the OpenFlow protocol. Furthermore, we check the viability of our proposal by designing a new business model for network operators and providers.

The paper is organized as follows. Section 2 describes the state of the art regarding the integration of SDN in PONs. Section 3 explains the experimental implementation of our SDN approach over GPONs and the description of the online service configuration policies over our network scenario. Section 4 presents the experimental results and discussion of our SDN-GPON proposal using a particular business model. Finally, in Section 5, the most relevant conclusions obtained in this experimental study are shown.

2. State of the Art

Many researches are focusing on the integration of SDN in PONs. On the one hand, some studies integrate OpenFlow switching paradigms in the PON control layer. In this way, authors in [8] propose a SDN-OLT solution in GPONs using virtual switches and the OpenFlow standard in order to emulate a SDN layer in the OLT. In addition, authors in [9] propose to implement virtual switches inside ONUs able to select OFDM channels for TWDM-PONs. Other proposals regarding virtual switches for PONs were proposed in [10] to improve the efficiency and flexibility for data center networks, especially for intra-datacenter communications. The papers [11] describe an SDN-controller to simultaneously manage several SDN-OLTs in PON infrastructures. Other researches are working on mapping OpenFlow messages to native PON configuration commands [12] and defining extensions of the OpenFlow standard for PONs technologies [13]. In addition, the Open Source Virtual OLT Hardware Abstraction (VOLTHA) project [14] proposes to abstract PONs as programmable Ethernet switches that can be controlled by an SDN controller and therefore, to manage its network configuration and devices (OLT, ONTs).

On the other hand, issues related to the Dynamic Bandwidth Allocation (DBA) or the registration and management of ONTs are difficult to cover with SDN in PONs, due to the latency between the SDN controller and the OLTs. In this line, authors in [15] propose the SIEPON architecture where the ONT registration process and some global DBA policies are moved to an external SDN controller, since they are not time-critical operations. Authors in this paper describe the simulation model and they analyze some network parameters such as the delay and the throughput performance by a simulation study. In addition, authors in [16] describe a novel SDN architecture for EPONs integrating a reprogrammable DBA module inside the SDN controller in contrast to the hardware DBA modules (inside OLTs) existing in traditional PONs. In this proposal, an SDN controller deals with several DBA algorithms and it deploys them in the OLT according to the network global status and the traffic demands of ONTs. This proposal and the results are supported with simulations. Authors in [17] experimentally demonstrate a novel SDN architecture for remote unified control with SA-FS to dynamically provision and adjust services according to the real-time Quality of Service (QoS) demand. Regarding TWDM-PONs, authors in [18] designed a novel software-defined solution to fulfil QoS requirements using a centralized SDN controller that dynamically allocates bandwidth and wavelengths to users. This model and the algorithm's performance are supported with simulations. In the same way, paper [19] describes a new SDN architecture and bandwidth allocation policy to guarantee QoS requirements in TWDM-PONs. Therefore, it can be noticed that most of the research related to DBA strategies or online service reconfiguration policies controlled by SDN are implemented using simulation models, but they have not been tested in real PON network environments with legacy equipment.

Although we do not focus on protection and energy savings in PONs, it is worthy to mention that other research works on the use of SDN in PONs have considered those aims. For instance, authors in [20] developed an experimental SDN architecture to allow protection and dynamic service provisioning in a multi-wavelength PON in coordination with a core SDN-based network, and authors in [21] propose moving the power control of OLTs and ONTs to the SDN controller to achieve energy savings. Finally, SDN solutions are also applied in Virtual PONs (VPON), such as authors in [22] that carry out a simulation analysis of a novel software-defined Hybrid Passive Optical Network (HPON) architecture proposed over multiple VPONs managed by a priority-based DBA mechanism called SPB-DBA. In addition, the Flex PON architecture proposed in [23] defines a full-service Software-Defined Solution to provide a flexible and a programmable service provisioning in VPONs. In the article, authors experimentally demonstrate flex link using Digital Signal Processing (DSP) technologies.

Therefore, it can be stated that the integration of SDN technologies in PONs could provide great advantages to network operators, since they can optimize the network operation and reduce the network complexity. However, some tasks regarding online service and network configuration or dynamic bandwidth allocation are difficult to move to external SDN controllers since they are time-critical operations, although some of these policies can be very interesting to be managed by SDN controllers to provide a more efficient control in PON architectures. As a consequence, we propose and experimentally test an SDN solution over legacy GPONs, which allows moving some global service configuration policies to an SDN controller by means of programming an external SDN management layer that dynamically adjusts these policies according to the stipulated dynamic QoS requirements of residential users. This proposal can enable new business models for network operators and providers, so the viability of our solution is demonstrated proposing a new possible business scenario for users and operators. However, other business models can be applied with our strategy to exploit the possibilities it offers. Finally, our proposal also demonstrates the feasibility of moving some QoS services and network configurations out of the PON network to an external SDN controller. To the best of our knowledge, this is the first time that a real-time reconfiguration service policy is experimentally tested in a legacy GPON using SDN techniques apart from the approach described in [17]. However, in the architecture presented in [17] they build a new OpenFlow-enabled OLT node with an embedded OpenFlow agent. To control this network architecture, authors propose extending the existing OpenFlow protocol to interact with the main characteristics of PONs (such as guard time, bandwidth, time slot, ONU identification) and the architecture permits dealing with the DBA algorithm inside the OLT. In contrast, in our proposal, we use the real OpenFlow standard that does not allow the control of PON parameters in its flow rules. Moreover, legacy GPON equipment does not permit modifying the DBA algorithm, since it is integrated in a chipset inside the OLT. Therefore, we propose a different solution using existing technologies to provide a realistic network scenario with legacy and commercial GPON equipment.

3. Experimental OpenFlow Proposal over Legacy GPONs to Allow Real-Time Service Reconfiguration Policies

3.1. Implementation of a SDN Approach over Legacy GPON Equipment

We propose integrating a SDN approach into commercial GPONs using the OpenFlow standard together with the ODL controller and a set of OVSs connected to GPON devices, OLTs and ONTs, to emulate a SDN layer on these devices. Among the different SDN controllers, we have deployed OpenDayLight (ODL), although other OpenFlow compatible controllers could be deployed. For instance, ODL and ONOS are quite featured SDN controllers, since they show high modularity, good productivity, effective for large scale networks and a good Graphical User Interface (GUI) [26,27]. Moreover, both are open source and benefit from a large developer and user communities under the Linux Foundation Networking. Moreover, ODL can support so many applications, such as new IoT southbound interfaces, that could make it one of the most important future SDN controllers [26]. As a consequence, we selected ODL in our proposal, although other high-performance SDN controllers, such as ONOS, could be deployed in the same way. In this network scenario, the ODL controller will be able to manage the GPON configuration and its services/profiles according to the stipulated QoS requirements by sending OpenFlow messages to every OVS, as it can be observed in Figure 1.

Therefore, we deploy an OVS (Central OVS, COVS) on a computer connected to the OLT to implement a SDN layer on the OLT to manage the GPON downstream channel (traffic coming from outside the GPON). In the same computer, a router and a DHCP server were installed to control the IP addresses of the connected ONTs. Moreover, we deployed OVSs (Residential OVS, ROVS) on mini computers connected to each ONT to implement and SDN layer on the ONTs to manage the GPON upstream channel (traffic coming from network subscribers). It can be noticed that in the future, the Openflow switches (COVS and ROVS) should be integrated inside OLTs and ONTs (SDN based OLT/ONT), but as commercial GPON equipment do not support SDN it is necessary to emulate a SDN layer.

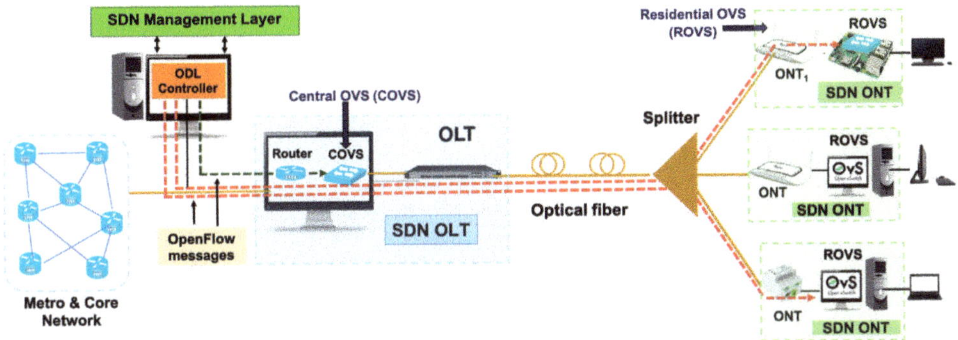

Figure 1. The software defined networking (SDN)-OpenFlow scenario implemented in the gigabit passive optical networks (GPONs) testbed.

In order to dynamically control our SDN-GPON approach, we have implemented an SDN management layer controlled by ISPs or Network Operators that communicates with SDN controllers (a single ODL controller in our case) to modify the GPON configuration according to the real-time QoS requirements and the traffic conditions. Therefore, the SDN controller/s and the management layer belong to ISPs/Network Operators that provide Internet connections and services to their residential subscribers, such as data (Internet), VoIP (Voice over IP), and HDTV (High-Definition TV). Therefore, the ODL controller sends OpenFlow messages to the different OVSs (OLT, ONTs) following the instructions of the SDN management layer to configure the QoS requirements of services (e.g., guaranteed bandwidth). These instructions are kept in entries (called flows) of the OpenFlow tables that are created and modified by the SDN controller, so that each entry corresponds to specific service QoS requirements. The flow tables are real-time modified and sent by the ODL by means of flows (one flow corresponds to one entry) encapsulated in OpenFlow messages to the different OVSs when QoS requirements or services should be updated in OLTs and ONTs. Each flow has several matches (match conditions), so each provides a condition that must be met in the packets to be part of the flow, that is, a filter condition for packets at each OVS. In our SDN approach, packets are filtered with the MAC address of the ONT to distinguish the traffic of each residential user. Furthermore, as one service in GPONs must be configured for both channels (upstream, downstream) two flows must be created in the ODL, one sent to the corresponding user through its COV (with the downstream QoS requirements) and another to the corresponding ROVS (with the upstream QoS requirements). In addition, since a maximum bandwidth is assigned to each service as an important QoS requirement (Internet, Video, VoIP) to control, for example, the guaranteed bandwidth, this parameter is reflected in the OpenFlow standard using meters [28] which are directly associated with the flows of the said service. Therefore, a meter measures the rate of packets and controls the maximum rate of its associated flow, so it controls the maximum bandwidth associated with the corresponding service of one residential user.

3.2. Design of a SDN Management Layer to Control Service Reconfiguration Policies on the SDN-GPON Model

As it was said before, we have deployed an OpenFlow-GPON scenario to propose moving some time-critical service reconfiguration policies to an external SDN controller that provide a full SDN control of the GPON network operation and configuration. In order to do it, we have programmed an SDN management layer that interacts with the ODL controller, which is responsible for periodically monitoring the demanded bandwidth of residential users (ONTs traffic) and to apply real-time policies to efficiently reconfigure the maximum allocated bandwidth of their contracted services, as it can be observed in

Figure 2. Indeed, our approach configures Internet services and their associated maximum bandwidth to users according to real-time bandwidth requirements of users by means of flows that are sent by the SDN controller following the instructions of the SDN management layer. Our OpenFlow-based proposal provides several advantages for ISPs and network operators. First, although legacy OLT/ONT devices are built compliant with the GPON standard, many times the management software and the Application Programming Interfaces (APIs) provided by vendors to configure the network are quite different, thus causing interoperability and compatibility issues between the equipment of different vendors. Therefore, our OpenFlow proposal permits controlling the legacy GPON equipment of different vendors since every device (OLT, ONT) understand the same protocol, that is OpenFlow, due to the emulated SDN layer implemented on each of them, as it can be observed in Figure 2.

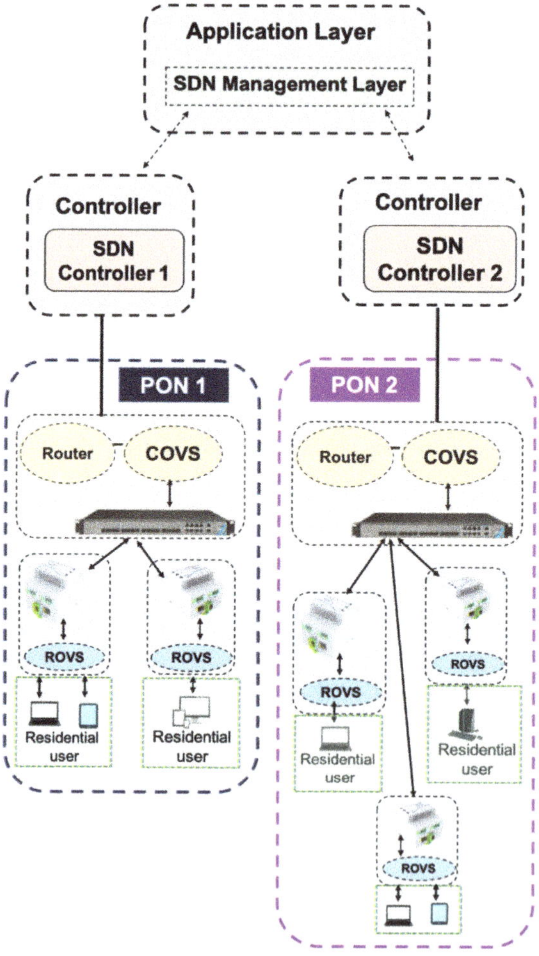

Figure 2. Block diagram of our OpenFlow approach over GPONs to apply external service configuration policies.

Secondly, our proposal allows simultaneously controlling several SDN controllers and GPONs using the same SDN management layer in a centralized, efficient, and dynamic way. Then, as it is shown in Figure 2, the SDN management layer can deal with a set of

GPONs sending configuration instructions to its corresponding SDN controller. Moreover, in our approach, any SDN controller capable of understanding the OpenFlow protocol can be deployed in the network scenario (ODL, ONOS), since the SDN management layer is able to interact with every controller. As a consequence, these advantages optimize the network operation and reduce its complexity, since our approach allows remotely controlling and allocating flows to provide and configure services in GPONs in a dynamic, programmable, centralized, and unified way.

Therefore, to carry out this dynamic control, the SDN management layer needs to know the packet statistics of every OVS (COVS, RVOS), so these statistics are periodically sent by every OVS to the ODL and the ODL to the SDN management layer. These statistics are sent in specific OpenFlow messages, called OFPM_METER messages, which contain the number of packets and bytes processed by the meters inside the OVS, the number of packets and bytes deleted, and the time the message is sent. Then, the SDN management layer is continually executing a dynamic bandwidth management program, which is listening to the messages sent by every OVS so when the program receives a message, it extracts the different statistics of the message together with the MAC address of the corresponding residential user (ONT). These data provide the real-time demanded bandwidth of each user, since one OVS is located at each residential home just before the ONT. In order to calculate the mean requested bandwidth using these OpenFlow statistics of the different residential OVSs (ROVSs), fixed sliding windows that store this information are used, one for each service of the residential subscriber. Each sliding window contains N samples (*win_size*), and its operation is represented in Figure 3. As it can be observed, each sample contains the total number of bytes sent by the ONT $\left(Bytes^{ONT_i,service_j}\right)$ of one specific service that passes over the ROVS, and the time in which that statistics was received $(t_N, t_{N+1}, t_{N+2}, \ldots)$. Each time a sample is inserted, the estimation of the mean requested bandwidth (in Mbps) for the specific residential user and service is updated with every sample of the window $\left(\overline{B_{request}^{ONT_i,service_j}}\right)$. To calculate the term $\left(\overline{B_{request}^{ONT_i,service_j}}\right)$, at a certain instant, the mean is done with all the $Bytes^{ONT_i,service_j}$ values contained in the samples and the time they were sent. Moreover, the first sample of the queue will be discarded when the maximum duration of the window is exceeded, that is, *win_size* samples. Sliding windows show a good performance, since they contain the updated requested bandwidth of residential users and their services.

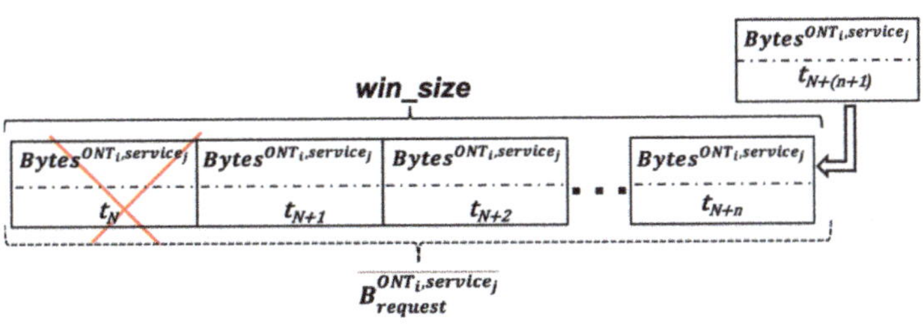

Figure 3. Operation of the sliding window implemented in the algorithm.

The process of assigning bandwidth to users (ONTs) and their services inside PON architectures is led by the OLT. Then, DBA algorithms (based on TDMA) are the most widespread methods, since the OLT allocates bandwidth to users in a dynamic way depending on the priority of the contracted services of users and its real-time demand. In this way, among the different bandwidth assignation disciplines, the DBA algorithm implemented in the OLT in our GPON testbed (as well as in legacy GPONs) follows a limited scheme, one of the most widespread policies in PONs due to its high efficiency and easy operation [29]. In this scheme, the OLT gives the required bandwidth to each service ($service_j$) of one ONT (ONT_i) in one cycle as long as its demand is lower than an established maximum bandwidth $\left(B_{max}^{ONT_i,service_j}\right)$ in Mbps, which consists of a minimum guaranteed bandwidth $\left(B_{guaranteed}^{ONT_i,service_j}\right)$ that has to be ensured plus an extra best effort or non-guaranteed bandwidth $\left(B_{effort}^{ONT_i,service_j}\right)$, which is optional and only depends on the configuration made by the operator or service provider inside the DBA (Equation (1)).

$$B_{max}^{ONT_i,service_j} = B_{guaranteed}^{ONT_i,service_j} + B_{effort}^{ONT_i,service_j} \tag{1}$$

In case the user asks for more than this maximum, the algorithm will allocate at most this maximum value. In this way, the DBA algorithm in the OLT must ensure the guaranteed bandwidth and it may provide non-guaranteed bandwidth in case GPON resources are available. In contrast, if the demanded bandwidth is lower than the guaranteed bandwidth, the algorithm will offer this demand. Therefore, a limited scheme makes the cycle time adaptive depending on the updated demand of every ONT.

Therefore, in our OpenFlow approach, the bandwidth is not only controlled by the DBA algorithm in the OLT but is also managed by the SDN layer to provide a tighter control of this process. The SDN layer limits the maximum bandwidth for each residential user (and services) and the global available bandwidth at both channels of the GPON (upstream, downstream). In this way, the DBA is configured to treat residential users equally, but allocating bandwidth considering real-time traffic demands, as the differentiation and the fulfilment of bandwidth guarantees is provided by the emulated SDN layer implemented on GPON devices (OVS). Therefore, the OVSs (ROVS, COVS) drops those packets exceeding the maximum assigned bandwidth at each channel. In this way, the maximum associated with each service $\left(B_{max}^{ONT_i,servicej}\right)$ can be dynamically modified according to specific bandwidth allocation policies implemented by ISPs/Network Operators, allowing a real-time bandwidth reconfiguration according to the real-time requested bandwidth of residential users. Consequently, we propose an online service reconfiguration policy, in which users contract a guaranteed bandwidth $\left(B_{guaranteed}^{ONT_i,service_j}\right)$ and it can be offered an extra maximum according to the priority and conditions of the contracted Service Level Agreement (SLA) up to a maximum limit $\left(B_{max}^{ONT_i,servicej}\right)$. Therefore, users do not need to request more bandwidth whenever they need it, but it is the external SDN management layer that monitors changes and decides to increase or decrease the bandwidth of residential users at a given time. Under this network scenario, the external SDN management layer by means of the ODL controller transparently gives a user some bandwidth in excess if it demands it and there is enough capacity in the GPON. Then, a bandwidth management program (located in the SDN management layer) is constantly listening for the arrival of OpenFlow messages (OFM_METER) and collecting the value of the demanded bandwidth of the sliding window that corresponds to one service of one specific ONT, calculating the average of the said window in real-time, as it can be observed in the flowchart of Figure 4. Every time the sliding window updates the mean requested bandwidth $\left(\overline{B_{request}^{ONT_i,service_j}}\right)$ an instance is launched to the service reconfiguration policy to detect if the average bandwidth of the window is higher than the current maximum bandwidth, called $B_{max,\ updated}^{ONT_i,servicej}$ in our algorithm. To do this, the algorithm first connects to the internal database that stores the maximum bandwidth $\left(B_{max}^{ONT_i,service_j}\right)$ and the guaranteed bandwidth of the contracted service. In the next step, the management program must assure the user an assigned rate always greater or equal than the basic contracted bandwidth $\left(B_{guaranteed}^{ONT_i,service_j}\right)$ in case the residential user demands it. Then, the program must verify that the requested bandwidth $\left(\overline{B_{request}^{ONT_i,service_j}}\right)$ does not exceed the maximum contracted bandwidth $\left(B_{max}^{ONT_i,service_j}\right)$. If all the conditions are met, the bandwidth management algorithm makes sure that the available resources are enough for the requested new bandwidth. In that case, the algorithm modifies this value $\left(B_{max,updated}^{ONT_i,service_j}\right)$, updates the database, and sends a request to the ODL controller to add new flows to the corresponding ONT (COVS, ROVS) with the new QoS requirements and discard the existing flows with the old QoS requirements. Otherwise, the maximum bandwidth $\left(B_{max,updated}^{ONT_i,service_j}\right)$ is not updated and no commands are sent to the ODL controller.

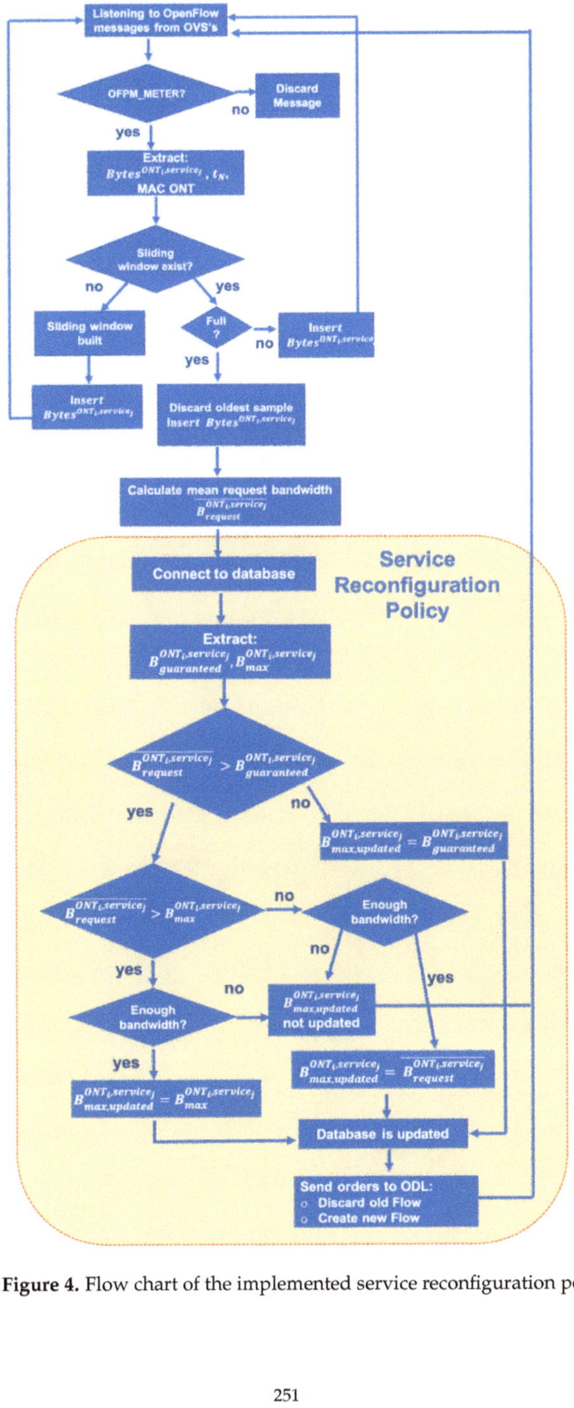

Figure 4. Flow chart of the implemented service reconfiguration policy.

Therefore, it could be very appealing that some online service management strategies can be moved out of the OLT to be centralized and controlled by the external SDN management layer. In addition, it avoids the synchronized time required in PON devices each time a service reconfiguration is made over the OLT and the ONTs. Then, we propose that the external SDN management layer could have implemented different service reconfiguration strategies. Finally, this approach avoids modifying the legacy GPON layer and it provides the full functionality of the integrated OpenFlow solution.

4. Experimental Setup and Results

To demonstrate the efficiency of our approach we have defined a new business model for ISPs/Network Operators and we check its viability using a legacy GPON. In fact, we have set up a network scenario which offers different priority SLAs based on flexible Internet plans so that users contract a fixed guaranteed bandwidth (basic bandwidth) plus an extra bandwidth, and when network subscribers demand this extra bandwidth and there are available resources, it can be given at the expense of a special pricing. Contrary to most of the proposed online service reconfiguration policies or DBA algorithms managed by SDN which are implemented with simulations, we have experimentally demonstrated our proposal using a GPON testbed with legacy equipment.

4.1. Description of the Legacy GPON Testbed

The commercial GPON (Figure 4) used to demonstrate the viability of our approach consists of an OLT (SmartOLT 350 of Telnet-RI vendor [30]), that supports 2.5 Gbps at the downstream channel and 1.25 Gbps at the upstream channel. The GPON testbed shows a tree topology configuration with a single splitting level. In this way, the OLT is connected to the optical splitter by means of three spools of Standard Single Mode Fiber (SMF) of different lengths (total length of 20 km). Then, the splitter (1:8) is connected to every ONT using SMF distribution fibers and the distance to each ONT is configurable, from 100 m up to 5 km, using a connection panel, as it can be noticed in Figure 5. This configurable network scenario allows setting up realistic situations where ONTs are located at different distances from the Central Office (OLTs). Finally, the GPON is equipped with L3 (with router functionalities) and L2 (without router functionalities) ONTs that comply with the ITU-T G.984.x/G.988 recommendations. The ONT models are the Wave Access 3021 (L3) and the Wave Access 512 (L2) [30]. Finally, in our particular approach, we attach computers (mini computers) to build routers with flexible capabilities by means of OVSs. For our experimental study, we connect five ONTs, two L2 ONTs, and three L3 ONTs.

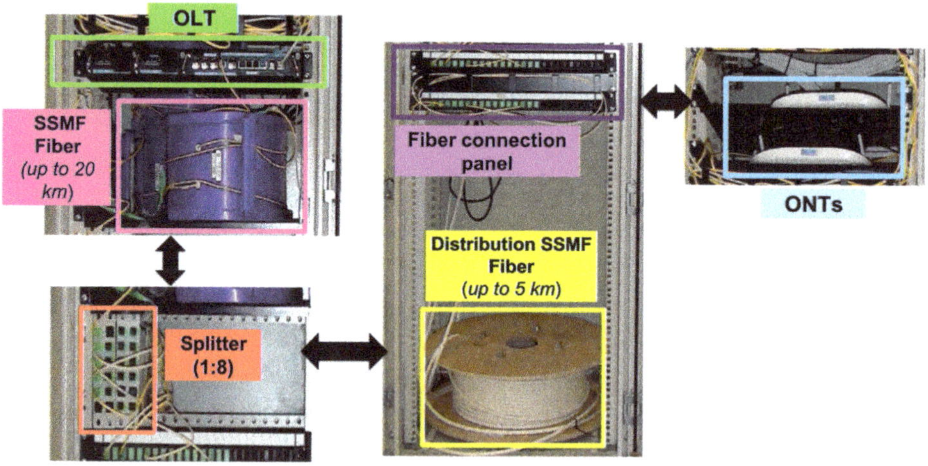

Figure 5. Real structure of the GPON testbed and its devices.

4.2. Description of the Case of Use

In our case of use, the policy that dynamically manages the maximum bandwidth allocated to services of residential users is implemented and demonstrated. As an example, we design Flexible Internet Plans differentiated by a basic bandwidth (guaranteed bandwidth) plus an excess bandwidth (Table 1). In case the basic bandwidth is below 100 Mbps, the excess bandwidth is considered as 20% of the basic (Basic Internet Plan), while if the basic bandwidth is higher than 100 Mbps, a 30% of excess bandwidth is considered (Premium Internet Plan). These maximum rates are calculated independently for the upstream and downstream bandwidth although they are symmetric. It should be noted, that although the proposed Internet plans in our case consider a certain amount of the basic bandwidth (20% and 30%), the operators and ISPs could stipulate other levels or conditions for the excess bandwidth. Then, these Flexible Internet plans ensure a fixed basic bandwidth to residential users plus an extra bandwidth, so that while users do not require this extra bandwidth it will be free to use transparently among other users who have greater bandwidth needs, so there is greater control over the total available bandwidth to distribute it in real-time and efficiently between all users, thus making the most of all network resources.

Table 1. Flexible Internet plans offered by Internet service providers (ISPs)/network operators.

Flexible Internet Plans	Basic Bandwidth	Excess Bandwidth
Basic Flexible Plan	≤100 Mbps	+20% Mbps
Premium Flexible Plan	>100 Mbps	+30% Mbps

4.3. Selection of Parameters in the Sliding Window

The first analysis regarding the sliding window used to calculate the average requested bandwidth of residential users, compared the response time of the dynamic bandwidth algorithm considering different window sizes. The optimal performance of the algorithm implies an optimum choice of some parameters, such as the size of the sliding window, which contains the last samples of the requested bandwidth of users. Then, if the number of samples is too large, the sliding window may contain quite old values of the requested bandwidth which can make the algorithm not react fast enough when changes in traffic conditions appear. In contrast, if the size of the sliding window is very short, the real-time estimation of the mean requested bandwidth cannot be reliable as the number of samples is not enough. Then, Figure 6a–d show the time that the algorithm takes to react when the associated bandwidth of one network subscriber (ONT) is being modified according to its real-time demands, considering different window sizes (5, 10, 20, and 30 samples). All the graphs are represented with Wireshark that can capture real- time network data [31]. As it can be observed in Figure 6c, the window of 20 samples has a response of approximately 16 s, while higher windows sizes such as 30 samples (Figure 6d) add a delay of approximately 20 s. In contrast, if low window sizes are considered, five and 10 samples (Figure 6a,b), the response time is lower, 8 and 12 s, respectively.

However, a too small window can be more sensitive to possible transmission peaks and fluctuations, as it can be seen in Figure 7. As a consequence, to test our dynamic service reconfiguration algorithm, a 10-samples window is chosen since the delay and the fluctuations are low and its performance is quite similar to high window sizes (20/30 samples).

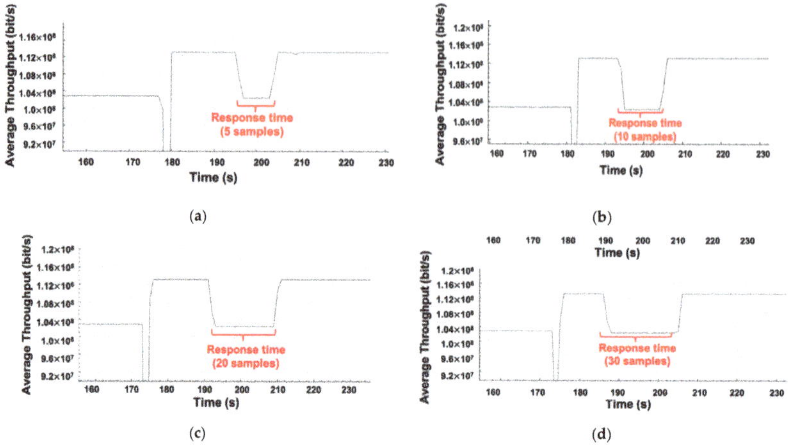

Figure 6. Test done on the size of the sliding window: (**a**) Five samples; (**b**) 10 samples; (**c**) 20 samples; (**d**) 30 samples.

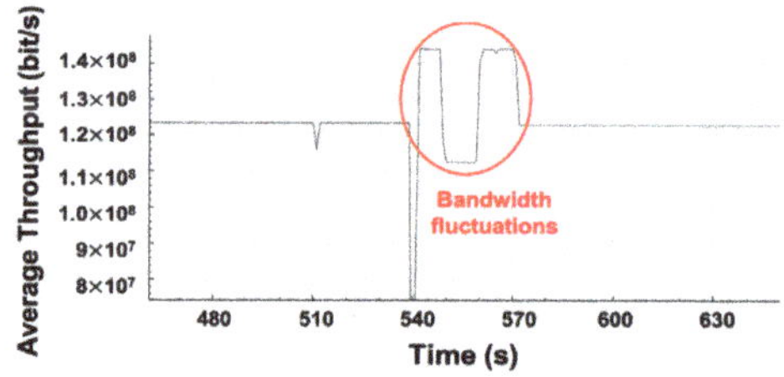

Figure 7. Real-time fluctuations on the average throughput due to a fall in the transmission rates.

4.4. Analysis and Results of the Algorithm Performance of the SDN Management Layer

To validate and analyze the performance of the algorithm experimentally, we have considered the availability of the two Flexible Internet Plans (Basic and Premium) previously mentioned, and we have assumed that a network subscriber (one ONT in the testbed) contracts a Flexible Internet Plan (Basic or Premium) and changes its real-time bandwidth demand, whereas the remaining four ONTs in the testbed contract a static Internet Plan. The experimental analysis was carried out at both channels, but in this paper, only the results of the upstream channel are presented for lack of space. However, the performance of the downstream channel is similar. This study should be considered as a proof of concept, since a typical PON will have a higher number of ONTs (e.g., 32–64) with varying demands than we have in our testbed. Nevertheless, to the best of our knowledge, this is the first time that a real-time reconfiguration service policy is experimentally tested in a legacy GPON using SDN techniques.

The Basic Flexible Internet plan consists of a guaranteed bandwidth of 100 Mbps plus an excess bandwidth of 20% (that is 20 Mbps), and the Premium Flexible Internet plan shows a guaranteed bandwidth of 150 Mbps with a 30% of excess bandwidth (45 Mbps), both symmetric services. Therefore, for the Basic Internet plan, as shown in Figure 8, the subscriber (ONT) that contracts this plan begins to transmit at 100 Mbps (60 s), and during the following intervals of 5 min (300 s) increases the transmission rate first to 110 Mbps, then to 120 Mbps, and finally to 130 Mbps, ending with a final period transmitting again at 100 Mbps. In addition, Table 2 indicates the transmitted bandwidth at the different interval times and the bandwidth expected to be offered to the residential user at each interval, according to the performance of the online service reconfiguration algorithm. Since the basic contracted bandwidth is equal to 100 Mbps (Table 1), the excess bandwidth is expected to increase up to 20% of this value (Basic Flexible Plan), so the maximum bandwidth assigned to the user will be 120 Mbps, whenever the user demands it and there is enough available bandwidth. Therefore, in the fourth interval (660–960 s), although the user demands 130 Mbps, the maximum assigned bandwidth is limited to 120 Mbps (Table 2).

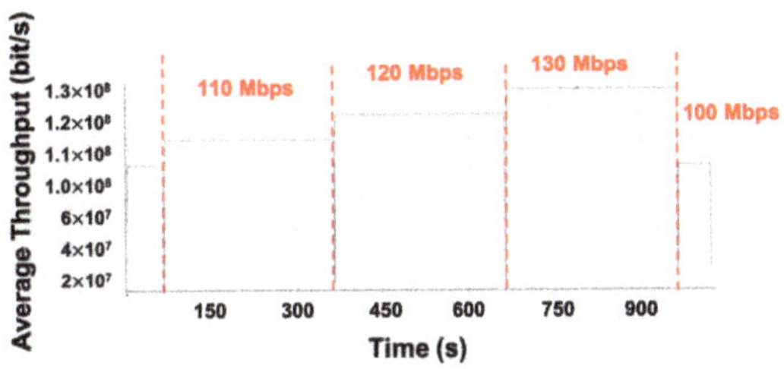

Figure 8. Real-time evolution of the requested bandwidth for the residential user that contract the basic flexible Internet plan.

Table 2. Demanded and received bandwidth for the basic flexible Internet plan.

Interval Time	User Transmission Rate (Mbps)	Expected Allocated Bandwidth (Mpbs)
0–60 s	100 Mbps	100 Mbps
60–360 s	110 Mbps	110 Mbps
360–660 s	120 Mbps	120 Mbps
660–960 s	130 Mbps	120 Mbps
960–1020 s	100 Mbps	100 Mbps

Then, Figure 9 shows the real-time evolution of the allocated bandwidth for the service of the ONT measured using Wireshark when the algorithm is working. Therefore, in the first Interval (0–60 s) the service is not modified, since the transmitted rate is equal to the basic contracted bandwidth (100 Mbps). In addition, it can be observed that at the beginning of each interval, a short step appears due to the reaction time of the OVS to a change in the user's transmission rate, as the OVS needs a few seconds (around 5–7 s) to notice an increase. If we analyze each interval one by one, in the second (60–360 s), where the residential user demands 110 Mbps, we observe the first step, followed by the time the algorithm needs to process the real-time demanded bandwidth of the sliding window, and then the instant where the algorithm provides the offered bandwidth to 110 Mbps (its bandwidth demand). The third Interval (360–660 s), where the user demands a rate of 120 Mbps, the result is similar and 120 Mbps are given (the maximum excess bandwidth). However, in the fourth interval (660–960 s) the residential user increases its demand up to 130 Mbps, but the algorithm limits this demand, since the maximum bandwidth associated with this Internet plan is 120 Mbps. Finally, in the last interval (960–1020 s), the user decreases its demand to 100 Mbps, so the algorithm assigns this bandwidth instantaneously. Consequently, it can be noticed that the algorithm implemented in the SDN layer shows a fast and stable response according to the contracted Internet plans of residential users. Moreover, it should be noted that DBA algorithms may need extra time to react to changes in the transmission rate of users as well as the OVS.

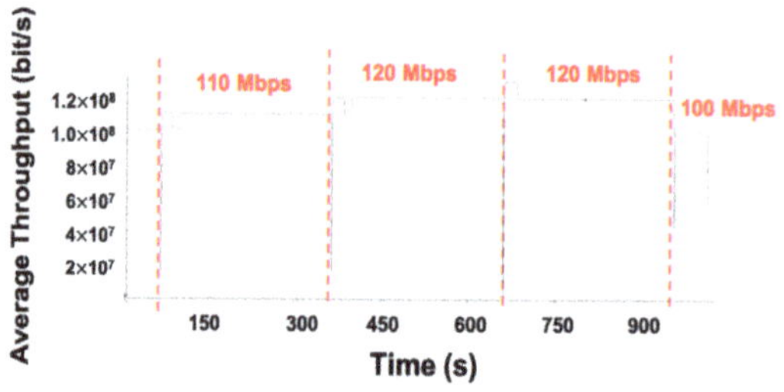

Figure 9. Real-time evolution of the allocated bandwidth for the residential user with the basic flexible Internet plan.

For the Premium Internet Plan, we consider a residential user that contracts an Internet service, which consists of a basic symmetric service of 150 Mbps plus an excess bandwidth of 30% (in that case 45 Mbps or more). However, in that case, the transmission rate of the residential user starts at 150 Mbps for 60 s and after that, it is increased in five min intervals (300 s). It has been considered that in each interval the bandwidth demand is increased by 20 Mbps, so that the evolution of the demanded bandwidth in the different period times is 150, 170, 190, 210, and finally 150 Mbps again. All these levels are reflected in Table 3, which represents the demanded bandwidth by the user in the different time intervals and the expected bandwidth that will be received according to the algorithm's performance.

Table 3. Demanded and received bandwidth for the premium flexible Internet plan.

Interval Time	User Transmission Rate (Mbps)	Expected Allocated Bandwidth (Mpbs)
0–60 s	150 Mbps	150 Mbps
60–360 s	170 Mbps	170 Mbps
360–660 s	190 Mbps	190 Mbps
660–960 s	210 Mbps	195 Mbps
960–1020 s	150 Mbps	150 Mbps

Unlike the previous case, now the residential user has a basic bandwidth higher than 100 Mbps, in which the corresponding Internet Plan may provide an excess bandwidth of 30%, so the maximum bandwidth that could be assigned is 195 Mbps, as observed in Table 3 in the interval 660–960 s. Therefore, the algorithm will create new flows based on the average bandwidth of the window, until reaching the maximum established of 195 Mbps. Figure 10 represents the real-time evolution of the allocated bandwidth for the service of the ONT measured using Wireshark. During the first period of change 60–360 s, in which the requested transmission rate is 170 Mbps, the algorithm shows two small steps and two transition zones between them just after offering this requested bandwidth. As it was mentioned before, the first step corresponds with the time required by the OVS to detect the change in the transmission rate of the user. Meanwhile, the algorithms process the data of the sliding window and send the order to the ODL controller to change the maximum bandwidth. Then, the second step corresponds with the time that the OVS needs to receive, process, and update the new flow sent by the ODL with the new transmission rate (meter) so the OVS becomes unstable during this period of time. Finally, the OVS updates the new maximum permitted bandwidth to 170 Mbps, that is the real- time demanded bandwidth. For the second period of change and the third, the performance of the algorithm is similar. Therefore, it can be noticed that the algorithm is able to dynamically evolve the allocated bandwidth according to the stipulated QoS requirements although the OVS needs a certain time to react to real-time changes in the transmission rate of users. In the same way, typical DBA algorithms inside OLTs probably also take time to deal with real-time changes in the transmitted traffic of residential users.

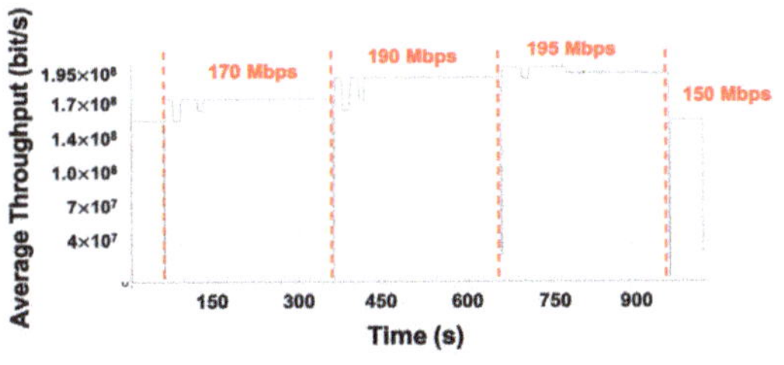

Figure 10. Real-time evolution of the allocated bandwidth for the residential user that contract the Premium Flexible Internet Plan.

5. Conclusions

In this paper, we have proposed and experimentally tested an OpenFlow-based approach over a legacy GPON to allow real-time service configuration policies applying SDN techniques. In this way, we have implemented an external SDN management layer able to communicate with SDN controllers, in this case ODL, to send orders through OpenFlow messages that configure and modify Internet services in residential users connected to the GPON. In particular, the SDN management layer can configure real-time Internet services and their associated maximum bandwidth to users according to their real-time demand and contracted bandwidth requirements. These service configuration policies do not require a cycle-by-cycle communication between the SDN management layer and the OLT, thereby not degrading performance despite the distance between the SDN infrastructure and the PON. Therefore, our approach provides a global network management and configuration of legacy GPONs using OpenFlow. Results have shown that the SDN management layer efficiently implements online service reconfiguration policies over residential users and their services demonstrate a good performance in terms of speed and efficiency, since it responds quite quick to real-time bandwidth requirements and fluctuations of bandwidth transitions are not very important.

This SDN proposal could provide great advantages for the GPON network and configuration. In fact, the approach allows ISPs or network operators to simultaneously control GPONs that belong to different providers, avoiding interacting with specific APIs of equipment from different manufacturers and minimizing compatibility problems. In our approach, every PON device can understand the OpenFlow protocol, since we have emulated a SDN layer over each device and consequently the SDN management layer interacts with SDN controllers (ODL in this case) in a transparent and automatic way. Even more, the implemented solution allows simultaneously controlling several SDN controllers and GPONs from the same SDN management layer in a centralized, efficient, and dynamic way. Finally, our approach allows moving global service configuration strategies out of the PON intelligence to an external SDN management layer which dynamically adjust these types of strategies according to the real-time QoS requirements, permitting ISPs/Network Operators to enable new business models for residential users. As a consequence, these advantages optimize the network operation and reduce its complexity, since it permits remotely controlling and configuring service profiles (Internet, VoIP, Video) in a dynamic, programmable, centralized, and unified way.

Author Contributions: Conceptualization, N.M., J.C.A., I.d.M., and R.J.D.; Methodology, N.M., J.C.A., I.d.M., R.J.D., P.F.; Software, D.d.P., N.M., and J.C.A.; Investigation, N.M., D.d.P., J.C.A.; Validation, I.d.M., R.J.D., P.F., R.M.L., and E.J.A., Writing—Original Draft Preparation, N.M., D.d.P., J.C.A., I.d.M., R.J.D., P.F.; Writing—Review & Editing, R.M.L. and E.J.A.; Supervision, R.M.L. and E.J.A.; Project Administration, N.M., R.J.D., I.d.M., and E.J.A. All authors have read and agreed to the published version of the manuscript.

Funding: This research has been funded by the Spanish Ministry of Science, Innovation and Universities (TEC2017-84423-C3-1-P and RTC2019-007043-7), Consejería de Educación de la Junta de

Castilla y León (VA085G19), and the European Union through INTERREG V-A España-Portugal (POCTEP) program (0677_DISRUPTIVE_2_E).

Institutional Review Board Statement: Not applicable.

Informed Consent Statement: Not applicable.

Data Availability Statement: Data is contained within the article.

Conflicts of Interest: The authors declare no conflict of interest.

References

1. Kerpez, K.J.; Cioffi, J.M.; Ginis, G.; Goldburg, M.; Galli, S.; Silverman, P. Software-defined access networks. *IEEE Commun. Mag.* **2014**, *52*, 152–159. [CrossRef]
2. Thyagaturu, A.S.; Mercian, A.; McGarry, M.P.; Reisslein, M.; Kellerer, W. Software Defined Optical Networks (SDONs): A Comprehensive Survey. *IEEE Commun. Surv. Tutorials* **2016**, *18*, 2738–2786. [CrossRef]
3. Open Networking Foundation (ONF). Available online: https://www.opennetworking.org/ (accessed on 3 October 2020).
4. NETCONF Configuration Protocol. Network Configuration Working Group. Available online: https://tools.ietf.org/wg/netconf (accessed on 3 October 2020).
5. RESTCONF Protocol. Internet Engineering Task Force (IETF). Available online: https://tools.ietf.org/html/rfc8040 (accessed on 3 October 2020).
6. State of Fibre. New Market Forecast 2020–2026 Revealed. Available online: https://www.ftthcouncil.eu/home/latest-news/state-of-fibre-new-market-forecasts-2020-2026-revealed?news_id=3863&back=/resources/key-publications (accessed on 10 December 2020).
7. Europe Broadband Status Market Forecast by 2020–2025, FTTH Council. Available online: https://www.ftthcouncil.eu/documents/Reports/2019/FTTH%20Council%20Europe%20%20Forecast%20for%20EUROPE%202020-2025.pdf (accessed on 1 October 2020).
8. Lee, S.S.W.; Li, K.-Y.; Wu, M.-S. Design and implementation of a GPON-based virtual Open Flow-enabled SDN switch. IEEE/OSA. *J. Lightwave Technol.* **2016**, *34*, 2552–2561. [CrossRef]
9. Rouskas, G.N.; Dutta, R.; Baldine, I. A New Internet Architecture to Enable Software Defined Optics and Evolving Optical Switching Models. In Proceedings of the International Conference Broadband Communications Network Systems (BROADNETS), London, UK, 8–11 September 2008; pp. 71–76.
10. Gu, R.; Ji, Y.; Wei, P.; Zhang, S. Software defined flexible and efficient passive optical networks for intra-datacenter communications. *Opt. Switch. Netw.* **2014**, *14*, 289–302. [CrossRef]
11. Lee, Y.; Kim, Y. *A Design of 10 Gigabit Capable Passive Optical Network (XG-PON1) Architecture Based on Software De-Fined Network (SDN)*; (ICOIN): Siem Reap, Cambodia, 2015.
12. Amokrane, A.; Hwang, J.; Xiao, J.; Anerousis, N. Software defined enterprise passive optical network. In Proceedings of the 10th International Conference on Network and Service Management (CNSM) and Workshop, Rio de Janeiro, Brazil, 17–21 November 2014; pp. 406–411.
13. Parol, P.; Pawlowski, M. Towards networks of the future: SDN paradigm introduction to PON networking for business applications. In Proceedings of the 2013 Federated Conference on Computer Science and Information Systems), Kraków, Poland, 8–11 September 2013.
14. The Open Source Project VOLTHA. Available online: https://www.opennetworking.org/voltha/ (accessed on 3 October 2020).
15. Khalili, H.; Sallent, S.; Piney, J.R.; Rincón, D. A proposal for an SDN-based SIEPON architecture. *Opt. Commun.* **2017**, *403*, 9–21. [CrossRef]
16. Lia, C.; Guoa, W.; Wanga, W.; Hua, W.; Xiab, M. Programmable bandwidth management in software-defined EPON architecture. *Opt. Commun.* **2016**, *370*, 43–48. [CrossRef]
17. Zhang, J.; Zhao, Y.; Jialin, W.; Ji, Y.; Yi, L.; Han, J.; Lee, Y. Experimental demonstration of remote unified control for Open-Flow-based software-defined optical access networks. *Photonic Netw. Commun.* **2016**, *31*, 568–577.
18. Wang, F.; Liu, B.; Zhang, L.; Jin, F.; Zhang, O.; Tian, Q.; Tian, F.; Rao, L.; Xin, X. Dynamic bandwidth allocation based on multiservice in software-defined wavelength-division multiplexing time-division multiplexing passive optical net-work. *Opt. Eng.* **2017**, *56*, 036104. [CrossRef]
19. Hwang, I.-S.; Rianto, A.; Pakpahan, A.F. Software-defined Peer-to-Peer file sharing architecture for TWDM PON. In Proceedings of the 2018 27th Wireless and Optical Communication Conference (WOCC), Hualien, Taiwan, 30 April–1 May 2018; pp. 1–4.
20. Ruffin, M.; Slyne, F.; Bluemm, C.; Kitsuwan, N.; McGettrick, S. Software defined networking for next generation con-verged metro-access networks. *Opt. Fiber Technol.* **2015**, *26*, 31–41. [CrossRef]
21. Yan, B.; Zhou, J.; Wu, J.; Zhao, Y. Poster: SDN based energy management system for optical access network. In Proceedings of the 9th International Conference on Communications and Networking in China, Shanghai, China, 14–16 August 2014; pp. 658–659.
22. Quian, C.; Li, Y.; Zhang, O.; Cao, B.; Li, Z.; Wang, M. Staged priority-based dynamic bandwidth allocation in soft-ware-defined hybrid passive optical network. *Opt. Eng.* **2018**, *57*, 126101.

23. Zhou, L.; Peng, G.; Chand, N. Demonstration of a novel software-defined Flex PON. *Photonic Netw. Commun.* **2015**, *29*, 282–290. [CrossRef]
24. Open vSwitch. Available online: https://www.openvswitch.org/ (accessed on 15 November 2020).
25. Open Day Light SDN Controller. Available online: https://www.opendaylight.org/ (accessed on 15 November 2020).
26. Salman, O.; Elhajj, I.H.; Kayssi, A.; Chehab, A. SDN Controllers: A Comparative Study. In Proceedings of the 18th Mediter-ranean Electrotechnical Conference (MELECON), Lemesos, Cyprus, 18–20 April 2016.
27. Paliwal, M.; Shrimankar, D.; Tembhurne, O. Controllers in SDN: A Review Report. *IEEE Access* **2018**, *6*, 36256–36270. [CrossRef]
28. OpenFlow Switch Specification and Meters, OpenFlow Switch Specification. Available online: https://www.opennetworking.org/wp-content/uploads/2014/10/openflow-switch-v1.5.1.pdf (accessed on 15 November 2020).
29. Kramer, G.; Mukherjee, B.; Dixit, S.; Ye, Y.; Hirth, R. Supporting differentiated classes of service in Ethernet passive optical networks. *J. Opt. Netw.* **2002**, *1*, 280–298.
30. Telnet Vendor Web Page, Telnet R-I. Available online: https://www.telnet-ri.es/ (accessed on 4 October 2020).
31. Wireshark: Network Protocol Analyzer. Available online: https://www.wireshark.org/ (accessed on 4 October 2020).

applied sciences

MDPI

Letter

Rearrangeable Nonblocking Conditions for Four Elastic Optical Data Center Networks

Bey-Chi Lin

Department of Applied Math, National University of Tainan, Tainan City 70005, Taiwan; beychi@mail.nutn.edu.tw; Tel.: +886-6-2133-111

Received: 18 September 2020; Accepted: 19 October 2020; Published: 22 October 2020

Abstract: Four variants of elastic optical data center network (DCN) architectures based on optical circuit switching were proposed in an earlier study. The necessary and sufficient values of frequency slot units (FSUs) per fiber required for these four DCNs in the sense of there being strictly nonblocking (SNB) were derived, but no results in the sense of being rearrangeable nonblocking (RNB) were presented. In reality, only limited bandwidths are available, and reducing the value of FSUs per fiber has become a critical task to realize nonblocking optical DCN architectures in practice. In this paper, we derive the sufficient value of FSUs per fiber required for the four DCNs to be RNB by two multigraph approaches. Our results show that the proposed RNB conditions in terms of FSUs per fiber for a certain two of the four DCNs reduce their SNB results down to at least half for most cases, and even down to one-third.

Keywords: data center networks; elastic optical networks; rearrangeable nonblocking (RNB); edge-coloring

1. Introduction

Recently, high transmission speed between the servers in data centers [1] has become an increasing requirement to meet the needs of current applications such as cloud computing and data mining. To support such high transmission speed, various data center network (DCN) architectures have been proposed [1–4]. One among them integrates an electronic packet switching (EPS) network and an optical circuit switching (OCS) network [1,2,4]. In such a hybrid electrical/optical architecture, both the EPS and OCS networks connect to each top-of-rack (ToR) switch simultaneously, where EPS serves small flows and OCS serves big flows. It has been shown that such a hybrid electrical/optical architecture reduces the power consumption and the operating expense [5,6].

An elastic optical network (EON) [7–11] is a candidate for being the OCS part of a DCN [2]. In EONs, flexible frequency grids proposed by ITU-T [12] are used, and a different number, say m, of *adjacent* frequency slot units (FSUs) are assigned to an optical connection, where m is usually upper bounded by a value, say $m_{\max}$. The bandwidth of an FSU is 12.5 GHz [12], and a connection is called an m-slot connection if it is assigned m adjacent FSUs. Four variants of optical DCN architectures based on elastic optical switches, called DCN1, DCN2, DCN3 and DCN4, were proposed in [13]. The four DCN architectures are similar to wavelength-space-wavelength (W-S-W) networks [8–10], which are Clos-like architectures, but they do not adopt costly tunable wavelength converters as W-S-W networks do. In addition, the maximum number of connections generated from each input fiber in the four DCNs and W-S-W networks is limited due to the different components used. This leads to the nonblocking conditions derived for these four DCNs being different from those derived for W-S-W networks.

When a network is called nonblocking, it is in reference to the nonblocking traffic assigned to the network [14]. To prevent excessive blocking of connections, the network should be nonblocking. A network is called strictly nonblocking (SNB) if a connection will never be blocked by existing connections, and a network is called rearrangeable nonblocking (RNB) if a new connection can be accommodated by rearranging some existing connections [14]. An RNB network is also defined as one where any set (or frame) of connections can be routed simultaneously. The necessary and sufficient number of FSUs per fiber required for these four DCNs in the sense of their being SNB were given in [13], but no results in the sense of being RNB were proposed.

The four DCNs usually require a great number of FSUs per fiber to be SNB, especially when $m_{\max}$ is growing. However, the resource of FSUs per fiber in practical systems is limited since for the EON switches, and the C-band has only around 350 available FSUs (1530–1565 nm). Reducing the value of FSUs per fiber is a challenging task, and this issue has been studied in various research on EONs [8–11]. In order to reduce the value of FSUs per fiber to realize nonblocking optical DCN architectures in practice, we studied the four DCNs in the sense of being RNB in this paper, and derived the sufficient number of FSUs per fiber by adopting two multigraph approaches. Our results show that two of the proposed RNB conditions reduced the SNB results significantly.

The rest of the paper is organized as follows: In Section 2, we give a brief review of the four DCN architectures and introduce the notations used in the paper. In Section 3, we prove the RNB conditions for the DCN1 and DCN3 networks. In Section 4, we prove the RNB conditions for the DCN2 and DCN4 networks. Section 5 concludes the paper.

2. Preliminaries and Notations

In this section, we will review the four elastic optical DCN architectures proposed in [13] and introduce the notations used in this paper. The four elastic optical DCN architectures require bandwidth-variable, waveband-selective switches (BV-WSSs) [15,16], bandwidth-variable space switches (BV-SSs), passive combiners (PCs) and ToR switches. Both BV-WSSs and BV-SSs, the latter of which consist of BV-WSSs and PCs, can switch wavebands with flexible bandwidths without spectrum conversion capabilities. Each ToR switch consists of q bandwidth-variable transponders (BVTs), which are divided into two parts: the transmission part, denoted by BVT-T, and the receiving part, denoted by BVT-R. The part of each ToR switch consisting of q BVT-Ts (or BVT-Rs) and a PC (or BV-WSS) is denoted by ToR-T (or ToR-R) (see Figure 1). A BVT-T can use any m consecutive FSUs of its output; i.e., the frequency of its output is arbitrarily tunable. In addition, a BVT-T is connected to a BVT-R in a strict one-to-one manner, and thus a BVT-T does not simultaneously send connections to two or more BVT-Rs. All connections generated from the same ToR switch occupy different FSUs, so that all of them can be sent through one fiber connecting the ToR-T (or ToR-R) to the OCS network.

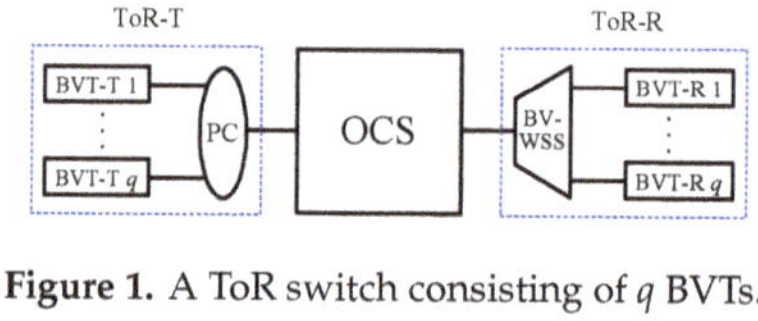

Figure 1. A ToR switch consisting of q BVTs.

The DCN1 architecture, denoted by DCN1(r, q, k), is given in Figure 2a. A DCN1(r, q, k) network contains one $r \times r$ BV-SS and r ToR switches, each of which consists of q BVT-Ts (or BVT-Rs) and is attached to an input (or output) fiber with k FSUs of the BV-SS. The DCN2 architecture, denoted by DCN2(s, r, q, k), is a variant of DCN1(r, q, k) and is given in Figure 2b. A DCN2(s, r, q, k) network contains one $r \times r$ BV-SS and r groups of s ToR switches, which are combined by one PC into (or directed from one BV-WSS to) one input (or output) fiber connecting to the BV-SS. We use ToR-T (or ToR-R) u-i to denote the ith ToR-T (or ToR-R) in group u, where $1 \leq u \leq r$ and $1 \leq i \leq s$. The DCN3 architecture is denoted by DCN3(r, q, k, p), and it contains p $r \times r$ BV-SSs and r ToR switches (Figure 3a). The output (or input) fiber of ToR-T u (or ToR-R v) is connected to one BV-WSS (or PC) which connects to the uth input (or vth output) of each BV-SS. Finally, the DCN4 architecture is denoted by DCN4(s, r,

q, k, p), and it is obtained from a DCN2(s, r, q, k) network by adopting p BV-SSs to connect ToR-Ts and ToR-Rs (Figure 3b).

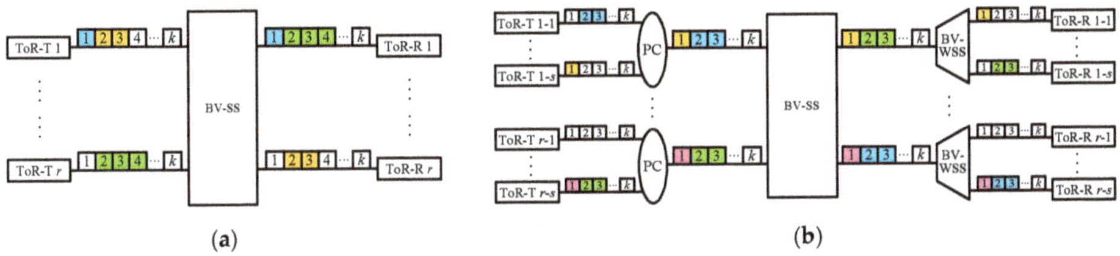

Figure 2. (a) A DCN1(r, q, k) network and (b) a DCN2(s, r, q, k) network, where each ToR-T (or ToR-R) consists of q BVT-Ts (or BVT-Rs), as given in Figure 1.

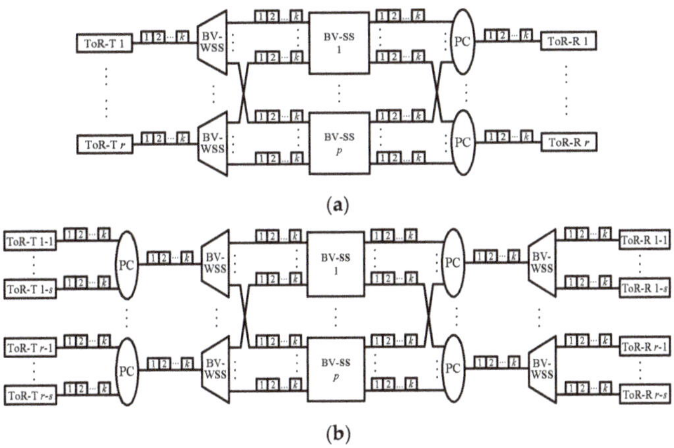

Figure 3. (a) A DCN3(r, q, k, p) network, and (b) a DCN4(s, r, q, k, p) network, where each ToR-T (or ToR-R) consists of q BVT-Ts (or BVT-Rs), as given in Figure 1.

The four DCN architectures serve m-slot connections with $m \leq m_{\max}$. To guarantee that each fiber occupying k FSUs is sufficient to carry connections served by all BVT-Ts, the value of k is assumed to be $k \geq qm_{\max}$ (or $k \geq sqm_{\max}$) for the DCN1 and DCN3 (or DCN2 and DCN4) architectures. An m-slot connection from a BVT-T in ToR-T u (or ToR-T u-i) to a BVT-R in ToR-R v (or ToR-R v-j) in a DCN1 or DCN3 (or a DCN2 or DCN4) is denoted by (u, v, m) (or (u-i, v-j, m)), where $1 \leq u, v \leq r$ and $1 \leq i, j \leq s$.

FSUs in each fiber are numbered from 1 to k. To set up a connection (u, v, m) (or (u-i, v-j, m)), the same sets of m adjacent FSUs must be found in both the fiber connecting ToR-T u (or ToR-T u-i) with one BV-SS and the fiber connecting this BV-SS with ToR-R v (or ToR-R v-j). If those sets do not exist, the connection is blocked. The necessary and sufficient values of k for DCN1 to DCN4 in the sense of being SNB were given in [13]. We quote the SNB results for the DCN1 and DCN2 networks in Lemmas 1 and 2 for further comparison in Sections 3 and 4.

Lemma 1. A DCN1(r, q, k) network for m-slot connections with $1 \leq m \leq m_{\max}$ is SNB if and only if

$$k \geq k_{\text{SNB}} = 2(q-1)\cdot(2m_{\max}-1) + m_{\max} \tag{1}$$

Lemma 2. A DCN2(s, r, q, k) network for m-slot connections with $1 \leq m \leq m_{\max}$ is SNB if and only if

$$k \geq k'_{\text{SNB}} = 2(sq - 1)\cdot(2m_{\max} - 1) + m_{\max} \quad (2)$$

3. RNB DCN1 and DCN3 Networks

In this section, we first consider the RNB DCN1 network and then the RNB DCN3 network. In order to derive the sufficient value of k for a DCN1(r, q, k) network in the sense of being RNB, we propose a multigraph approach and a routing algorithm in the following.

3.1. Multigraph Approach and Routing Algorithm

Given a DCN1(r, q, k) network and a frame F of connections, we propose Multigraph Approach A, given below, to convert the DCN1(r, q, k) network for frame F into a multigraph G_F.

Multigraph Approach A:

Let each left vertex u (or right vertex v) in multigraph G_F be ToR-T u (or ToR-R v) of the DCN1(r, q, k) network. In multigraph G_F, there is an edge connecting vertexes u and v if there is an m-slot connection from a BVT-T in ToR-T u and it is destined to a BVT-R in ToR-R v, i.e., (u, v, m) (see Figure 4a). Note that we call G_F a multigraph [17] because multiple connections between ToR-T u and ToR-R v are allowed, and thus there could be more than one edge connecting vertexes u and v in G_F.

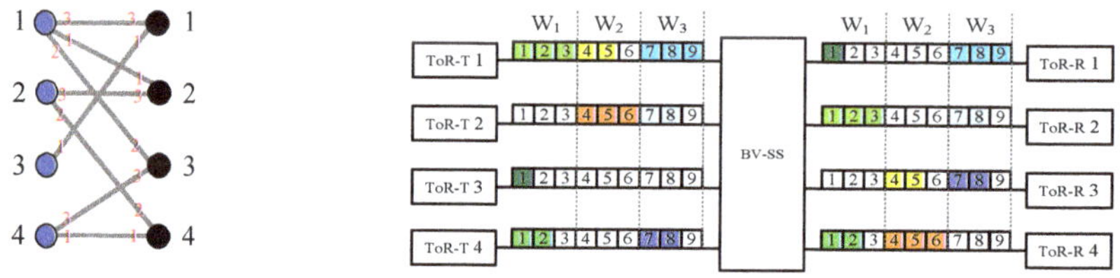

(**a**) (**b**)

Figure 4. Given a DCN1(4, 3, 9) network and a frame F of connections (1, 1, 3), (1, 2, 3), (1, 3, 2), (2, 2, 2), (2, 4, 3), (3, 1, 1), (4, 3, 2) and (4, 4, 2): (**a**) The corresponding multigraph G_F constructed by Multigraph Approach A. Note that G_F is edge-colored by colors 1, 2, and 3 (marked in red), and left and right vertices are marked in blue and black, respectively. (**b**) A routing of connections in frame F according to Routing Algorithm A in association with the edge-coloring of G_F.

In Property 1, we show that G_F is q-edge-colorable.

Property 1. Given a DCN1(r, q, k) network and a frame F of connections, let G_F be the corresponding multigraph constructed by Multigraph Approach A. Multigraph G_F is q-edge-colorable.

Proof. Let $\Delta(G_F)$ be the maximum degree of G_F. Since each ToR switch consists of q BVT-Ts and q BVT-Rs, at most q m-slot connections can be generated from a ToR-T (or destined to a ToR-R). Thus, we have $\Delta(G_F) \leq q$. From the construction of G_F, we can see that G_F is a bipartite multigraph. In addition, G_F is q-edge-colorable according to graph theory [17] if G_F is a bipartite multigraph with $\Delta(G_F) \leq q$. □

In a DCN1(r, q, k) network, we use I_u (or O_v) to denote the fiber connecting ToR-T u (or ToR-R v) and the BV-SS, where $k \geq qm_{\max}$ and $1 \leq u, v \leq r$. In addition, we partition each fiber with k FSUs into q parts, each of which consists of $m_{\max}$ consecutive FSUs. Each part is called a window, and these q windows, denoted by W_l for $1 \leq l \leq q$, are numbered from 1 from left to right. We use $|W_l|$ to represent the size of window W_l, and also use $I_{u,l}$ (or $O_{v,l}$) to represent the lth window in fiber I_u (or O_v) for $1 \leq u, v \leq r$ and $1 \leq l \leq q$.

Recall that G_F is q-edge-colorable (Property 1). Let colors 1, 2, ... , q be adopted to edge color G_F. We route each (u, v, m) for the RNB condition [14] using Routing Algorithm A given below.

Routing Algorithm A:
Connection (u, v, m) is routed in windows $I_{u,c}$ and $O_{v,c}$ if color c is assigned to the corresponding edge of (u, v, m) in G_F (see Figure 4b).

3.2. RNB Sufficient Conditions

A sufficient value of k for a DCN1(r, q, k) network in the sense of being RNB is derived in Property 2.

Property 2. A DCN1(r, q, k) network for m-slot connections with $1 \leq m \leq m_{\max}$ is RNB if

$$k \geq k_{\text{RNB}} = q \cdot m_{\max} \tag{3}$$

Proof. This property holds if Routing Algorithm A is feasible, and Routing Algorithm A is feasible if each m-slot connection can be carried by the corresponding windows. Since $m \leq m_{\max}$, each m-slot connection can be carried by any window W_l if $|W_l| = m_{\max}$ for $1 \leq l \leq q$, which implies that each fiber has $k = q \cdot m_{\max}$ FSUs. Therefore, when Routing Algorithm A is applied, a DCN1(r, q, k) network with $k \geq q \cdot m_{\max}$ is RNB. □

Comparing Equation (3) with Equation (1), we have $k_{\text{RNB}}/k_{\text{SNB}} \leq 1/2$ for $m_{\max} \geq 2$ and $q \geq 3$. Property 2 implies that k_{RNB} reduces the SNB DCN1 result given in [13], namely, k_{SNB}, down to at least half for most cases. In addition, numerical results are given in Table 1 which show that k_{RNB} can reduce k_{SNB} down to as low as one third, for example, the cases with $m_{\max} \geq 6$ and $q = 4$, and the cases with $m_{\max} \geq 4$ and $q \geq 8$.

Table 1. Numerical results of k required for being an SNB or RNB DCN1(r, q, k) network for m-slot connections with q = 4, 8, 10 and $1 \leq m \leq m_{\max}$, where k_{SNB} and k_{RNB} are given in Equations (1) and (3), respectively.

$m_{\max}$	$q = 4$		$q = 8$		$q = 10$	
	k_{SNB}	k_{RNB}	k_{SNB}	k_{RNB}	k_{SNB}	k_{RNB}
2	20	8	44	16	56	20
4	46	16	102	32	130	40
6	72	24	160	48	204	60
8	98	32	218	64	278	80
10	124	40	276	80	352	100

The sufficient condition for being an RNB DCN1(r, q, k) network (Property 2) is also the necessary condition if only one connection rate $m_{\max}$ is considered (Property 3).

Property 3. Suppose only one connection rate, $m_{\max}$, is considered. Then, the DCN1(r, q, k) network is RNB if and only if $k \geq k_{\text{RNB}} = q \cdot m_{\max}$.

Proof. The sufficient condition of this property is true since it is a special case with one connection rate of Property 2. In addition, the necessary condition holds when q connections $(u, u, m_{\max})$ for $1 \leq u \leq r$ are generated from each ToR-T u. □

From the architectures of the DCN1 and DCN3 networks (see Figures 2a and 3a), we can see that a DCN1(r, q, k) network for $k \geq q{\cdot}m_{\max}$ functions the same as a DCN3(r, q, k, p) network with $p = 1$. Thus, we derive Property 4 immediately.

Property 4. A DCN3(r, q, k, p) network for m-slot connections with $1 \leq m \leq m_{\max}$ and $k \geq q{\cdot}m_{\max}$ is RNB if $p \geq 1$.

Proof. This property is true for two reasons: i) a DCN3(r, q, k, p) network with $p = 1$ functions as well as a DCN1(r, q, k) network, and ii) a DCN1(r, q, k) network for $k \geq q{\cdot}m_{\max}$ is RNB (Property 2). □

For a DCN1(r, q, k) (or DCN3(r, q, k, p)) network, recall that the resource of FSUs per fiber in practical systems is limited, namely, $k \leq 350$. This implies that to have an RNB DCN1(r, q, k) (or DCN3(r, q, k, p)) network for m-slot connections with $1 \leq m \leq m_{\max}$ in the real word, we also need $q{\cdot}m_{\max} \leq 350$ due to Property 2 (or Property 4).

4. RNB DCN2 and DCN4 Networks

Similar to Section 3, we first consider the RNB DCN2 network and then the RNB DCN4 network. For the DCN2 network (Figure 2b), $I_{u\text{-}i}$ (or $O_{v\text{-}j}$) is used to represent the fiber connecting ToR-T u-i (or ToR-R v-j) and the uth PC (or vth BV-WSS), and I'_u (or O'_v) is used to represent the fiber connecting the uth PC (or vth BV-WSS) and the BV-SS for $1 \leq u, v \leq r$ and $1 \leq i, j \leq s$. Next, we will propose Multigraph Approach B and Routing Algorithm B for the DCN2 network in the sense of being RNB by modifying Multigraph Approach A and Routing Algorithm A, respectively.

Multigraph Approach B:

Given a DCN2(s, r, q, k) network and a frame F of connections, multigraph G'_F is constructed in the following way. Let each left vertex u (or right vertex v) in G'_F be the uth (or vth) group of s ToR-Ts u-i (or ToR-Rs v-j) for $1 \leq i, j \leq s$. An edge is added between two vertexes u and v in G'_F if there is an m-slot connection from the uth ToR-T group destined to the vth ToR-R group (see Figure 5a).

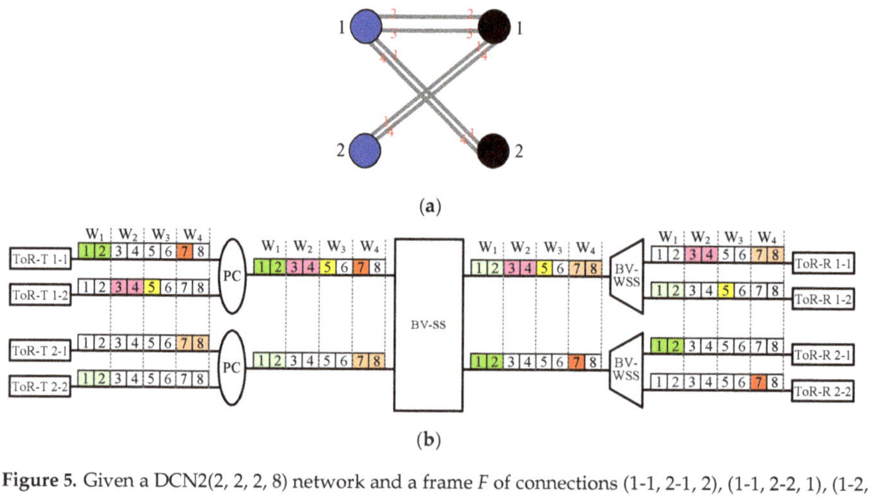

Figure 5. Given a DCN2(2, 2, 2, 8) network and a frame F of connections (1-1, 2-1, 2), (1-1, 2-2, 1), (1-2, 1-1, 2), (1-2, 1-2, 1), (2-1, 1-1, 2) and (2-2, 1-2, 2): (**a**) The corresponding multigraph G'_F generated by Multigraph Approach B. Note that G'_F is edge-colored by colors 1, 2, 3 and 4 (marked in red font), and left and right vertices are marked in blue and black, respectively. (**b**) A routing of connections in frame F according to Routing Algorithm B in association with the edge-coloring of G'_F.

Since each group of ToR switches can generate at most sq m-slot connections, we derive that $\Delta(G'_F) \leq sq$, and thus G'_F is sq-edge-colorable [17]. Let colors 1, 2, … , sq be used to edge-color G'_F. We adopt Routing Algorithm B, shown below, to route each (u-i, v-j, m) in association with the edge-coloring of G'_F for the RNB condition.

Routing Algorithm B:

Connection (u-i, v-j, m) is routed in windows $I_{u\text{-}i,c}$, $I'_{u,c}$, $O'_{v,c}$ and $O_{v\text{-}j,c}$ if color c is assigned to the corresponding edge of (u-i, v-j, m) in G'_F (see Figure 5b).

Similar to Properties 2–4, we have Properties 5–7, as follows.

Property 5. A DCN2(s, r, q, k) network for m-slot connections with $1 \leq m \leq m_{\max}$ is RNB if

$$k \geq k'_{\text{RNB}} = sq \cdot m_{\max} \tag{4}$$

Proof. The proof is similar to that of Property 2. □

Property 6. Suppose only one connection rate, $m_{\max}$, is considered. Then a DCN2(s, r, q, k) network is RNB if and only if $k \geq k'_{\text{RNB}} = sq \cdot m_{\max}$.

Proof. The proof is similar to that of Property 3. □

Property 7. A DCN4(s, r, q, k, p) network for m-slot connections with $1 \leq m \leq m_{\max}$ and $k \geq sq \cdot m_{\max}$ is RNB if $p \geq 1$.

Proof. From the topology of the DCN4 architecture (see Figure 3b), we can see that a DCN4(s, r, q, k, p) network with $p = 1$ and $k \geq sq \cdot m_{\max}$ functions as well as a DCN2(s, r, q, k) network. According to Property 5, the property holds immediately. □

Comparing Equation (2) with Equation (4), we have $k'_{\text{RNB}}/k'_{\text{SNB}} \leq 1/2$ for $m_{\max} \geq 2$ and $sq \geq 3$. Property 5 implies that k'_{RNB} reduces the SNB DCN2 result given in [13], namely, k'_{SNB}, down to at least half for most cases, and even down to one third. In addition, numerical results are given in Table 2, which shows that k'_{RNB} can reduce k'_{SNB} down to as low as one third, for example, all the cases with $m_{\max} \geq 4$, $s = 3$ and $q \geq 4$. Again, due to the limited resource of FSUs per fiber in practical systems, to have an RNB DCN2(s, r, q, k) (or DCN4(s, r, q, k, p)) network for m-slot connections with $1 \leq m \leq m_{\max}$ in the real word, we need $sq \cdot m_{\max} \leq 350$ due to Property 5 (or Property 7).

Table 2. Numerical results of k required for being an SNB or RNB DCN2(s, r, q, k) network for m-slot connections with $s = 3$, $q = 4, 8, 10$ and $1 \leq m \leq m_{\max}$, where k'_{SNB} and k'_{RNB} are given in Equations (2) and (4), respectively.

$m_{\max}$	$s = 3, q = 4$		$s = 3, q = 8$		$s = 3, q = 10$	
	k'_{SNB}	k'_{RNB}	k'_{SNB}	k'_{RNB}	k'_{SNB}	k'_{RNB}
2	68	24	140	48	176	60
4	158	48	326	96	410	120
6	248	72	512	144	644	180
8	338	96	698	192	878	240
10	428	120	884	240	1112	300

5. Conclusions

Four variants of elastic optical DCN architectures, called DCN1, DCN2, DCN3 and DCN4, were proposed in [13]. The four DCNs in the sense of being SNB usually require a large number of FSUs per fiber. To reduce the value of FSUs, we considered the four DCNs in the sense of their being RNB in this paper. We proposed two multigraph approaches to firstly prove the sufficient number of FSUs per fiber for these four DCNs in the sense of there being RNB. Our results show that the proposed RNB conditions in term of FSUs per fiber for the DCN1 and DCN2 networks reduce their SNB results down to at least half in most scenarios, and even down to one third. In addition, we show that the sufficient condition for an RNB DCN3 (or DCN 4) network is exactly the same as that derived for an RNB DCN1 (or DCN 2) network. The proposed multigraph approaches can be applied to all Clos-like architectures for studying RNB conditions.

Funding: This work was supported by the Ministry of Science and Technology, Taiwan, under Contract MOST 108-2221-E-024-002-MY2.

Conflicts of Interest: The authors declare no conflict of interest.

References

1. Kachris, C.; Tomkos, I. A survey on optical interconnects for data centers. *IEEE Commun. Surv. Tutor.* **2012**, *14*, 1021–1036. [CrossRef]
2. Yoo, S.J.B. Integrated photonic-electronic technologies for next generation data centers and the future internet. In Proceedings of the 2012 International Conference on Photonics in Switching (PS), Ajaccio, France, 11–14 September 2012.
3. Fiorani, M.; Aleksic, S.; Casoni, M. Hybrid optical switching for data center networks. *J. Electr. Comput. Eng.* **2014**, *2014*, 1–13. [CrossRef]
4. Hamza, A.S.; Deogun, J.S.; Alexander, D.R. Wireless communication in data centers: A survey. *IEEE Commun. Surv. Tutor.* **2016**, *18*, 1572–1595. [CrossRef]
5. Ji, P.N.; Kachris, C.; Tomkos, I.; Wang, T. Energy efficient data center network based on a flexible bandwidth MIMO OFDM optical interconnect. In Proceedings of the 4th IEEE International Conference on Cloud Computing Technology and Science Proceedings, Taipei, Taiwan, 3–6 December 2012.
6. Kachris, C.; Ji, P.N.; Wang, T.; Tomkos, I. Energy efficient flexible bandwidth OFDM-based data center network. In Proceedings of the 2012 IEEE 1st International Conference on Cloud Networking (CLOUDNET), Paris, France, 28–30 November 2012.
7. Lopez, V.; Velasco, L. *Elastic Optical Networks: Architectures, Technologies, and Control*; Springer International Publishing: Cham, Switzerland, 2016.
8. Kabaciński, W.; Al-Tameemi, A.; Rajewski, R. Necessary and sufficient conditions for the rearrangeability of WSW1 switching fabrics. *IEEE Access* **2019**, *7*, 18622–18633. [CrossRef]
9. Lin, B.-C. Rearrangeable W-S-W elastic optical networks generated by graph approaches. *IEEE OSA J. Opt. Commun. Netw.* **2018**, *10*, 675–685. [CrossRef]
10. Kabaciński, W.; Abdulsahib, M. Wide-sense nonblocking converting-space-converting switching node architecture under XsVarSWITCH control algorithm. *IEEE ACM Trans. Netw.* **2020**, *28*, 1550–1561. [CrossRef]
11. Danilewicz, G. Asymmetrical space-conversion-space SCS1 strict-sense and wide-sense nonblocking switching fabrics for continuous multislot connections. *IEEE Access* **2019**, *7*, 107058–107072. [CrossRef]
12. Spectral Grids for WDM Applications: DWDM Frequency Grid. In *ITU-T Standard G.694.1*; International Telecommunications Union (CCIR) and (CCITT): Geneva, Switzerland, 2012.
13. Kabaciński, W.; Michalski, M.; Rajewski, R.; Żal, M. Optical datacenter networks with elastic optical switches. In Proceedings of the 2017 IEEE International Conference on Communications, Paris, France, 21–25 May 2017.
14. Hwang, F. *The Mathematical Theory of Nonblocking Switching Networks*; World Scientific: Singapore, 2004.
15. 1×9/1×20 Flexgrid®Wavelength Selective Switch (WSS). 2015. Available online: https://www.finisar.com/sites/default/files/downloads/1$\times$9_1$\times$20_flexgrid_wss_pb_v3.pdf (accessed on 18 September 2020).

16. Xie, D.; Wang, D.; Zhang, M.; Liu, Z.; You, Q.; Yang, Q.; Yu, S. LCoS-based wavelength-selective switch for future finer-grid elastic optical networks capable of all-optical wavelength conversion. *IEEE Photon. J.* **2017**, *9*, 7101212. [CrossRef]
17. West, D.B. *Introduction to Graph Theory*; Pearson: London, UK, 2001.

applied sciences

MDPI

Article

Low-Loss and Broadband Silicon Photonic 3-dB Power Splitter with Enhanced Coupling of Shallow-Etched Rib Waveguides

Vinh Huu Nguyen, In Ki Kim and Tae Joon Seok *

School of Electrical Engineering and Computer Science, Gwangju Institute of Science and Technology, Gwangju 61005, Korea; vinhnguyen@gist.ac.kr (V.H.N.); inkikim94@gist.ac.kr (I.K.K.)

* Correspondence: tjseok@gist.ac.kr

Received: 10 June 2020; Accepted: 25 June 2020; Published: 29 June 2020

Featured Application: Silicon photonics, High-density integrated photonic devices, Photonic switches, 1 × N power splitters.

Abstract: A silicon photonic 3-dB power splitter is one of the essential components to demonstrate large-scale silicon photonic integrated circuits (PICs), and can be utilized to implement modulators, 1 × 2 switches, and 1 × N power splitters for various PIC applications. In this paper, we reported the design and experimental demonstration of low-loss and broadband silicon photonic 3-dB power splitters. The power splitter was realized by adiabatically tapered rib waveguides with 60-nm shallow etches. The shallow-etched rib waveguides offered strong coupling and relaxed critical dimensions (a taper tip width of 200 nm and gap spacing of 300 nm). The fabricated device exhibited an excess loss as low as 0.06 dB at a 1550-nm wavelength and a broad operating wavelength range from 1470 nm to 1570 nm. The relaxed critical dimensions (≥200 nm) make the power splitter compatible with standard fabrication processes of existing silicon photonics foundries.

Keywords: silicon photonics; 3-dB power splitter; low-loss; broadband; shallow-etched rib waveguides

---

1. Introduction

Silicon photonics has emerged as a powerful technology for optical networks in datacom and computercom with the commercial success of silicon photonic transceivers [1–3]. In the past decade, silicon photonics has been rapidly matured with the process standardization and the process design kit (PDK) development of silicon photonic foundries [4,5]. Large-scale silicon photonic integrated circuits have been demonstrated for a variety of applications, such as photonic switches [6–8], optical phased arrays (OPAs) [9–12], and programmable photonic processors [13–17]. A 3-dB optical power splitter is one of the key components to build high-density integrated photonic devices with multi-channel and parallel processing. A 3-dB optical power splitter can be employed in integrated Mach-Zehnder interferometers (MZI) to demonstrate intensity modulators or 1 × 2 switches [18–20], where the light traveling in a waveguide is split into two arms, the sensing and the reference arms. The optical power uniformity between two arms is extremely significant, which has motivated us to design a compact, low-loss, and excellent power uniformity optical power splitter. The 3-dB power splitters can be also utilized to demonstrate 1 × N splitters [9,12,21,22]. Notably, low-loss and uniform 1 × N power splitters are essential devices for large-scale silicon photonic OPAs. Various silicon photonic 3-dB power splitters have been demonstrated with multimode interferometers [9,12,22–24], directional couplers (DCs) [25], asymmetric waveguide-based phase control [26], adiabatic couplers with S-bend based Y-branches [27], and parameterized Y-junctions [28], which typically exhibit non-negligible excess losses or limited bandwidths.

Recently, low-loss and broadband power splitters based on adiabatically tapered waveguides have been reported [29,30]. However, these power splitters were designed with silicon strip waveguides, which cause tight mode confinement and weak coupling strength between waveguides. As a result, the designed critical dimensions (CDs) such as tapered tips and waveguide-gaps (30~50 nm), are beyond the deep ultraviolet (DUV) lithography limit of typical silicon photonics foundries. In this paper, we reported the experimental demonstration of power splitters based on adiabatically tapered silicon rib waveguides. The shallow-etched rib waveguides offered relatively strong coupling and relaxed critical dimensions (CDs ≥ 200 nm), which allowed the device dimensions to be compatible with standard processes of existing silicon photonic foundries. The demonstrated device had a low excess loss of 0.06 dB and a wide operating wavelength range over 100 nm (1470~1570 nm). The fabrication tolerance analysis confirmed that the proposed power splitter is fabrication insensitive within a large deviation range of ±50 nm for both the width and gap.

2. Device Design

As the proposed power splitter is based on the adiabatic coupling of tapered waveguides, the coupling strength of the waveguides is a key parameter to achieve a compact device design. The strong coupling can readily be obtained by reducing the gap spacing between the waveguides. However, small gap spacing beyond the lithography limit prohibits a wide spread of applications of the device. An alternative way to achieve strong coupling is to utilize shallow-etched rib waveguides for enhanced mode overlap. Figure 1 shows the coupling strength of two identical waveguides for different waveguide types: (1) Silicon strip waveguides and (2) silicon rib waveguides. The thickness and width of the waveguides were chosen to be 220 nm and 500 nm, which well supported the single TE_0 mode for both types of waveguides. The partial etch depth of the rib waveguide was 60 nm. The result clearly shows that the rib waveguides had stronger couplings than the strip waveguides. The gap spacing dimension can be relaxed for the rib coupled-waveguide. Therefore, we chose rib waveguides for our power splitter design to achieve DUV lithography-compatible CDs. Shallow-etched waveguides also provided low propagation loss, owing to the small mode overlap with the sidewall roughness.

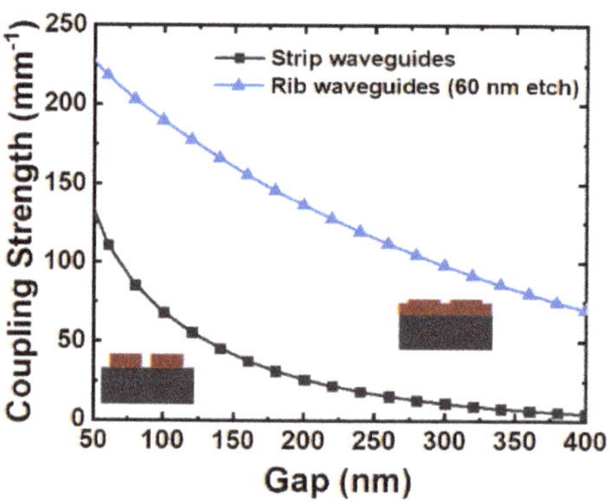

Figure 1. Coupling strength versus gap of two types of waveguide structure. Structures in red and grey are silicon (Si) and buried oxide (BOX), respectively.

Figure 2 shows the schematic of the proposed 3-dB power splitter, consisting of one input waveguide and two output waveguides. A 2-μm-thick buried oxide (BOX) layer (n = 1.444 at 1550 nm) is in grey. The silicon rib waveguides (n = 3.476 at 1550 nm), with a thickness of 220 nm and a partial etch depth of 60 nm, are in red. The input waveguide and the two output waveguides had identical dimensions. Using eigenmode expansion (EME) solver, the device design was optimized.

The waveguides were linearly tapered from the tip width (w_1) of 200 nm to the waveguide width (w_2) of 700 nm. The gap spacing between waveguides was designed to be 300 nm. In order to design the taper length, we investigated the transmission with the taper length sweep, as shown in Figure 3a. The taper length was chosen to be 40 μm, at which the transmission converged to −3.03 dB. Although our current design employed the linear taper, the taper length can be significantly reduced by utilizing the optimal curvature of the taper [31]. The spectral responses of the designed device were also simulated using the FDTD (Finite-Difference Time-Domain) solver. Thanks to the adiabatically tapered waveguides, the broad operation bandwidth over the 200-nm wavelength range was confirmed, as shown in Figure 3b. The normalized E-field intensity profile of the device is presented in Figure 4, which confirms symmetric 3-dB power splitting characteristic.

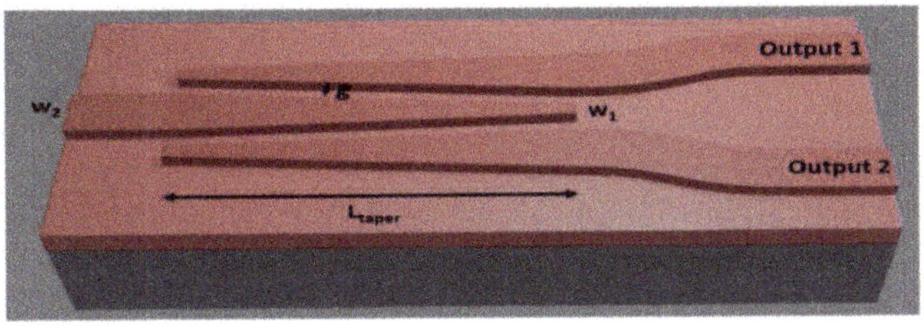

Figure 2. Schematic of 3-dB tapered rib waveguide optical power splitter. Structures in red and grey are silicon (Si) and buried oxide (BOX), respectively.

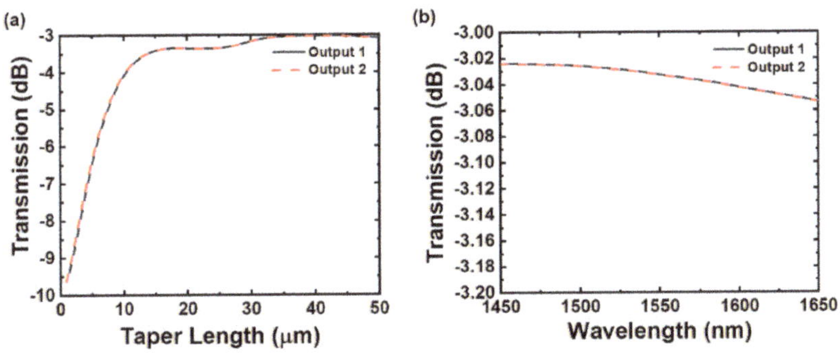

Figure 3. Simulated transmission result. (**a**) Taper length span; (**b**) wavelength sweep over 200-nm bandwidth.

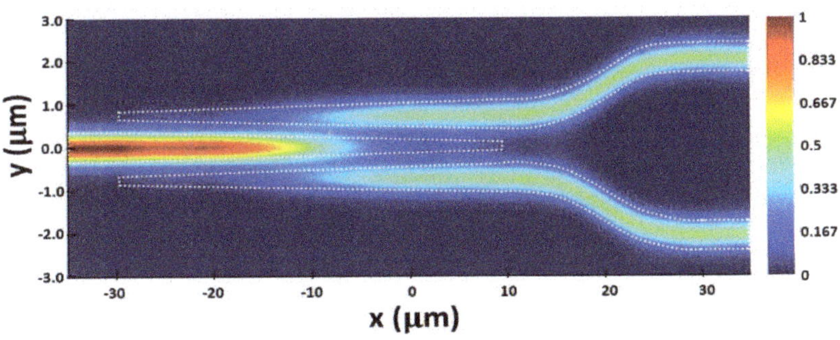

Figure 4. The E-field intensity profile of 3-dB optical power splitter. The field values in the legend are normalized values.

3. Experimental Results

The silicon photonic 3-dB power splitters were fabricated in an 8-inch wafer-scale silicon process facility using deep-UV (DUV) lithography. The designed power splitters were implemented on an 8-inch silicon-on-insulator (SOI) wafer with a 220-nm-thick device layer and a 2-µm-thick buried oxide (BOX) layer. First, we deposited a 70-nm-thick layer of Tetraethyl Orthosilicate (TEOS) by low-pressure chemical vapor deposition (LPCVD) as a hard mask layer. Then, we deposited patterned rib waveguides with DUV lithography, and a 60-nm-deep shallow etch on the silicon device layer. Strip waveguides were also implemented by subsequent DUV lithography and silicon full etch. Finally, the remaining oxide hard mask layer was removed by wet etching.

In order to characterize the fabricated power splitter, we used cascaded six stages of the 1 × 2 3-dB power splitter, as shown in Figure 5a. We employed vertical grating couplers to couple light into the fabricated device. The coupling loss of the grating couplers was measured to be ~6.0 dB from a back-to-back grating coupler pair on the same chip. The input power is P, and the cascaded output powers were then P/2, P/4, P/8, P/16, P/32, and P/64, respectively, as indicated in the Figure 5a. Figure 5b shows the scanning electron microscope (SEM) images of the fabricated power splitter, which confirm a good agreement with the device design. The nominal waveguide width and gap were observed to be 740 nm and 340 nm. The fabrication deviation from the design was below 50 nm. For the experimental characterization, we used a linear array of polarization-maintaining (PM) fibers with 127-µm pitch to guarantee consistent coupling losses from optical ports of the cascaded power splitters. TE-polarized light was coupled to the fabricated device through grating coupler array. The transmission of each stage was measured at the wavelength of 1550 nm and plotted as a function of the number of splitters, as shown in Figure 6a. The transmission of the single power splitter was measured to be −3.06 dB from the linear regression. The excess loss was estimated to be 0.06 dB (= 3.06 dB − 3 dB). The spectral response of each stage is shown in Figure 6b, exhibiting the broad operating bandwidth over the wavelength range of 100 nm from 1470 nm to 1570 nm. The transmission of each stage (one splitter, two splitters, etc.) was normalized to the transmission of a back-to-back grating coupler pair.

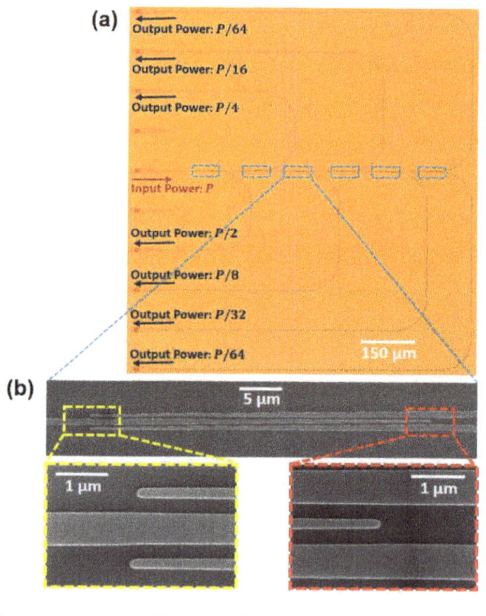

Figure 5. (**a**) Cascaded six-stage structure for excess loss measurement. (**b**) Scanning electron microscope (SEM) images of fabricated optical power splitter.

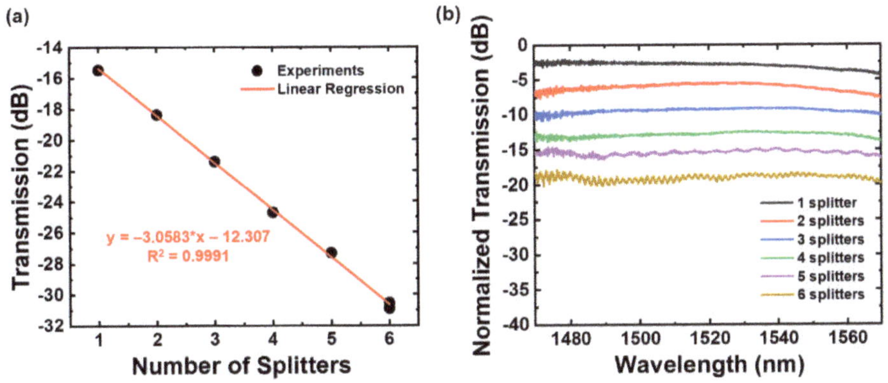

Figure 6. Experimental results (**a**) for excess loss at 1550 nm and (**b**) normalized transmission in broad bandwidth.

4. Discussion

The strong coupling of the rib waveguides enabled the device design with large CD (200 nm), which was compatible with typical existing silicon photonics foundries. It was noted that the top oxide cladding could be added without significant changes in the device's performance. We also investigated the fabrication tolerance of the device using EME simulations. Figure 7a shows the calculated transmissions of the power splitter with the deviations of the waveguide width and the gap from the optimum designs, confirming good fabrication tolerance of our power splitter design. The calculated excess loss of the device exhibited an excess loss as low as 0.07 dB, even with considerable fabrication variations (dark and light purple regions). For example, the excess loss of the power splitter can be below 0.07 dB as long as the waveguide width and gap are controlled within the fabrication variations of ±50 nm, which can be readily achieved from commercial silicon photonics foundries. For our fabricated device (the corresponding dimensions for waveguide width and gap spacing of 740 nm and 340 nm, respectively), the excess loss was expected to be lower than 0.07 dB. Our measured excess loss of 0.06 dB agrees well with this fabrication tolerance study. In addition, the tolerance simulation for the tip width was also performed. The excess loss was below 0.05 dB as long as the tip width was controlled to be smaller than 300 nm, as shown in Figure 7b.

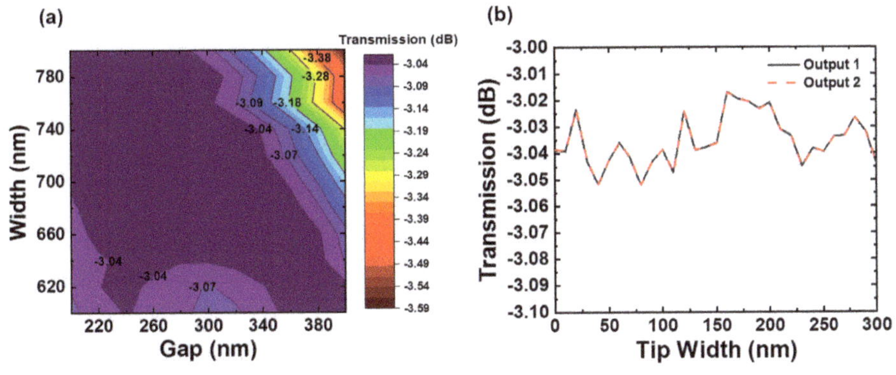

Figure 7. Fabrication tolerance to deviation (**a**) of the waveguide width and the gap and (**b**) of the tip width of the power splitter using EME simulation.

Table 1 shows the summary of previously reported experimental power splitters and this work. Power splitters based on adiabatically tapered waveguides exhibited relatively low excess losses compared to other types. Whereas adiabatic power splitters with strip waveguides suffered from their small CDs, adiabatic power splitters with shallow-etched rib waveguides are promising in large-scale integrated photonic devices, considering their low loss (0.06 dB) and relaxed CD (200 nm), which are compatible with typical existing silicon photonics foundries.

Table 1. Summary of parameters and performances of reported power splitters and this work.

References	Dimension	Excess Loss (at 1550 nm)	Wavelength Bandwidth	Critical Dimension	Splitter Type
[21]	L = 200 µm	0.15~0.6 dB	100 nm	150 nm	Arc-Shaped
[22]	L = 10 µm	0.6 dB	60 nm	200 nm	Tapered Branch
[23]	2 µm × 3.6 µm	2.5 dB	300 nm	200 nm	MMI
[24]	1.2 µm × 2 µm	0.27 dB	80 nm	200 nm	MMI
[25]	L_{Taper} = 100 µm	1 dB	100 nm	200 nm	Adiabatic Tapers
[26]	L_{DC} = 31.4 µm	1 dB	88 nm	200 nm	DC
[28]	1.4 µm × 2.3 µm	0.36 dB	40 nm	200 nm	Y-junction
[30]	L_{Taper} = 5 µm	0.12 dB	70 nm	30 nm	Adiabatic Tapers
This work	L_{Taper} = 40 µm	0.06 dB	100 nm	200 nm	Adiabatic Tapers

5. Conclusions

We demonstrated low loss, broadband, and fabrication-insensitive silicon photonic 3-dB power splitters based on adiabatically tapered rib waveguide. The shallow-etched rib waveguides were employed to make the critical dimensions of the designed device larger than 200 nm. The compact power splitter with the taper length of 40 µm exhibited a low excess loss of 0.06 dB and a broad bandwidth over a 100-nm wavelength. The design compatibility with existing silicon photonics foundries and good fabrication tolerance enable the wide usage of the power splitter on a variety of silicon photonics applications. Thanks to its low loss, the proposed power splitter can favorably be integrated to demonstrate large-scale photonic integrated circuits.

Author Contributions: Design, V.H.N.; Simulation and Layout, V.H.N. and I.K.K.; writing—original draft preparation, V.H.N.; writing—review and editing, V.H.N. and T.J.S.; supervision, T.J.S. All authors have read and agreed to the published version of the manuscript.

Funding: This research was supported by National Research Foundation of Korea (NRF) grant funded by the Korean government (MSIT) (No. 2018R1C1B6005302); Nano·Material Technology Development Program through the National Research Foundation of Korea (NRF) funded by the Ministry of Science, ICT and Future Planning. (2009-0082580); GIST Research Institute (GRI) grant funded by the GIST in 2020.

Acknowledgments: The authors would like to thank Jong-Bum You for discussion on the fabrication process at NNFC.

Conflicts of Interest: The authors declare no conflict of interest.

References

1. Wang, H.; Chai, H.; Lv, Z.; Zhang, Z.; Meng, L.; Yang, X.; Yang, T. Silicon photonic transceivers for application in data centers. *J. Semicond.* **2020**, *41*, 1–16.
2. Abrams, N.; Cheng, Q.; Glick, M.; Jezzini, M.A.; Morrissey, P.E.; O'brien, P.; Bergman, K. Silicon Photonic 2.5D Multi-Chip Module Transceiver for High-Performance Data Centers. *J. Lightwave Technol.* **2020**. [CrossRef]
3. Pasricha, S.; Nikdast, M. A Survey of Silicon Photonics for Energy Efficient Manycore Computing. *IEEE Des. Test* **2020**. [CrossRef]
4. Thomson, D.; Zilkie, A.; Bowers, J.E.; Komljenovic, T.; Reed, G.T.; Vivien, L.; Marris-Morini, D.; Cassan, E.; Virot, L.; Fédéli, J.-M.; et al. Roadmap on silicon photonics. *J. Opt.* **2016**, *18*, 073003. [CrossRef]
5. Chen, H.; Zhang, Z.; Huang, B.; Mao, L.; Zhang, Z. Progress in complementary metal-oxide-semiconductor silicon photonics and optoelectronic integrated circuits. *J. Semicond.* **2015**, *36*, 121001. [CrossRef]

6. Tu, X.; Song, C.; Huang, T.; Chen, Z.; Fu, H. State of the Art and Perspectives on Silicon Photonic Switches. *Micromachines* **2019**, *10*, 51. [CrossRef]
7. Seok, T.J.; Kwon, K.; Henriksson, J.; Luo, J.; Wu, M.C. Wafer-scale silicon photonic switches beyond die size limit. *Optica* **2019**, *6*, 490. [CrossRef]
8. Dumais, P.; Goodwill, D.J.; Celo, D.; Jiang, J.; Zhang, C.; Zhao, F.; Tu, X.; Zhang, C.; Yan, S.; He, J.; et al. Silicon Photonic Switch Subsystem with 900 Monolithically Integrated Calibration Photodiodes and 64-Fiber Package. *J. Lightwave Technol.* **2018**, *36*, 233–238. [CrossRef]
9. Kim, S.-H.; You, J.-B.; Ha, Y.-G.; Kang, G.; Lee, D.-S.; Yoon, H.; Yoo, D.-E.; Lee, D.-W.; Yu, K.; Youn, C.-H.; et al. Thermo-optic control of the longitudinal radiation angle in a silicon-based optical phased array. *Opt. Lett.* **2019**, *44*, 411. [CrossRef]
10. Chung, S.; Abediasl, H.; Hashemi, H. A Monolithically Integrated Large-Scale Optical Phased Array in Silicon-on-Insulator CMOS. *IEEE J. Solid State Circuits* **2018**, *53*, 275–296. [CrossRef]
11. Sun, J.; Timurdogan, E.; Yaacobi, A.; Su, Z.; Hosseini, E.S.; Cole, D.B.; Watts, M.R. Large-Scale Silicon Photonic Circuits for Optical Phased Arrays. *IEEE J. Sel. Top. Quantum Electron.* **2014**, *20*, 264–278. [CrossRef]
12. Acoleyen, K.V.; Bogaerts, W.; Jágerská, J.; Thomas, N.L.; Houdré, R.; Baets, R. Off-chip beam steering with a one-dimensional optical phased array on silicon-on-insulator. *Opt. Lett.* **2009**, *34*, 1477–1479. [CrossRef] [PubMed]
13. Teng, M.; Fathpour, S.; Safian, R.; Zhuang, L.; Honardoost, A.; Alahmadi, Y.; Polkoo, S.S.; Kojima, K.; Wen, H.; Renshaw, C.K.; et al. Miniaturized Silicon Photonics Devices for Integrated Optical Signal Processors. *J. Lightwave Technol.* **2020**, *38*, 6–17. [CrossRef]
14. Zhou, H.; Zhao, Y.; Wei, Y.; Li, F.; Dong, J.; Zhang, X. All-in-one silicon photonic polarization processor. *Nanophotonics* **2019**, *8*, 2257–2267. [CrossRef]
15. Harris, N.C.; Carolan, J.; Bunandar, D.; Prabhu, M.; Hochberg, M.; Baehr-Jones, T.; Fanto, M.L.; Smith, A.M.; Tison, C.C.; Alsing, P.M.; et al. Linear programmable nanophotonic processors. *Optica* **2018**, *5*, 1623. [CrossRef]
16. Shen, Y.; Harris, N.C.; Skirlo, S.; Prabhu, M.; Baehr-Jones, T.; Hochberg, M.; Sun, X.; Zhao, S.; Larochelle, H.; Englund, D.; et al. Deep learning with coherent nanophotonic circuits. *Nat. Photonics* **2017**, *11*, 441–446. [CrossRef]
17. Pérez, D.; Gasulla, I.; Crudgington, L.; Thomson, D.J.; Khokhar, A.Z.; Li, K.; Cao, W.; Mashanovich, G.Z.; Capmany, J. Multipurpose silicon photonics signal processor core. *Nat. Commun.* **2017**, *8*, 636. [CrossRef]
18. Yuan, D.; Dong, Y.; Liu, Y.; Li, T. Mach-Zehnder Interferometer Biochemical Sensor Based on Silicon-on-Insulator Rib Waveguide with Large Cross Section. *Sensors* **2015**, *15*, 21500–21517. [CrossRef]
19. Dante, S.; Duval, D.; Sepúlveda, B.; González-Guerrero, A.B.; Sendra, J.R.; Lechuga, L.M. All-optical phase modulation for integrated interferometric biosensors. *Opt. Express* **2012**, *20*, 7195. [CrossRef]
20. Chu, T.; Yamada, H.; Ishida, S.; Arakawa, Y. Compact 1 × N thermo-optic switches based on silicon photonic wire waveguides. *Opt. Express* **2005**, *13*, 10109. [CrossRef]
21. Tao, S.H.; Fang, Q.; Song, J.F.; Yu, M.B.; Lo, G.Q.; Kwong, D.L. Cascade wide-angle Y-junction 1 × 16 optical power splitter based on silicon wire waveguides on silicon-on-insulator. *Opt. Express* **2008**, *16*, 21456. [CrossRef] [PubMed]
22. Fan, G.; Li, Y.; Han, B. A Wide Wavelength Range of 1 × 8 Optical Power Splitter with an Imbalance of Less than ±1.0 dB on Silicon-on-Insulator Technology. *IEEE Photonics J.* **2017**, *9*, 1–5. [CrossRef]
23. Sun, C.; Zhao, J.; Wang, Z.; Du, L.; Huang, W. Broadband and high uniformity Y junction optical beam splitter with multimode tapered branch. *Optik* **2019**, *180*, 866–872. [CrossRef]
24. Zhang, Y.; Yang, S.; Lim, A.E.-J.; Lo, G.-Q.; Galland, C.; Baehr-Jones, T.; Hochberg, M. A compact and low loss Y-junction for submicron silicon waveguide. *Opt. Express* **2013**, *21*, 1310. [CrossRef]
25. Yun, H.; Shi, W.; Wang, Y.; Chrostowski, L.; Jaeger, N.A.F. 2 × 2 adiabatic 3-dB coupler on silicon-on-insulator rib waveguides. In Proceedings of the Photonics North 2013, International Society for Optics and Photonics, Ottawa, ON, Canada, 3–5 June 2013; Volume 8915, p. 89150V.
26. Lu, Z.; Yun, H.; Wang, Y.; Chen, Z.; Zhang, F.; Jaeger, N.A.F.; Chrostowski, L. Broadband silicon photonic directional coupler using asymmetric-waveguide based phase control. *Opt. Express* **2015**, *23*, 3795. [CrossRef]
27. Han, L.; Kuo, B.P.-P.; Alic, N.; Radic, S. Ultra-broadband multimode 3dB optical power splitter using an adiabatic coupler and a Y-branch. *Opt. Express* **2018**, *26*, 14800. [CrossRef]

28. Lin, Z.; Shi, W. Broadband, low-loss silicon photonic Y-junction with an arbitrary power splitting ratio. *Opt. Express* **2019**, *27*, 14338. [CrossRef] [PubMed]
29. Kim, H.; Shin, H. Tailorable and Broadband On-Chip Optical Power Splitter. *Appl. Sci.* **2019**, *9*, 4239. [CrossRef]
30. Wang, Y.; Gao, S.; Wang, K.; Skafidas, E. Ultra-broadband and low-loss 3 dB optical power splitter based on adiabatic tapered silicon waveguides. *Opt. Lett.* **2016**, *41*, 2053. [CrossRef] [PubMed]
31. Sun, X.; Liu, H.-C.; Yariv, A. Adiabaticity criterion and the shortest adiabatic mode transformer in a coupled-waveguide system. *Opt. Lett.* **2009**, *34*, 280. [CrossRef]

MDPI
St. Alban-Anlage 66
4052 Basel
Switzerland
www.mdpi.com

Applied Sciences Editorial Office
E-mail: applsci@mdpi.com
www.mdpi.com/journal/applsci